DESIGN AND TECHNOLOGY OF HEAT PIPES FOR COOLING AND HEAT EXCHANGE

DESIGN AND TECHNOLOGY OF HEAT PIPES FOR COOLING AND HEAT EXCHANGE

Calvin C. Silverstein
CCS Associates
Bethel Park, Pennsylvania

HEMISPHERE PUBLISHING CORPORATION
A member of the Taylor & Francis Group
Washington Philadelphia London

USA	Publishing Office:	Taylor & Francis 1101 Vermont Avenue, N.W. Suite 200 Washington, DC 20005-3521 Tel: (202) 289-2174 Fax: (202) 289-3665
	Distribution Center:	Taylor & Francis Inc. 1900 Frost Road Suite 101 Bristol PA 19007-1598 Tel: (215) 785-5800 Fax: (215) 785-5515
UK		Taylor & Francis Ltd. 4 John St. London WC1N 2ET Tel: 071 405 2237 Fax: 071 831 2035

DESIGN AND TECHNOLOGY OF HEAT PIPES FOR COOLING AND HEAT EXCHANGE

1 2 3 4 5 6 7 8 9 0 B R B R 9 8 7 6 5 4 3 2

This book was set in Times Roman by Princeton Editorial Associates. Cover design by Michelle Fleitz. Printing and binding by Braun-Brumfield, Inc.

A CIP catalog record for this book is available from the British Library.
♾ The paper in this publication meets the requirements of the ANSI Standard Z39.48-1984 (Permanence of Paper)

Library of Congress Cataloging-in-Publication Data

Silverstein, Calvin C.
Design and technology of heat pipes for cooling and heat exchange/Calvin C. Silverstein.
p. cm.
Includes bibliographical references and index.
1. Heat pipes. I. Title
TJ264.S55 1992
621.402′5—dc20 92-4393
ISBN 0-89116-859-1 CIP

CONTENTS

PREFACE

The heat pipe is a device for transporting heat from one point to another with quite extraordinary properties. Heat transport occurs via evaporation and condensation, and the heat transport fluid is recirculated by capillary forces which automatically develop as a consequence of the heat transport process. With this mode of heat transfer, the heat pipe has the capability to transport heat at high rates over appreciable distances, virtually isothermally and without any requirement for external power. It may be regarded as a quasipassive heat transport system with an extremely high effective thermal conductivity in the direction of heat transport.

Heat pipes have been considered for a wide variety of applications, ranging from heat rejection in space to cooling of gas turbine vanes. Heat pipe fluids range from liquid hydrogen and nitrogen at cryogenic temperatures, to water, ammonia, and alcohol at near-ambient temperatures, to liquid metals such as sodium and lithium at elevated temperatures.

In this book, the characteristics of heat pipes under steady-state and transient operating conditions are described. Liquid metal heat pipes are emphasized, reflecting the author's experiment, although nonliquid metal heat pipes have by no means been neglected. The physical aspects of heat pipe behavior have been given considerable emphasis. Design formulas have been developed on the basis of both mathematical models and empirical observations.

The primary purpose of the book is to impart an appreciation of heat pipe behavior, technology, design methods, and application potential to managers and specialists in heat transfer and related fields. Because of its pronounced tutorial approach, it may also be suitable as a primary or reference text in the classroom.

While the book content reflects the work of many heat pipe researchers, the contributions of Joseph E. Kemme warrant special acknowledgment. Much of the current body of heat pipe knowledge is the result of Mr. Kemme's extensive experimental investigations at Los Alamos National Laboratory

Calvin C. Silverstein

CHAPTER
ONE

INTRODUCTION

This book presents information on the application of heat pipe technology, design methods, and data to heat pipe cooling and heat exchange requirements. It is intended to provide the nonspecialist with sufficient understanding of heat pipe technology to appreciate and assess its application potential, while also meeting the needs of the experienced heat pipe designer and researcher.

1.1. HISTORICAL PERSPECTIVE

In 1964, Grover, Cotter, and Erickson of Los Alamos National Laboratory described heat transfer experiments with three structures of unique internal construction, which they referred to as "heat pipes" (Ref. 1.1). Each heat pipe consisted of a sealed tube whose inner surface was lined with a thin layer of porous material. The pores were filled with a liquid, and the remaining internal volume of the tube was occupied by the vapor of the liquid. One of the heat pipes contained water and the other two contained sodium.

Heat was added to each pipe over a portion of its length, and removed over the remaining length by natural convection and/or radiation. It was found that the temperatures of the heat pipes remained essentially constant over most of their length, even at relatively high heat addition rates. One of the sodium heat pipes is shown under test in a vacuum chamber in Figure 1.1. This heat pipe was 0.75 in. (0.019 m) in diameter and 35.4 in. (0.90 m) long, and was fabricated from stainless steel. It operated at a power level of about 3413 Btu/hr (1000 W) at a temperature of about 1520°F (827°C).

Following these early heat pipe experiments, research and development were

Figure 1.1. Original sodium heat pipe under test. *(Courtesy of Los Alamos National Laboratory, Los Alamos, New Mexico.)*

expanded at Los Alamos, and heat pipe programs were established at other government, industrial, and academic organizations in the United States and abroad. The succeeding years have been marked by an ever-increasing flow of literature on heat pipe research and applications. At the Sixth International Heat Pipe Conference, which was held in May 1987 in Grenoble, France, participants represented nineteen countries on five continents.

Initial efforts were directed toward space heat transport applications because of the potential high reliability of the heat pipe, its capability for operation under the weightless conditions of space, and its capability for isothermal operation without the need for external power. It was soon realized that the unique characteristics of the heat pipe would be equally useful here on earth, and terrestrial applications began to receive considerable interest as well.

Heat pipes have been considered for such widely diverse applications as heat removal from nuclear space power plants, cooling of stagnation regions on hypersonic vehicles, and use as cooking pins to promote more rapid, more uniform cooking of food. They have been operated at cryogenic temperatures, and at temperatures in excess of 3000°F (1649°C). A wide variety of heat pipe fluids have been used, including liquid nitrogen, ammonia, water, methanol, the alkali liquid metals, and molten silver.

One may wonder why, in view of its conceptual simplicity and exceptional heat transport capability without external power, the heat pipe concept was not given serious consideration earlier. One reason may have been that, since the surface tension forces on which capillarity depends are quite weak, heat transfer systems based on capillary force were thought to have inconsequential heat transport capability. An awareness that substantial capillary force could be developed in a fine-pored structure, and that heat transfer via latent heat would greatly reduce the fluid flow requirement, was a key ingredient in arriving at the heat pipe concept.

Another, possibly more cogent, reason was that, prior to the inception of the space program, there really was no compelling need for a heat transport device based on capillary recirculation of the heat transfer fluid. The advent of the space program, however, created the need for a heat transport device for use in space power systems that would be extremely reliable, self-actuating, relatively isothermal, and capable of operation under the weightless conditions of space. The heat pipe, as developed by Grover and his associates at Los Alamos, admirably filled this need while spurring keen interest in earthbound heat pipe applications as well.

1.2. HEAT PIPE CHARACTERISTICS

A heat pipe is a device used to transport heat from one location to another by means of the evaporation and subsequent condensation of an appropriate fluid, in which circulation of the fluid is maintained by capillary forces. The basic elements of a heat pipe are shown in the cross section of Figure 1.2.

The interior surface of the closed container is lined with a thin layer of porous material, usually referred to as a wick. The container may have a cylindrical shape, or

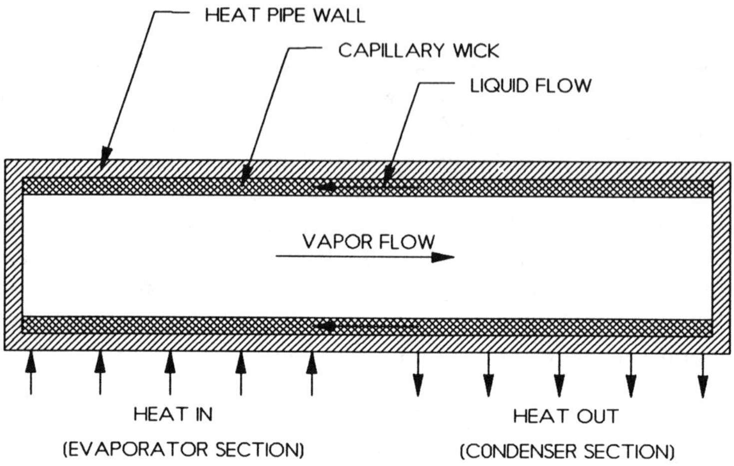

Figure 1.2. Basic elements of a heat pipe.

any other shape that can reasonably be fabricated. The pores of the wick are filled with a liquid appropriate to the application, and the vapor of the liquid occupies the remaining internal volume.

Since the liquid and its vapor coexist in equilibrium, the pressure inside the container is equal to the vapor pressure corresponding to the heat pipe temperature. Thus, a water heat pipe at 212°F (100°C) would have an internal pressure of 14.7 psi (101,400 N/m^2).

The addition of heat along a portion of the container surface causes liquid to evaporate from the wick, while the removal of heat from some other portion of the surface causes vapor to condense on the wick. The pressure gradient resulting from the accumulation of vapor at one end of the heat pipe and its depletion at the other end causes the vapor to flow through the vapor space.

As liquid evaporates, the liquid retreats into the wick pores. In the process, the liquid surface becomes curved and the liquid pressure drops below the pressure of the adjacent vapor. At the condensing end of the heat pipe, liquid accumulates on the wick, tending to maintain a flat surface without any depression of pressure in the liquid. The resulting pressure gradient in the liquid causes the liquid to flow through the wick toward the evaporator end, in a direction counter to that of the adjacent flowing vapor.

Capillary pressure develops in the wick pores to balance the pressure difference that exists along the entire heat pipe length between the flowing vapor and the adjacent flowing liquid.

An important heat pipe feature is the ability to transport heat along the direction of fluid flow with a very small temperature drop. Heat transfer through a heat pipe is virtually isothermal because the vapor pressure drop is usually on the order of 1% or less. The associated change in temperature of the saturated vapor is then quite small. The effective thermal conductivity along the direction of heat transport is generally at least four or five orders of magnitude greater than the thermal conductivity of copper. Consequently, most of the temperature drop associated with heat pipe heat transport occurs as heat is transferred through the wall and wick into and out of the vapor space.

Still another interesting heat pipe property is the capability of converting or transforming the surface heat flux (heat transfer rate per unit surface area) at the heat input zone to a higher or lower heat flux at the heat output zone. Since the heat transport rate is the same at both zones, the transformed heat flux varies inversely as the ratio of the surface areas. Thus, to reduce the heat flux by a factor of ten, the area of the heat output zone should be ten times that of the heat input zone.

The property of heat flux transformation is quite valuable when the heat flux associated with a heat source is too high to be accommodated by the device that is to utilize the heat, such as an energy converter. Placement of a properly designed heat pipe between the heat source and the heat utilization device can overcome this incompatibility.

1.3. THE HEAT PIPE AS A NATURAL CONVECTION DEVICE

The capability of the heat pipe of transporting heat over appreciable distances without any need for external power to circulate the heat transfer fluid is one of its most useful properties. Elimination of the fluid pump and power supply leads to greater reliability of the heat transport system and reduced weight, in addition to the saving in power consumption.

Self-actuated heat transport is characteristic of heat transport systems that rely on natural convection, and the heat pipe falls into this category. In natural convection systems, the forces necessary to maintain circulation of the heat transfer fluid arise naturally as a consequence of the heat transfer process itself.

Natural convection usually refers to the process in which heat transferred to a fluid raises its temperature and reduces its density, giving rise to buoyant forces that lift the fluid and transport the absorbed heat to some other location, where the heat can be removed. Natural convection occurs in a similar manner in two-phase systems. Here, the application of heat to the liquid phase produces a low-density vapor that is free to rise through the liquid and condense at some other location.

In either case, continuous circulation of the heat transfer fluid is maintained as the cooled fluid or condensate returns to the heat addition zone under the influence of gravity.

When the liquid and vapor phases are separated, with the liquid forming a pool at the bottom of a closed container and a film of condensate flowing down the wall, the natural convection system is referred to as a "thermosyphon." (See Figure 1.3.)

Natural convection heat transport can be accomplished with forces other than gravitational force. In the heat pipe, capillary force is employed in combination with two-phase flow to produce an extremely efficient mode of natural convection heat transport. Reliance on capillary force for fluid circulation means that heat pipe operation is not dependent on buoyant forces generated in a gravitational field, although heat transport capability is influenced by gravitational as well as external acceleration.

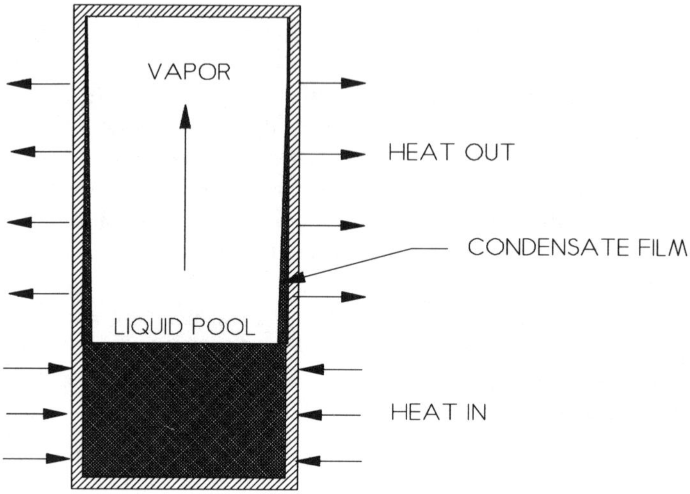

Figure 1.3. Thermosyphon natural convection heat transport.

As shown in Figure 1.4, if the heat input end of a heat pipe is at a lower elevation than the heat output end, gravity will facilitate the return of condensing liquid, enhancing heat transport. If the orientation of the heat pipe is reversed so that the heat input end is at the higher elevation, gravity impedes the return of condensate and heat transport capability is reduced.

A heat pipe deliberately designed to take advantage of the heat transport–enhancing property of gravity is called a "gravity-assisted heat pipe." The gravity-dependent nature of a heat pipe that is inclined to the horizontal may be used to advantage as a

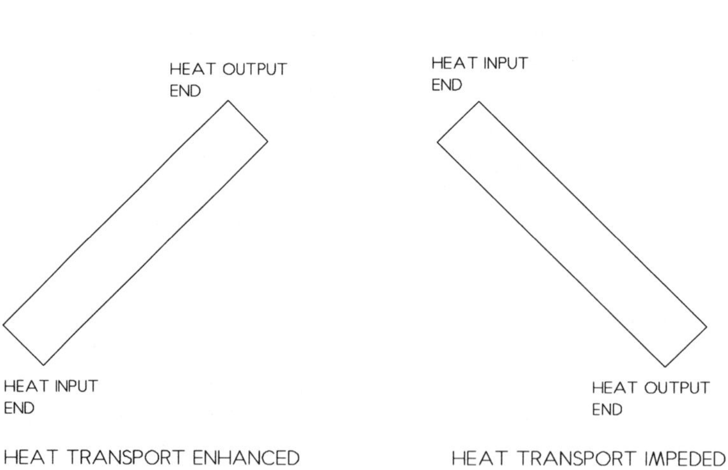

Figure 1.4. Effect of heat pipe orientation on heat transport capability.

directional thermal valve, enhancing the flow of heat in one direction along the heat pipe axis while limiting heat flow in the opposite direction.

The heat transport capability of a heat pipe is also affected by external acceleration. If the acceleration acts to drive condensate back to the heat input end, heat transport will be enhanced. If the acceleration acts to impede the return flow of condensate, heat transport will be reduced.

1.4. HEAT PIPE VARIATIONS

The basic heat pipe concept may be modified to yield characteristics that are particularly well adapted to specialized heat transport requirements. Variations on the basic heat pipe concept include the variable conductance heat pipe, the capillary-pumped heat transfer loop, and the rotating heat pipe.

1.4.1. Variable Conductance Heat Pipe

Although the temperature distribution through a heat pipe will remain fixed for given heat input and heat output conditions, any variation in either or both of these conditions will generally result in a temperature change. For some heat pipe applications, such as the cooling of experimental equipment or the conversion of heat into electricity, precise temperature control is required.

The variable conductance heat pipe has the capability of maintaining its temperature within narrow limits despite wide variations in the heat input rate or heat sink characteristics. Temperature control is achieved by automatic variation of the heat rejection length in response to heat input and/or output changes.

As shown in Figure 1.5, the control mechanism is an inert gas reservoir with a large volume and a relatively constant pressure connected to the condensing end of the heat pipe. The heat input (evaporator) length is constant, with a heat input rate that may vary with time. The internal pressure, and hence temperature, of the saturated liquid-vapor system then remains relatively constant as the effective length of the heat output

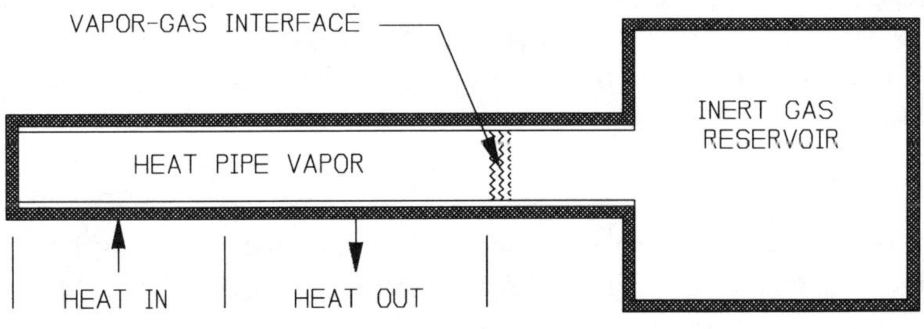

Figure 1.5. Variable conductance heat pipe concept.

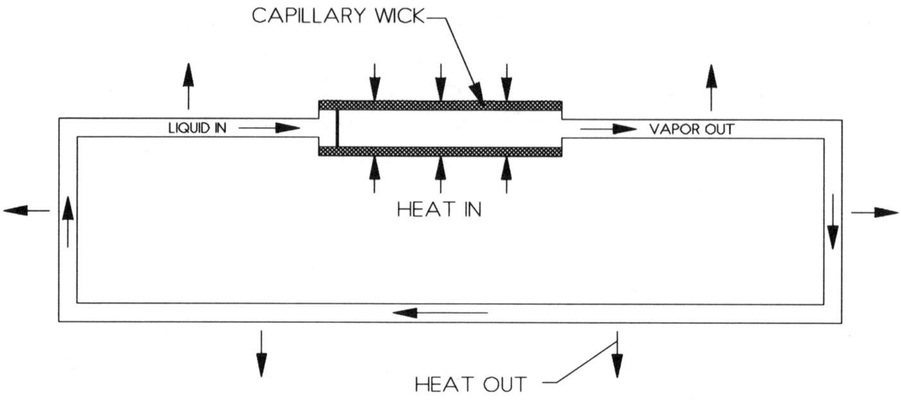

Figure 1.6. Capillary-pumped heat transfer loop.

(condenser) section varies in response to changing heat input rates and/or external heat sink conditions.

1.4.2. Capillary-Pumped Heat Transfer Loop

Surface tension forces can be used to recirculate automatically the heat transport fluid in a heat transfer loop, as well as in a heat pipe. The heat input section of such a loop is similar to that of a heat pipe, with a capillary wick structure. The heat output section is fabricated from ordinary tubing, with no wick structure present. The resulting "capillary-pumped heat transfer loop" is a hybrid device, incorporating elements of both the heat pipe and a conventional heat transfer loop. (See Figure 1.6.)

Vapor generated in the wicked heat addition section condenses as it traverses the heat removal section. The condensate, at a lower pressure than the initially generated vapor, reenters the heat input section and is evaporated. In the process of reevaporation, the pressure losses that have been incurred as the heat transfer fluid traverses the loop are restored by capillary forces in the wick pores.

The capillary-pumped heat transfer loop is useful in situations in which the heat input zone is a relatively small portion of the total heat transport length and the heat output section must follow a tortuous path. In contrast to the heat pipe, only the heat input zone need be wicked. Since the heat output section consists of ordinary tubing, it may readily be formed into complex shapes without concern for wick damage.

1.4.3. Rotating Heat Pipe

When the need exists for heat transfer adjacent to a rotating shaft, centrifugal force associated with the rotation may be tapped to create a natural convection heat transport device known as the rotating heat pipe.

As shown in Figure 1.7, the rotating heat pipe consists of a closed cavity extending

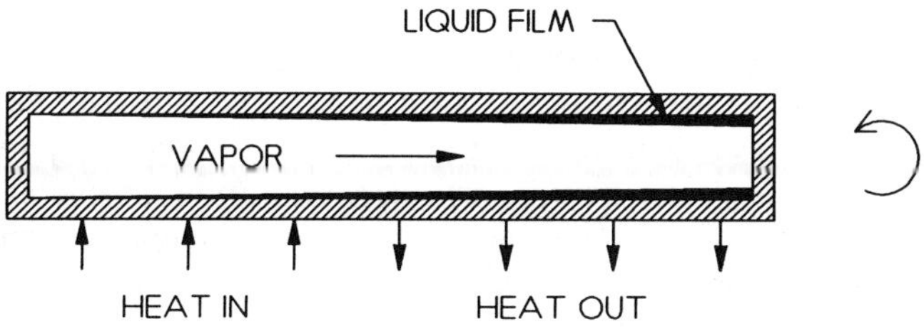

Figure 1.7. Rotating heat pipe.

along the shaft axis, which is occupied by a liquid-vapor mixture. When the shaft is rotating, the liquid forms a film along the cavity wall. The addition of heat at some point, as from a bearing, for example, causes the liquid to boil or evaporate. The resulting vapor travels through the cavity, condensing at a point where heat is being removed. The liquid film becomes thicker in the condensing zone and thinner in the evaporating zone. The resulting centrifugal head pumps the condensate back to the heat input zone, assuring automatic recirculation of the heat transfer fluid.

As in the basic heat pipe, natural convection heat transfer within the rotating heat pipe is not dependent on the presence of gravity. The heat transport process proceeds via evaporation, condensation, and the countercurrent flow of vapor and liquid phases, precisely as occurs in a conventional heat pipe. Unlike the heat pipe, however, there is no need for an internal wick structure. For this reason, the rotating heat pipe is also called the "wickless heat pipe." (This term is also sometimes used in referring to the thermosyphon.)

The heat transport capability of a rotating heat pipe will vary with changes in the rotational speed.

1.5. THE HEAT PIPE AS AN ELEMENT OF A HEAT TRANSFER SYSTEM

In considering the use of heat pipes for cooling or heat exchange applications, it is important to keep in mind that the heat pipe (or group of heat pipes) is but one link in an overall heat transfer system. Means must be provided for thermally coupling the heat pipe to the source of heat and the heat sink. The design of the heat source and heat sink, the integration of heat pipes with these system elements, the thermal isolation of the heat transfer system from adjacent components, and the impact of the heat pipe system on the performance of the device of which it is a part can be important aspects of the total heat pipe design activity.

Of particular significance is the fact that the operating temperature of a given heat pipe system, or conversely the size of the system for a given operating temperature, is determined from the requirement that the heat input and output rates must balance. These determinations can generally be made independently of heat pipe internal heat transport requirements.

In cases in which the heat pipe system is to operate continuously at a constant heat transfer rate, the system usually can be brought up to rated power gradually, and the startup process is of little concern. In cases in which the system must be brought up to rated power fairly rapidly, the startup process must be considered as part of the overall heat pipe system design effort. In cases in which the heat rate varies continuously with time, as for example during the reentry of a spacecraft, the design of the heat pipe system must be based on its complete transient history.

1.6. CONTENT OF THIS BOOK

In this book, considerable emphasis has been given to liquid metal heat pipe technology, although the technology of heat pipes with nonmetallic fluids has by no means been neglected. This emphasis is reflective of the author's design experience, as well as his desire to impart an appreciation for the excellent heat transport capabilities of liquid metal heat pipes. The commonality in heat transport characteristics of widely disparate heat pipe fluids when operating at comparable vapor pressures has also been stressed.

Heat transport devices such as the variable conductance heat pipe, the capillary-pumped heat transfer loop, and the rotating heat pipe are interesting and significant extensions of heat pipe technology. However, design aspects of these devices have not been discussed in this book. This decision was made by the author so that the basic heat pipe concept could be given maximum emphasis, and to limit the size and scope of the book to reasonable proportions.

The first three chapters constitute a descriptive, largely nonmathematical treatment of heat pipe characteristics and technology. These chapters are intended to provide useful background information for nonspecialists as well as heat transfer practitioners.

In this introductory chapter, an attempt has been made to impart to the reader some feeling for the unique characteristics of heat pipes, how these characteristics may be used to enhance heat transport capability, and how the basic heat pipe concept may be modified to accommodate special heat transfer situations. In Chapter 2, heat pipe operating characteristics are examined in some detail. Important aspects of heat pipe technology are reviewed in Chapter 3, including heat pipe fluids and materials, wick properties, fabrication techniques, environmental factors affecting heat pipe operation and life, and design aspects of heat transfer systems that incorporate heat pipes.

Important aspects of the fluid flow process in heat pipes are treated in Chapter 4. Factors that limit heat transport rates in heat pipes are discussed in Chapter 5. An account of heat pipe design methods, including the design of heat pipe heat exchangers, is given in Chapter 6. Chapter 7 presents methods for predicting the startup and transient behavior of heat pipes. Selected examples of design applications for heat pipes are given in Chapter 8, including heat dissipation in space, heat pipe transient behavior, and heat exchange.

Appendix A explains the nomenclature used throughout the book. Useful conversion factors are summarized in Appendix B. Selected material properties useful in the design and analysis of heat pipe systems are included in Appendix C. Important heat exchanger relationships are derived in Appendix D. Liquid flow in the capillary pumping layer of two-layer heat pipe wicks is treated in Appendix E.

1.7. A WORD ABOUT UNITS

In this book, the magnitude of physical quantities is expressed in both English or engineering units and SI (International System) units. In the text and illustrations, engineering units are given first, followed by SI units in parentheses. In graphs, separate scales are provided for engineering and SI units. When experimental data are being referenced, the data are usually given in the originally specified form, even if this does not correspond to standard notation. For example, a length would be given in cm if originally so specified, rather than in the standard SI length unit of m. Also, unless the text is specifically referring to absolute units, temperatures are usually presented as °F and °C.

In the English system of units, the term lb designates both force and mass, an ambiguous situation that must be handled carefully to avoid inadvertent numerical errors. When force and mass both appear in the same term of an equation, mass should always be expressed in slugs. The mass in slugs is equal to the mass in lb divided by the acceleration of gravity (32.17 ft/sec^2), and has the units lb-sec^2/ft.

In order to make the necessary conversion from lb mass to slugs, it must be known whether a given "lb" designation refers to force or mass. Frequently, this distinction is made by expressing lb force as lb_f and/or lb mass as lb_m.

In some cases, the distinction is evident from the nature of the physical quantity whose dimensions are being described. Thus, a density (mass/unit volume) of 10 lb/ft^3 clearly indicates that lb ≡ lb mass, and a pressure (force/unit area) of 10 lb/ft^2 clearly indicates that lb ≡ lb force. In other cases, such as a viscosity of 1 lb/sec-ft, the fact that lb refers to lb mass rather than lb force is not at all clear.

In this book, the somewhat tedious accounting for lb mass and lb force is avoided by reliance on dimensional consistency. If the dimension of a term in an equation has the desired units, whether lb refers to lb mass or lb force is of no concern. If the desired units have not been obtained, then the numerator or denominator of the term is multiplied by 32.17 ft/sec^2. The following examples make this clear.

Reynolds number equation:

$$\mathrm{Re} = \frac{\rho V D}{\mu}$$

where ρ is density in lb/ft^3, V is velocity in ft/sec, D is diameter in in., μ is viscosity in lb/ft-sec, and Re is the Reynolds number (dimensionless). After inserting the indicated units for the terms in the equation and converting in. in the D term to ft, all the dimensions cancel out, leaving the desired dimensionless expression for the Reynolds number without concern for whether the lb units refer to force or mass. The correct dimensionless form is obtained in this case because both lb units are actually units of mass.

Sonic velocity equation:

$$V_s^2 = k_s R_g T_v$$

where V_s is the sonic velocity in ft/sec, k_s is the ratio of specific heats at constant pressure and volume (dimensionless), R_g is the gas constant in lb-ft/lb-°R, and T_v is the vapor temperature in °R.

When the units for the various terms are inserted into the equation, V_s^2 has the unit ft. The correct dimension of ft^2/sec^2 is obtained by multiplying the equation by 32.17 ft/sec^2, without consideration of whether the lb units in R_g refer to mass or force. The correction actually converts lb mass in the denominator of R_g to slugs, but was done solely on the basis of dimensional consistency.

The preceding examples illustrate the general requirement for dimensional consistency in all terms of an equation, regardless of whether or not a correction is needed for the acceleration of gravity.

1.8. REFERENCE

1.1. Grover, G. M., Cotter, T. P., and Erickson, G. F., "Structures of Very High Thermal Conductance," *Journal of Applied Physics,* Vol. 35, No. 6, June 1964, pp. 1990–1991.

CHAPTER
TWO

HEAT PIPE OPERATING CHARACTERISTICS

In this chapter, various phenomena associated with heat pipe operation are identified, and their influence on heat pipe operational characteristics is explained. The treatment is primarily descriptive in nature, although some key mathematical relationships involved in heat pipe design and analysis are introduced. Topics to be covered include the role of capillarity in heat transport, pressure and temperature distributions, heat transport limits, and heat pipe startup.

2.1. THE ROLE OF CAPILLARITY IN HEAT TRANSPORT

Capillarity refers to the capability of the interface between a liquid and a gas or vapor in a fine-pored structure to withstand a pressure difference across the interface. Capillarity plays an essential role in the transport of heat through heat pipes by providing the mechanism for automatic recirculation of the heat pipe fluid. The manner in which this is accomplished will be described shortly, but first the phenomenon of capillarity itself will be reviewed.

Before discussing capillarity, the concepts of surface tension and wettability must be introduced.

2.1.1. Surface Tension (Refs. 2.1, 2.2)

All liquid surfaces are in a state of tension. Surface tension arises because the molecules at the liquid surface are subjected to a net force by adjacent molecules within the liquid

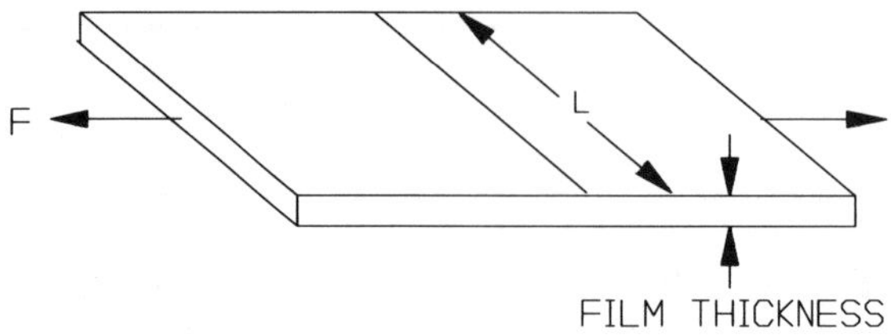

Figure 2.1. Surface tension in a liquid surface film.

interior. Since the interior molecules are all surrounded by other molecules, the net force on an interior molecule is zero.

Tensile stress is usually defined as the force at which a substance will be pulled apart per unit area normal to the direction of the force. In the case of a liquid surface film, the tensile force is applied normal to any arbitrary line lying in the surface. The area over which the force acts is the product of the surface length and the thickness of the surface film. (See Figure 2.1.) The film thickness is not known precisely, but is probably on the order of one to two molecular dimensions. By expressing surface tension as a force per unit length of the surface, uncertainty about the actual area over which the force acts is avoided.

Surface tension is a function of temperature, decreasing as the temperature rises. Its magnitude varies slightly with the nature of the gas or vapor with which the liquid surface is in contact. The surface tension for a number of heat pipe liquids is given in Table 2.1.

2.1.2. Wettability (Ref. 2.3)

Wettability refers to the tendency of a liquid to adhere to and spread over the surface of a solid. In general, a drop of liquid placed on a solid surface will partially deform, producing a contact angle ϕ_{ls} with that surface (see Figure 2.2). The size of the contact

Table 2.1. Surface tension of some heat pipe liquids

Liquid	Temperature, °F(°C)	Surface tension, 10^{-3} lb/ft (N/m)
Methyl alcohol	122 (50)	1.38 (0.0201)
Ammonia	52 (11)	1.61 (0.0235)
Water	68 (20)	4.99 (0.0728)
Sodium	1500 (816)	8.3 (0.121)
Lithium	2200 (1204)	17.8 (0.260)

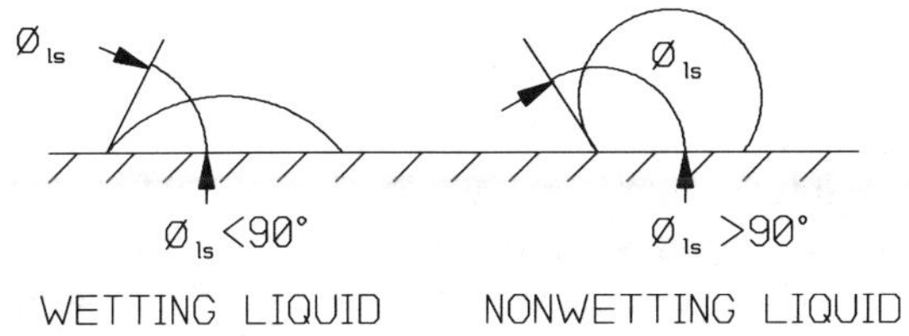

Figure 2.2. Wettability of liquids.

angle depends on the relative magnitude of forces of cohesion and adhesion. Cohesion refers to the tendency of the liquid phase molecules to remain together, while adhesion refers to the tendency of the liquid phase molecules to bind to the solid phase molecules.

The liquid is said to wet the solid when the adhesive forces predominate over the forces of cohesion. The contact angle is then less than 90 deg. When the cohesive forces predominate, the liquid is said to be nonwetting to the solid. The contact angle is then greater than 90 deg.

As will be seen shortly, capillarity effects are greatest when a liquid is either strongly wetting (ϕ_{ls} is close to 0 deg) or strongly nonwetting (ϕ_{ls} is close to 180 deg). Capillarity effects are minimal when the contact angle is in the vicinity of 90 deg, and nonexistent if ϕ_{ls} is exactly equal to 90 deg.

In a heat pipe, since capillarity is responsible for circulation of the heat pipe fluid, strongly wetting or nonwetting liquids are preferred. Although in theory a nonwetting liquid can be used in a heat pipe (see Section 2.2.3), heat pipe liquids are generally of the wetting type.

Generally, a liquid that acts as an effective solvent will have good wettability characteristics on hard surfaces such as glass or metal. Alcohol, ammonia, and the liquid metals, all of which are viable heat pipe liquids, fall into this category. Water, another heat pipe liquid, also wets such surfaces effectively, provided that they have been adequately cleaned. The contact angle for water has been reported to be 42 deg on a clean nickel surface, and zero deg on a clean nickel oxide surface (Ref. 2.4).

The wettability of a given surface by different liquids varies with their surface tension. As the surface tension increases, the contact angle increases and wettability decreases. Thus, water, at 0.00499 lb/ft (0.0728 N/m), effectively wets steel and glass surfaces whereas mercury, with a surface tension of 0.0326 lb/ft (0.475 N/m) and a contact angle of 154 deg, does not.

Wettability is also influenced by surface roughness, in a manner that depends on whether a smooth surface of the same material is wetting or nonwetting. If the contact angle for a smooth surface is less than 90 deg (a wettable surface), roughening that surface will decrease ϕ_{ls}, increasing wettability. If the contact angle for a smooth surface exceeds 90 deg (a nonwettable surface), ϕ_{ls} will increase with surface roughness, reducing wettability further. For example, if the surface of paraffin wax is roughened, the contact angle for water increases from 110 deg to 132 deg.

2.1.3. Capillarity

Capillarity refers to the capability of a curved liquid surface to sustain a pressure difference across that surface. The magnitude of the pressure difference that can be sustained is called the capillary pressure. It depends on the radius of curvature of the liquid surface, the surface tension of the liquid, and its contact angle (i.e., wettability). The capillarity of the wick in a heat pipe is a major factor in establishing heat transport capability.

Capillarity is responsible for the rise in elevation of a wetting liquid when a tube is inserted into the liquid, and for the depression in elevation of a nonwetting liquid on insertion of a tube into that liquid (see Figure 2.3). An expression for the capillary pressure that develops in these tubes will now be derived (Ref. 2.2). The result can also be used to find the capillary pressure in the pores of a heat pipe wick.

Let us consider the wetting liquid of Figure 2.3. The pressure at the base of the liquid column inside the tube is the same as that on the pool surface, which is just the atmospheric pressure P_a. The liquid pressure at the top of the column P_l is less than that at the column base by the hydrostatic head, which is equal to the product of the liquid weight density $\rho_l g$ and the column height h, where ρ_l is the liquid mass density and g is the acceleration of gravity. Therefore, $P_l = P_a - \rho_l g h$.

Since the pressure over the top of the column is also P_a, a net downward pressure difference equal to $P_a - P_l$ exists across the liquid surface in the tube. The net downward force is equal to the product of the pressure difference and the projected liquid surface area πr^2, where r is the inner tube radius.

When the liquid column has reached its maximum height, the downward force is just balanced by the vertical component of the surface tension force at the top of the column. This force is equal to the product of the surface tension σ, the inner tube periphery $2\pi r$, and the cosine of the contact angle ϕ_{ls} between the liquid and the tube surface.

Thus, we have

$$\pi r^2 (P_a - P_l) = 2\pi r \sigma \cos \phi_{ls}$$

or

$$P_a - P_l = (2\sigma / r) \cos \phi_{ls} \tag{2.1}$$

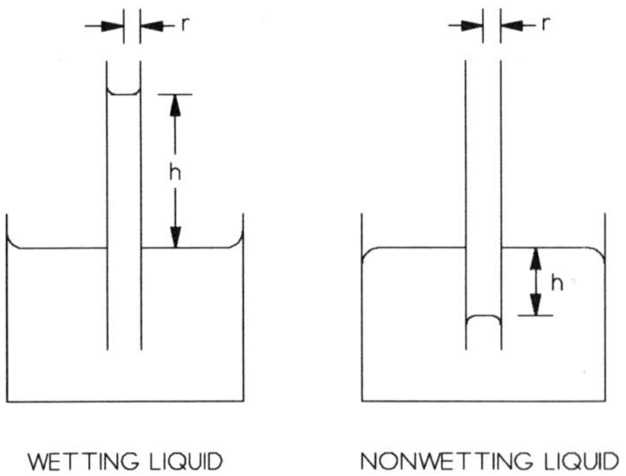

Figure 2.3. Capillarity in tubes with wetting and nonwetting liquids.

Equation (2.1) defines the maximum capillary pressure available in a tube of radius r with a liquid of surface tension σ and contact angle ϕ_{ls}.

Noting that $P_a - P_l = \rho_l gh$, and using Eq. (2.1), the liquid column height h can be expressed as

$$h = (2\sigma/\rho_l gr) \cos \phi_{ls} \tag{2.2}$$

In the foregoing example, it was shown that the hydrostatic pressure drop in a capillary tube can be offset by a corresponding rise in the capillary pressure. Capillary pressure can also be used to offset the dynamic pressure drop due to liquid flow through a wick structure. For example, if a horizontal porous wick is immersed in liquid at one end and heated at the other end, liquid will flow through the wick toward the heated end at the same rate at which liquid is being vaporized.

The pressure drop incurred by the liquid as it flows through the wick will be offset by the capillary pressure that develops at the liquid-vapor interface of the heated end. When the liquid flow rate resulting from heat addition produces a pressure drop equal to the maximum capillary pressure of the wick pores, the flow rate has reached a maximum value. Any further increase in the heat addition rate cannot be matched by an increase in the liquid flow rate, the heated part of the wick will dry out, and vaporization will cease.

The above interaction among heat addition, liquid and vapor flow, and capillary pressure in a wick is duplicated in the evaporator section of a heat pipe. In a heat pipe, however, the source of liquid is the vapor that condenses on the wick at the heat rejection end.

Equations (2.1) and (2.2) also apply in the case of a nonwetting liquid, for which the contact angle ϕ_{ls} lies between 90 and 180 deg. Then, cos ϕ_{ls} is negative, the liquid pressure inside the tube is greater than atmospheric, the surface tension force is now downward, and the liquid column height is negative (i.e., the liquid inside the tube is at a lower elevation than that of the surrounding pool, as shown in Figure 2.3).

In Table 2.2, the capillary pressure and the liquid column height obtainable in a steel tube are given for water ($\phi_{ls} = 0$ deg) and mercury ($\phi_{ls} = 154$ deg) and two different tube radii. The data apply at room temperature.

Table 2.2 shows that significant capillary pressure can be developed in fine-pored tubes. For the tube with a radius of 0.002 in. (0.0000508 m), water will rise to a height of more than 11 in. (0.279 m), and mercury will be depressed to a depth of almost 5 in. (0.126 m).

Equation (2.1), more general than Eq. (2.2), gives the maximum pressure difference that can be sustained across a liquid-vapor interface in any capillary structure. Therefore, Eq. (2.1) can be used to predict the maximum available capillary pressure in a heat pipe wick, and hence the maximum permissible pressure difference between the heat pipe vapor and its liquid.

Equation (2.1) can also be written in the form

$$\Delta P_c = 2\sigma/(r/\cos \phi_{ls}) = 2\sigma/r_p = 4\sigma/D_p \tag{2.3}$$

where ΔP_c is the maximum capillary pressure, r_p is the effective pore radius, and D_p is the effective pore diameter. Equation (2.3) is usually preferred in heat pipe capillary

Table 2.2. Capillary pressure and liquid column height

Liquid	Tube Radius, in. (m)	Capillary Pressure, psia (N/m^2)	Column Height, in. (m)
Water	0.200 (0.00508)	0.00416 (28.7)	0.115 (0.00293)
	0.002 (0.0000508)	0.416 (2870)	11.51(0.293)
Mercury	0.200 (0.00508)	−0.0244 (−168)	−0.050 (−0.00126)
	0.002 (0.0000508)	−2.44 (−16,800)	−4.97(−0.126)

pressure calculations because the contact angle ϕ_{ls} is not known with precision and r_p is more readily measured than r.

More generally, Eq. (2.3) can be expressed as

$$\Delta P_c = 4\sigma / D_c \tag{2.4}$$

where D_c is the diameter of curvature of the liquid-vapor interface in the wick pores, ΔP_c is the associated capillary pressure, and $D_c \geq D_p$. ΔP_c takes on its maximum value when $D_c = D_p$.

The effective pore radius for heat pipe wick structures can be on the order of 0.002 in. (0.0000508 m = 50.8 microns) or less. (In this book, the effective pore radius will be expressed in microns, in accordance with general practice.) For example, the effective pore radius for fully wetted 400 mesh screen, a not uncommon wick choice, is around 60 microns.

2.1.4. Tensile Stress in Liquids

With a wetting liquid, the liquid pressure in a capillary structure is equal to the vapor pressure over the liquid minus the pressure difference across the liquid-vapor interface (i.e., the capillary pressure). If the pressure difference should be greater than the vapor pressure, the pressure in the bulk liquid will be negative. That is, the liquid will be under tension rather than compression. Ordinarily, other than in the surface film, a liquid would not be considered capable of sustaining tensile stress. However, appreciable tensile stress can be sustained by liquids confined within fine capillary pores, including the liquid in a heat pipe wick.

For confirmation of this phenomenon, one need only look to the process whereby tree tissue is nourished with water (Ref. 2.5). Water is supplied to the upper reaches of a tree by capillary action through very fine pores called xylem. When water has been lifted to a height of 33.9 ft (10.33 m), a water head equal to 1 atm of pressure, by capillary action of the xylem, the liquid pressure at the upper water surface will have fallen to zero.

For trees that are taller than 33.9 ft (10.33 m), the water pressure at the upper reaches of the xylem is negative (i.e, the water is under tension), and the pressure drop across the liquid-vapor interface is greater than 1 atm. (See Figure 2.4.) In the xylem of giant redwoods, which can grow to a height or more than 300 ft (91.4 m), the tensile stress of water can exceed 115 psia (793,000 N/m^2).

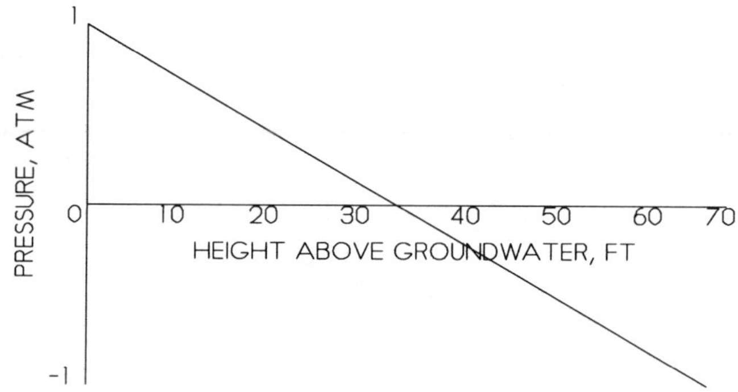

Figure 2.4. Liquid pressure in the xylem of trees.

In the laboratory, liquid tensions as high as 20 atm have been achieved under static conditions, and as high as 300 atm in liquids subjected to centrifugal force (Ref. 2.6).

In a heat pipe, the liquid in the wick can similarly be placed under tension for some operating conditions (Ref. 2.7). A state of tensile stress may exist if the vapor pressure is low and the liquid in the wick is subjected to a high pressure drop due to friction, gravity, and/or external acceleration. This condition is most likely to occur when a heat pipe is being started up.

2.2. PRESSURE AND TEMPERATURE DISTRIBUTIONS

The pressure distribution within a heat pipe takes on a form that is the result of the variable vapor and liquid flow rates along the heat pipe length, along with the counter-current flow of the adjacent vapor and liquid streams. The pressure distribution is needed in order to determine the maximum vapor-liquid pressure difference to be offset by capillary pressure, the minimum vapor and liquid pressures, and the heat pipe temperature distribution.

This section includes discussions of the pressure distribution within a heat pipe, the vapor and liquid pressure drop components, minimum vapor and liquid pressures, the heat pipe temperature distribution, and the pressure distribution for a nonwetting or inverted heat pipe.

2.2.1. Pressure Distribution in a Heat Pipe

The distribution of vapor and liquid pressures within a heat pipe plays a key role in the determination of heat transport capability. A typical pressure distribution is shown in Figure 2.5. The pressure scale has been exaggerated for clarity. It is evident that, except for the end of the condenser, the vapor pressure is higher than the liquid pressure over the entire heat pipe length. The point where the vapor and liquid pressures are equal has been called the "wet point" (Ref. 2.8). The pressure distribution is shown for the

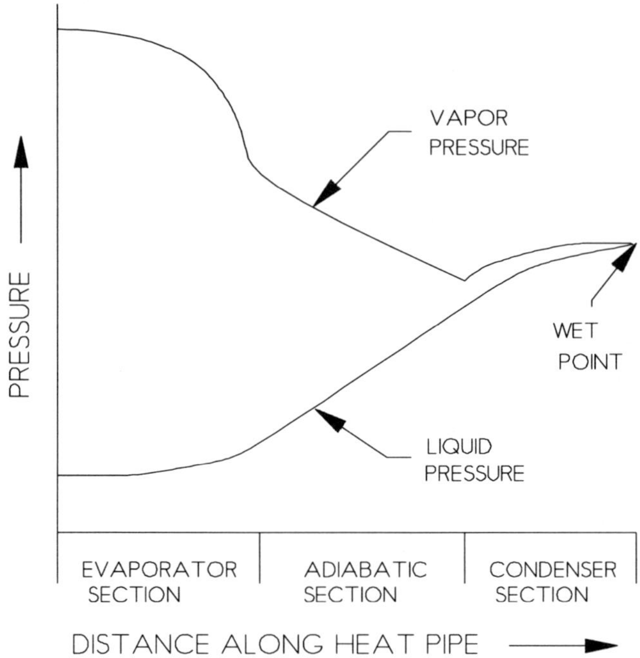

Figure 2.5. Liquid and vapor pressure distributions in a heat pipe (wet point at condenser end).

evaporator, adiabatic, and condenser sections of a heat pipe. The vapor and liquid pressure drop components will now be considered.

Vapor Pressure Drop. In the evaporator region, the vaporization of liquid produces a vapor whose mass flow rate and velocity continuously increase from zero at the start of the evaporator to maximum values at the evaporator exit. The frictional pressure drop per unit length then increases with length, producing a concave downward shape in the vapor pressure curve.

There is also a net increase of momentum (mass flow rate times velocity) in the vapor between the beginning of the evaporator and its exit. The momentum increase produces an additional pressure drop in the evaporator region, known as the momentum pressure drop.

Since there is no heat transfer into or out of the adiabatic region, the vapor mass flow rate remains constant. Velocity and momentum changes are due solely to friction, and are generally small. The vapor pressure drop may then be considered as having only a frictional component, and will be relatively constant per unit length. The vapor pressure curve in the adiabatic region of the heat pipe is then essentially linear.

When the vapor enters the condenser, the mass flow rate and velocity decrease continuously as the vapor condenses on the wick, finally becoming zero at the end of the condenser. The frictional pressure drop per unit length undergoes a similar decrease, with the result that the frictional vapor pressure curve is now concave up.

The momentum that was generated in the evaporator also drops to zero, producing a momentum-related pressure rise in the condenser. However, complete pressure recovery is not usually achieved, so that there is some net pressure loss associated with the vapor momentum generation and dissipation process in a heat pipe. If the momentum-related pressure rise exceeds the frictional pressure drop, there will be a net pressure rise in the condenser, as is indicated in Figure 2.5.

If the vapor mass flow rate generated in the evaporator becomes sufficiently large, the vapor flow regime may change from laminar to turbulent at some point along the evaporator length. The flow then remains turbulent over the remainder of the evaporator length, in the adiabatic section, and in the condenser until the mass flow has decreased sufficiently to enable the flow to revert to the laminar state.

Summing up, the vapor pressure drop ΔP_v has two components: the frictional pressure drop ΔP_{vf} and the momentum pressure drop ΔP_{vm}. Thus,

$$\Delta P_v = \Delta P_{vf} + \Delta P_{vm} \tag{2.5}$$

The vapor flow may be entirely laminar or, if the mass flow rate is sufficiently large, partly laminar and partly turbulent.

Liquid Pressure Drop. Liquid in the wick structure flows counter to the vapor flow, and the liquid flow rate varies in a manner opposite to that of the vapor. Thus, the liquid flow rate continuously increases along the condenser length, and continuously decreases with length in the evaporator.

The liquid frictional pressure drop curve is similar to that for the heat pipe vapor. The curve is concave down in the condenser, linear in the adiabatic section, and concave up in the evaporator. Although momentum changes accompany the changing liquid flow rate, these changes are usually negligible because of the low liquid velocity, and can be neglected for design purposes.

The liquid frictional pressure drop for a given flow rate is strongly dependent on the nature of the liquid flow channel through the wick. If the flow channel is open with no internal structure, resistance to flow will be minimal and the liquid pressure drop will be relatively small. If the flow channel is porous, with internal elements such as wires, particles, or fibers, there will be a substantial resistance to flow and a relatively large liquid pressure drop.

With porous flow channels, the liquid flow regime will generally remain laminar even at flow rates corresponding to extremely high heat transport rates. For open flow channels, the liquid flow may be laminar over the entire heat pipe length or, for sufficiently large flow rates, partly laminar and partly turbulent.

If the heat pipe is in a gravitational field or subject to external acceleration, a hydrostatic pressure difference will develop over the length of the liquid in the heat pipe wick. This pressure difference is manifested as a liquid pressure drop or pressure rise. If gravity or acceleration acts to impede the flow of condensate back to the evaporator, there will be a pressure drop. If the return flow of condensate is facilitated, there will be

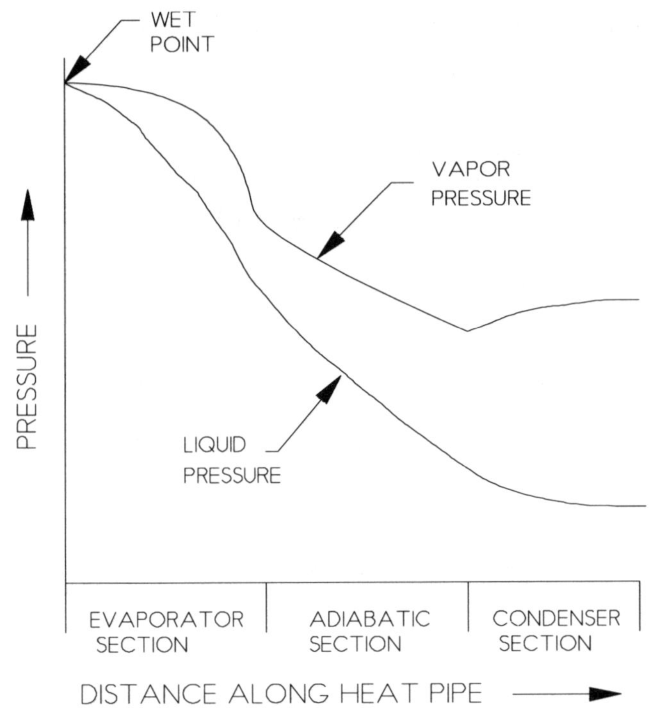

Figure 2.6. Liquid and vapor pressure distributions in a heat pipe (wet point at evaporator end).

a pressure rise. The resulting net pressure change depends on the magnitude of the acceleration or gravity component along the end-to-end heat pipe length.

Summing up, the liquid pressure drop ΔP_l is equal to the sum of the frictional pressure drop ΔP_{lf} and the acceleration pressure drop ΔP_{la}, or

$$\Delta P_l = \Delta P_{lf} + \Delta P_{la} \tag{2.6}$$

As defined here, the acceleration pressure drop includes both gravitational and external acceleration components, and may be negative (i.e., there may be a pressure rise) as well as positive. If the liquid flow channel is porous, flow may be considered as laminar. For an open flow channel, the liquid flow may be all laminar, or partly laminar and partly turbulent.

Minimum Vapor and Liquid Pressures. We consider here the minimum vapor and liquid pressures that can develop in a heat pipe.

Since a vapor cannot exist under negative pressure (i.e., tension), the lower limit on vapor pressure is zero. Under normal operating conditions, the vapor pressure drop is on the order of a few percent or less, and excessively low vapor pressure is not a problem. However, the vapor pressure may drop to zero at some point along the heat pipe during the initial phase of startup, when the temperature and vapor pressure are very low. The heat transport rate when this occurs has been called the "viscous limit" (Ref. 2.9). This is a transient condition that can be expected to disappear upon continued heating, and is usually not of concern in heat pipe design.

The minimum liquid pressure usually occurs at the evaporator end of a heat pipe (see Figure 2.5), but can occur at other locations. As indicated in Figure 2.6, the location of the wet point (i.e., the point where the vapor and liquid pressures are essentially equal) can shift when there is a large liquid pressure rise due to gravity or external acceleration.

Depending on the actual shape of the pressure distribution, the wet point can also occur at the interface of the evaporator and adiabatic sections, the adiabatic and condenser sections, or (if there is no adiabatic section) the evaporator and condenser sections.

When the wet point lies at the condenser end, and there is a net liquid pressure drop, the minimum liquid pressure $P_{l\min}$ is given by

$$P_{l\min} = P_v - (\Delta P_v + \Delta P_l) \tag{2.7}$$

where P_v is the maximum vapor pressure (at the beginning of the evaporator).

When the wet point is at the condenser end, and there is a net liquid pressure rise,

$$P_{l\min} = P_v - \Delta P_v \tag{2.7a}$$

When the wet point occurs at the evaporator end, the minimum liquid pressure is given by

$$P_{l\min} = P_v - \Delta P_l \tag{2.8}$$

where now ΔP_l is the liquid pressure rise.

From Eq. (2.7), it is evident that the liquid pressure will become negative (i.e., be in a state of tension) if the liquid pressure drop should exceed the difference between the maximum vapor pressure and the vapor pressure drop. For the situation in which Eq. (2.8) is applicable, the liquid will be under tension if the liquid pressure rise should be greater than the maximum vapor pressure.

As has been explained in Section 2.1.4, liquids in fine-pored structures can withstand significant tensile stress. A sodium heat pipe has been operated under conditions for which the liquid in the wick was under a tensile stress of 1–2 psia (6900–13,800 N/m^2) (Ref. 2.7). A heat pipe liquid is most likely to experience tensile stress when a hydrostatic head due to gravity or external acceleration produces a large liquid pressure drop.

2.2.2. Temperature Distribution in a Heat Pipe

During normal operation, the temperature along the length of a heat pipe is practically constant. This isothermal operating characteristic is one of the outstanding features of the heat pipe. However, significant temperature drops can be associated with the flow of heat into and out of a heat pipe, as illustrated in the schematic of Figure 2.7. In this figure, radial distances and the axial distance are both plotted along the abscissa, with the radial scale exaggerated for clarity.

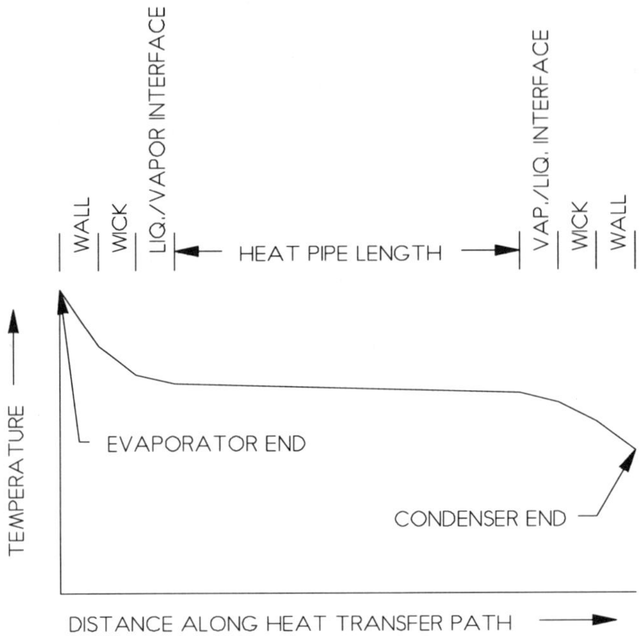

Figure 2.7. Temperature distribution along the heat pipe heat transfer path.

The flow of heat into the heat pipe interior is accompanied by temperature drops across the wall, the wick, and the liquid-vapor interface. The peak temperature occurs on the external wall surface in the evaporator region. As the heat pipe vapor flows through the vapor space, a temperature drop develops in the axial direction. As vapor condenses, a further temperature drop occurs across the liquid-vapor interface, the wick, and the wall of the condenser. The minimum heat pipe temperature is reached on the external wall surface in the condenser region.

As was pointed out in Section 1.2, the effective thermal resistance along the heat pipe length is extremely low. The thermal resistance across the liquid-vapor interface is also very low, and can usually be neglected unless the heat flux at the evaporator wall is very large.

Since the heat pipe vapor is nominally saturated, the vapor temperature is a unique function of the vapor pressure. Therefore, the axial temperature drop can be found from the drop in vapor pressure along the heat pipe.

In a properly designed heat pipe, the axial temperature drop will usually be isothermal within a few degrees. This level of isothermality can generally be assured by limiting the vapor pressure drop to 1%. The situation can be quite different during startup, when pressure drops in excess of 60% can occur. The axial temperature drop can then be in excess of 100°F (56°C). (See Section 2.4.)

The dominant thermal resistance in a heat pipe, and hence almost the entire temperature drop, occurs across the wall and wick thicknesses. If the condenser is much longer than the evaporator, the radial heat flux at the condenser will be much smaller than that at the evaporator. In this event, which is not uncommon in heat pipe design, most of the temperature drop will take place in the evaporator section.

If the wall and wick thicknesses are a significant fraction of the heat pipe radius, curvilinear geometry should be used in calculating temperature drops. If the thicknesses do not exceed 10% of the radius, rectilinear geometry is usually sufficient for design purposes.

2.2.3. Pressure Distribution in an Inverted Heat Pipe

In Section 2.1.2, the wettability of a surface to a liquid was defined in terms of the contact angle ϕ_{ls} between the liquid and the surface. For contact angles between 90 deg and 180 deg, the surface is said to be nonwetting to the liquid.

Virtually all heat pipe liquids wet the wick structure. The assumption of a wetting heat pipe liquid leads to the pressure distribution of Figure 2.5, in which the vapor pressure is greater than the liquid pressure over the entire heat pipe length (except at the wet point, where the liquid and vapor pressures are equal).

If a liquid is nonwetting to a surface, the liquid surface in a capillary structure assumes a convex shape with respect to the adjacent gaseous or vapor phase, and the liquid pressure exceeds that of the gas or vapor. (See Figure 2.3.) In a heat pipe with a nonwetting liquid, the pressure distribution would then be expected to look like that of Figure 2.8.

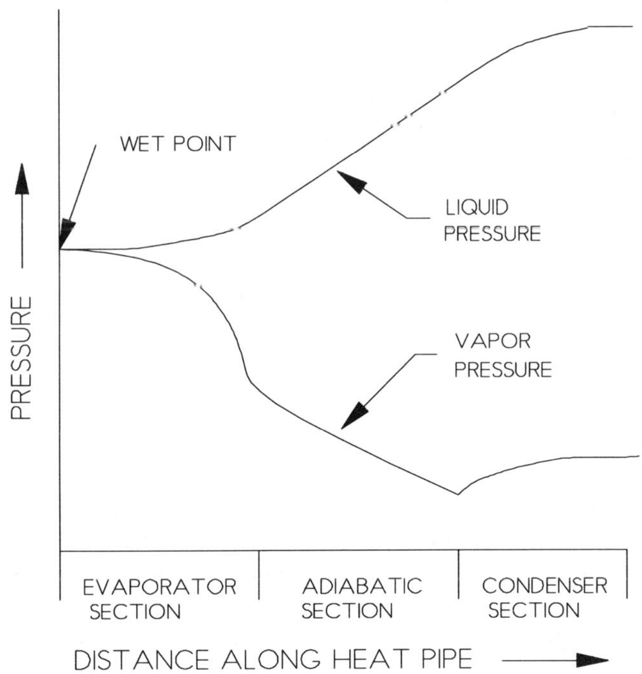

Figure 2.8. Liquid and vapor pressure distributions in an inverted heat pipe (wet point at evaporator end).

The liquid pressure is now greater than the vapor pressure over the entire heat pipe length. The wet point is located at the evaporator end, and the maximum liquid-vapor pressure difference occurs at the condenser end. Since this pressure distribution is essentially the inverse of that for a wetting liquid, a heat pipe with a nonwetting liquid will be referred to as an inverted heat pipe.

Mercury, which is highly nonwetting to steel, would seem to be an appropriate candidate for an inverted heat pipe. Mercury–stainless steel heat pipes have been operated successfully, but with special additives to promote wetting by the mercury liquid (Ref. 2.10).

Plastic wick structures in combination with high-surface-tension liquids will also exhibit nonwetting characteristics. An example of this combination is polytetrafluoroethylene (PTFE) and water. A nonwetting wick structure can be formed from a thin sheet of porous PTFE.

Possible advantages of the inverted heat pipe include lower cost (plastic porous sheets are quite inexpensive) and greater capillary pumping capability. This increased capability could permit higher heat transport rates and/or operational capability under larger hydrostatic liquid heads. It is a consequence of the very small effective pore radii that are obtainable in porous plastic materials. For example, effective pore radii as small as 0.060 microns are available with porous PTFE. In porous polypropylene, the effective pore radius is on the order of 0.035 microns.

A water heat pipe with an effective wick pore radius of 3 microns would be functional with the evaporator end elevated as much as 16 ft (4.9 m) above the condenser end. With an effective pore radius of 0.3 microns, the heat pipe could function with the evaporator end as much as 160 ft (49 m) above the condenser end. Such small wick pore sizes and associated capillary pumping capability are extremely difficult if not impossible to obtain with metallic wick structures.

With nonwetting wick structures, special procedures may be necessary to prime the wick properly (fill the wick pores with liquid prior to operation), since liquid will not flow through the pores unless pressurized in excess of the maximum pore capillary pressure. Also, the low thermal conductivity of many plastics could limit the permissible heat flux (heat transfer rate per unit area) through the wick structure.

2.3. HEAT TRANSPORT LIMITS

As was stated in Section 2.1.3, capillarity is responsible for heat transport within a heat pipe. Since there is a limit on the maximum capillary pressure that can be generated in the pores of a heat pipe wick, as calculable from Eq. (2.3), there exists an associated limit on the maximum heat transport capability. Heat transport capability can also be limited by other physical phenomena, including the onset of boiling in the heat pipe liquid, the entrainment of liquid from wick pores by the heat pipe vapor, and the attainment of sonic velocity in the heat pipe vapor.

In this section, physical aspects of the various limiting heat transport phenomena are emphasized. Mathematical methods for the prediction of heat transport limits are presented in Chapter 5.

2.3.1. Capillary Pumping Limit

The pressure distribution in a typical heat pipe is shown in Figure 2.5. Except for a single point at the condenser end, the vapor pressure is higher than the liquid pressure over the entire heat pipe length. In order for such an arrangement to be stable, another pressure must be present to balance the vapor-liquid pressure difference. That pressure is the capillary pressure that develops at the liquid-vapor interface in the wick pores. Its magnitude is given by Eq. (2.4).

As long as the heat pipe liquid wets the wick structure, the liquid pressure can never be larger than the vapor pressure. If such a situation were to develop, the liquid-vapor interface would become unstable and rapidly deteriorate.

The vapor-liquid pressure difference, and hence the capillary pressure, vary along the heat pipe length. From the standpoint of heat pipe design, the maximum capillary pressure is of primary concern. In Figure 2.5, the maximum capillary pressure occurs at the start of the evaporator section, since the pressure difference between the vapor and liquid is a maximum there. It can also be seen from Figure 2.5 that the maximum pressure difference, and hence the maximum capillary pressure encountered in the heat pipe, is equal to the sum of the vapor and liquid pressure drops. Thus,

$$\Delta P_{vl} = \Delta P_v + \Delta P_l \tag{2.9}$$

where ΔP_v is the vapor pressure drop, ΔP_l is the liquid pressure drop, and ΔP_{vl} is the maximum vapor-liquid pressure difference. The pressure drops of Eq. (2.9) are expressible as functions of the maximum mass flow rate of the heat pipe fluid.

Equations (2.4) and (2.9) are the fundamental equations governing capillary-driven heat transport in a heat pipe. As long as ΔP_{vl} as determined from Eq. (2.9) does not exceed the maximum capillary pressure capability of the wick pores, as determined from Eq. (2.4) when $D_c = D_p$, heat transport will proceed normally.

The maximum mass flow rate W in the heat pipe is the same for both liquid and vapor. The product of W and the heat of vaporization κ is equal to the rate Q at which heat is transferred into, through, and out of the heat pipe. Thus,

$$Q = \kappa W \tag{2.10}$$

The mass flow rate W obtained from Eq. (2.10) may be substituted in the pressure drop expressions of Eq. (2.9) to yield an expression for the maximum vapor-liquid pressure difference ΔP_{vl} as a function of the heat transfer rate Q.

If the calculated ΔP_{vl} exceeds the maximum capillary pressure ΔP_c that can be developed in the wick pores as determined from Eq. (2.4) when $D_c = D_p$, the liquid-vapor interface will be destroyed and the evaporator region will dry out and overheat. If the calculated ΔP_{vl} is equal to or less than ΔP_c, then the specified heat transport rate can be handled by the heat pipe without adverse consequences. When ΔP_{vl} is just equal to ΔP_c, the heat pipe will be operating at its maximum heat transport capability.

The maximum heat transport rate is called the capillary pumping limit or the wicking limit. The term *capillary pumping limit* is used throughout this book. The capillary pumping limit is expressible either as the maximum heat transport rate per unit of vapor space cross-sectional area, or simply as the maximum heat transport rate.

If the capillary pumping limit should be too small, it can be increased through appropriate design changes. For example, the liquid and/or vapor flow areas could be enlarged to reduce the frictional resistance to flow. Alternatively, smaller wick pores could be used at the liquid-vapor interface to increase the maximum available capillary pressure.

Effect of Gravity and Acceleration. The pressure distribution shown in Figure 2.5 and the pressure drop relationship of Eq. (2.9) are based on the assumption that the heat pipe liquid undergoes a net pressure drop. This situation does prevail under any of the following conditions: there is no gravitational or external acceleration head present, a gravitational or acceleration head acts to increase the liquid pressure drop, or gravitation or acceleration causes a pressure rise that is less than the sum of the vapor and liquid frictional pressure drops.

When gravitation or acceleration produces a pressure rise that exceeds the sum of the vapor pressure drop and liquid frictional pressure drop, the pressure distribution in the heat pipe changes radically. (See Figure 2.6.) Now the wet point (point where the liquid and vapor pressures are equal) shifts to the evaporator end of the heat pipe, and the maximum difference between vapor and liquid pressures shifts to the condenser end. The magnitude of this pressure difference is now equal to the difference between the liquid pressure rise and the vapor pressure drop.

If a liquid pressure rise is taken to be negative, then the liquid-vapor pressure difference can be expressed more generally as the absolute value of the algebraic sum of the vapor and liquid pressure drops. Thus,

$$\Delta P_{vl} = |\Delta P_v + \Delta P_l| \tag{2.11}$$

If a net liquid pressure rise should be encountered, Eq. (2.11) should used in the capillary pumping limit calculation instead of Eq. (2.9).

It is possible that the wet point could occur at some location within the heat pipe length, such as at the intersection of the evaporator and adiabatic sections. In this case, the pressure drop terms on the right side of Eq. (2.11) would be evaluated between the wet point and the end of the condenser, rather than over the entire heat pipe length.

2.3.2. Boiling Limit

During normal operation of a heat pipe, heat added to the evaporator flows through the wall and wick, vaporizing the heat pipe liquid at the liquid-vapor interface. If the heat flux should exceed a certain critical value called the boiling limit, the liquid in the wick will boil. For reasons that are cited later in this section, it is advisable to operate a heat pipe below the boiling limit.

The various factors that influence the boiling limit will now be examined. These include the concepts of saturation and superheat, temperature and pressure distributions across the wick thickness, nucleation sites, and interface heat and mass transfer. The consequences of boiling in a heat pipe wick will also be discussed.

Saturation and Superheat. When the liquid and vapor phases of a fluid are in contact, there is a continuous interchange of molecules between the phases. If the rate at which vapor molecules condense on the liquid surface is just equal to the rate at which liquid molecules evaporate from the surface, the two phases are in equilibrium at the same temperature and pressure, and are said to be saturated. The common temperature is referred to as the saturation temperature, and the associated pressure as the saturation or vapor pressure.

The liquid is said to be superheated if its actual temperature is higher than the saturation temperature. The amount of superheat is equal to the difference between the liquid and saturation temperatures. In a superheated liquid, the vapor pressure corresponding to the actual liquid temperature is greater than the actual liquid pressure.

All surfaces, no matter how smooth, have microscopic cavities called nucleation sites where vapor and noncondensable gas (if present) can accumulate. Therefore, liquid-vapor interfaces exist wherever a liquid is in contact with a solid surface.

A superheated liquid will be stable in the presence of its vapor at nucleation sites as long as the difference between the vapor and liquid pressures is balanced by capillary pressure. This pressure difference increases with the liquid superheat. If the vapor-liquid pressure difference should exceed the maximum available capillary pressure, vapor bubbles will issue forth from the nucleation sites into the surrounding liquid. The liquid is then said to be boiling.

The liquid superheat at incipient boiling is called the superheat for boiling. Although nucleation sites also exist on the internal surfaces of a porous wick structure, boiling will generally be initiated at the wall, where the superheat and hence the vapor-liquid pressure difference is highest.

Temperature and Pressure Distributions. Temperature and pressure distributions in the wick liquid and the adjacent vapor are shown in Figure 2.9. The temperature drops continuously through the wick, and then undergoes a further abrupt drop across the liquid-vapor interface to the temperature of the adjacent vapor. The abrupt temperature drop is a result of the thermal resistance that exists across the interface between two phases, as will be discussed subsequently.

The pressure is relatively constant through the wick, and then undergoes an abrupt rise across the liquid-vapor interface to the pressure of the adjacent vapor. The abrupt pressure rise is equal to the capillary pressure that develops to offset the pressure difference between the heat pipe vapor and its adjacent liquid. (See Section 2.1.3.)

Also shown in Figure 2.9 are the saturation temperature corresponding to the liquid pressure and the vapor pressure corresponding to the liquid temperature. It is evident that the liquid is superheated over the entire wick thickness, and that the maximum superheat occurs at the heat pipe wall.

The boiling limit q_b (heat flux at the outer wick surface at which boiling occurs) is obtained from the following equation:

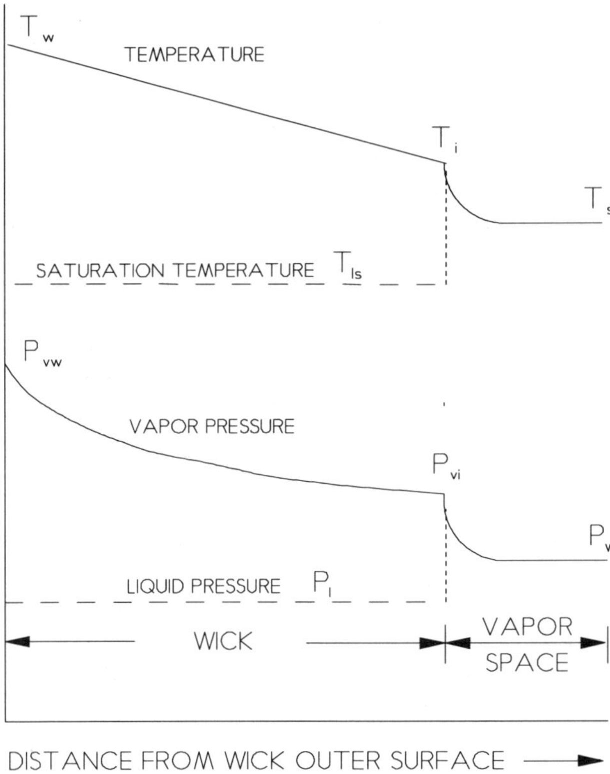

Figure 2.9. Pressure and temperature distributions across the thickness of a heat pipe wick.

$$q_b = C\Delta T_{wb}/A_{\text{wick}e} \tag{2.12}$$

where C is the thermal conductance of the wick in the evaporator, $A_{\text{wick}e}$ is the evaporator area at the outer wick surface, and ΔT_{wb} is the wick temperature drop corresponding to the superheat required for boiling.

The thermal conductance is defined as follows:

$$C = k_e A_{\text{wick}e}/t_{\text{wick}} \tag{2.12a}$$

where k_e is the effective thermal conductivity of the wick and t_{wick} is the wick thickness. In Eq. (2.12a), $A_{\text{wick}e}$ is interpreted as the mean wick surface area.

The boiling limit will be high if the superheat for boiling and the thermal conductance per unit area are high. High thermal conductance is attained with a large effective wick thermal conductivity and a small wick thickness.

The wick is a composite of two materials—the solid structure and the heat pipe liquid. Its effective thermal conductivity is a function of the wick geometry, the proportion of liquid and solid in the wick, and the ratio of their thermal conductivities.

If the wick geometry is cylindrical or spherical, the boiling limit at a given superheat for boiling will be smaller than that for rectilinear wick geometry. This is

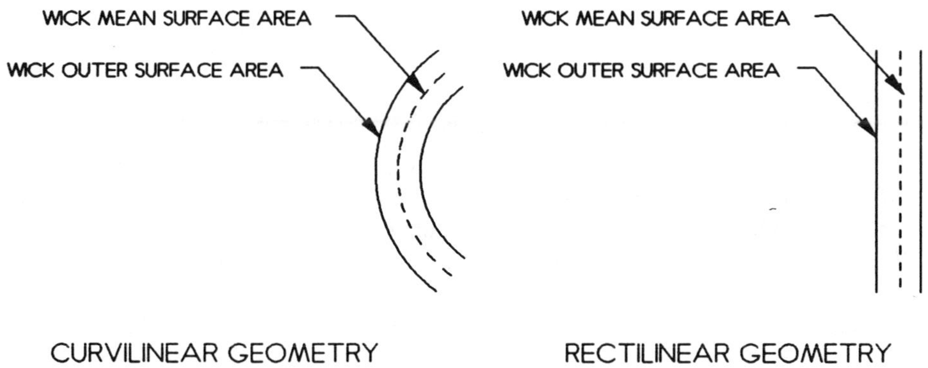

Figure 2.10. Effect of wick geometry on mean surface area.

because the mean wick area over its thickness is smaller than the wick outer surface area. Hence, the thermal conductance is smaller for a wick with a finite curvature. (See Figure 2.10.)

The boiling limit is quite large for liquid metal heat pipes. For all heat pipes, the boiling limit decreases with an increase in the heat pipe vapor temperature. Quantitative aspects of the boiling limit are treated in Chapter 5.

Nucleation Site Radius. The boiling limit is a strong function of the effective radius of the nucleation sites where vapor bubbles form. The larger the nucleation site radius, the smaller the heat flux at which boiling will be initiated.

The size of the nucleation sites depends on how effectively the heat pipe liquid can wet the wall surface and penetrate the microscopic cavities that exist in any surface. Wettability in turn is influenced by the presence of surface and liquid impurities, the chemical reactiveness of the liquid and the surface material, and the pressure-temperature history of the liquid (Ref. 2.11).

The alkaline liquid metals effectively wet most metal surfaces at temperatures above 900°F (482°C) (Ref. 2.12). The wetting action is believed to result from the reduction of metallic oxides on the surfaces, yielding a pure metal surface at the expense of the formation of liquid metal oxides.

As was the case for wick pore radii, nucleation site radii will also be expressed in microns in this book. An effective nucleation site radius of 0.81 micron has been found to correlate with incipient boiling experiments using sodium in a columbium–1% zirconium tube at saturation temperatures in the range of 1700 to 1950°F (927 to 1066°C). The measured distribution of radii for the actual cavities in the tube wall ranged from 0.13 to 2.54 microns. Similar results have been reported for incipient boiling of potassium in an as-received 316 stainless steel tube (Ref. 2.13).

In theory, nucleation site radii in heat pipes should be comparable to those encountered in all-liquid systems. However, the presence of the wick is a complicating factor that can add considerably to the uncertainty of predicting effective nucleation site radii.

If vapor or noncondensable gas should become trapped in the wick pores, the radius of the wick pores rather than the surface cavities could represent the effective nucleation site radius.

For example, the effective nucleation site radius for a potassium-nickel heat pipe was estimated from boiling limit data to lie in the range of 24 to 67 microns. The approximate wick pore radius, taken to be one-half the spacing between wires of the screen wick, was 37 microns. For a sodium–stainless steel heat pipe, an upper limit of 3 to 7 microns was estimated for the nucleation site radius (Ref. 2.14).

Since the effective nucleation site radius may be specific to a particular heat pipe design and fabrication technique, it can be difficult to predict with accuracy in the absence of experimental incipient boiling data for the heat pipe in question. For preliminary design purposes, a nucleation site radius in the range of 1 to 7 microns is recommended.

Interface Heat and Mass Transfer. Since there is a net transfer of molecules between the liquid and the vapor phase in the evaporator and condenser of a heat pipe, the heat pipe fluid is not in a true state of equilibrium. However, the departure from equilibrium is generally small. The vapor may then be considered as saturated at the liquid-vapor interface, with its temperature uniquely determined by the vapor pressure. As indicated in Figure 2.9, the heat pipe liquid is superheated and hotter than the vapor.

If there is a net transfer of molecules between phases, the phase from which molecules are being lost must be hotter. Thus, there is a temperature difference associated with mass transfer between phases. Since energy is also added or removed as molecules undergo a phase change, interphase mass transfer is always accompanied by heat transfer.

The heat transfer process across the interface between phases may be characterized by an interface heat transfer coefficient h_i, defined by the equation

$$q_i = h_i \Delta T_i \tag{2.13}$$

In Eq. (2.13), ΔT_i is the temperature difference across the interface and q_i is the interface heat flux.

For vapor pressures up to about 0.1 psia (690 N/m^2), interactions between evaporating and condensing molecules are negligible, and the interface heat transfer coefficient can be predicted from gas kinetics considerations (Ref. 2.15). At higher vapor pressures, such as are encountered in heat pipes, molecular interactions become increasingly frequent, and there is greater resistance to mass and heat transfer across the interface. The interface heat transfer coefficient may then be substantially smaller than values predicted from gas kinetics theory (Ref. 2.14).

For example, for sodium at a mean interface temperature of 1600°F (871°C), the predicted interface heat transfer coefficient based on negligible molecular interactions is 163.8 Btu/ft^2-sec-°F (3.35×10^{10} W/m^2-°K). Applicable experimental data, although quite imprecise, range from 2.05 to 16.38 Btu/ft^2-sec-°F (4.19×10^8 to 3.35×10^9 W/m^2-°K) (Refs. 2.14, 2.15).

If exposed to an incipient heat flux of 100 Btu/ft^2-sec (1.14×10^6 W/m^2), a sodium heat pipe at 1600°F (871°C) would then, using Eq. (2.13), experience a temperature drop across the liquid-vapor interface of between 6.1 and 48.8°F (3.4 and 27.1°C) based on experimental data. Using the coefficient obtained from gas kinetics theory, the predicted interface temperature drop is 0.6°F (0.3°C). At a heat flux of 1000 Btu/ft^2-sec (1.14×10^7 W/m^2), the interface temperature drop would be a factor of 10 higher.

For heat pipe design situations involving heat fluxes on the order of 20 Btu/ft^2-sec (2.3×10^5 W/m^2) or less, the interface temperature drop is generally small enough to be neglected.

Substantially higher heat fluxes may be encountered at the stagnation regions of hypersonic vehicles (vehicles traveling at speeds greater than Mach 3). If heat pipe cooling is contemplated for such regions, the temperature difference at the liquid-vapor interface of the heat pipe could be significant. However, its actual magnitude may be difficult to establish in the absence of test data because the interface heat transfer coefficient is not known very precisely.

Consequences of Boiling in the Wick. Boiling in a heat pipe wick can lead to an interruption in heat transport, and is generally to be avoided. However, the consequences of boiling will depend in no small measure on the specific design features of a particular heat pipe. The relationship between boiling effects and heat pipe design will now be discussed.

If the superheat for boiling is large, the onset of boiling can be explosive in nature, driving most of the liquid out of the wick. The resulting dryout of the wick will interrupt capillary pumping, causing the evaporator section to overheat and possibly leading to meltdown if the heat input is not terminated.

The superheat for boiling in liquid metal heat pipes is characteristically quite large (on the order of several hundred degrees). However, the heat flux required to achieve this level of superheat is generally so high that boiling in liquid metal heat pipes has rarely been observed.

If the wick should be replenished with liquid by capillary action prior to heat pipe failure, there could be a short period of cooling before the reinitiation of violent boiling. In such a situation, the heat pipe could be subjected to a "chugging" action, characterized by excessive temperature and pressure fluctuations.

If the boiling process should commence at low or moderate superheat, vapor bubble generation may proceed without displacement of the liquid from the wick. However, the presence of the bubbles will increase the resistance to liquid flow through the wick, raising the liquid pressure drop and lowering the capillary pumping limit. Under these circumstances, a heat pipe may continue to function at heat transport rates greater than the boiling limit, until capillary pumping is finally disrupted or until film boiling develops. In the latter case, a high-thermal-resistance vapor film will completely envelop the outer wick surface, resulting in immediate and severe overheating of the heat pipe wall.

If the heat source is located below or at the same elevation as the wick, the buoyant vapor bubbles produced during boiling can rise through the wick structure and into the vapor space. In the process, the liquid-vapor interface will be repeatedly distorted, broken, and reformed. At high boiling rates the interface may be disrupted to the point that surface tension can no longer sustain the vapor-liquid pressure differential. Capillary pumping will then cease and dryout of the wick will occur.

If the heat source is located above the wick, or in a gravity-free environment, buoyant forces can no longer be relied upon to carry the vapor bubbles into the vapor space. In this case, the bubbles will accumulate in the evaporator section. The bubbles will tend to coalesce, forming a large block of vapor that can impede the flow of liquid through the wick and reduce the capillary pumping limit. When the vapor block is large enough to bridge the wick completely, capillary pumping will be disrupted immediately. Alternatively, the vapor block may form a high-resistance vapor film that will immediately overheat the wall of the heat pipe evaporator section.

Vapor blockage is also a possibility in a two-layer wick consisting of an open or coarse-pored liquid flow layer next to the heat pipe wall and an adjacent fine-pored capillary pumping layer. Bubbles moving through the coarse-pored wick layer may not be able to penetrate the fine-pored layer.

For a cylindrical heat pipe in which heat is added over the entire circumference, vapor blockage at the upper section of the evaporator may lead to localized dryout and hot spots at heat fluxes below that at which widespread destructive dryout and overheating would be expected.

A special case arises when boiling is initiated by a heat source impinging on the evaporator surface whose lateral dimensions are comparable to or less than the wick thickness. If such a heat source is located above the wick, the vapor block formed will introduce a large localized thermal resistance, causing heat to flow around the vapor block along a relatively low-thermal-resistance path to the vapor space. If the temperature drop along the low-resistance path is less than the critical superheat, boiling at the wall-wick interface will cease. The heat pipe may then continue to function at a heat flux well beyond the normal boiling limit with no additional vapor generation.

From the preceding discussion, it is apparent that the consequences of boiling in a heat pipe depend on numerous design and performance factors, and are difficult to predict. In some instances, the boiling process itself may cause dryout of the evaporator. In other cases, the heat pipe may continue to function at heat transfer rates above the boiling limit, but with an adverse effect on the capillary pumping limit.

Since the limits of operation beyond the boiling limit are difficult to predict, a heat pipe should generally be designed to operate at a heat transport rate below the boiling limit. In situations in which operation of a heat pipe in the boiling mode may be feasible, test data will usually be needed to establish the permissible upper limit of the boiling heat flux.

Boiling in a Heat Pipe versus a Liquid Pool. A substantially higher heat flux may be required to initiate boiling in a heat pipe than in a static liquid pool, even though the superheat for boiling is comparable for the two cases.

For example, a sodium heat pipe at 1400°F (760°C) has operated at a surface heat flux (based on the inner wick diameter) of 900,000 Btu/ft^2-hr (2.84×10^6 W/m^2) without boiling (Ref. 2.16). In contrast, boiling was initiated at a heat flux of 300,000 Btu/ft^2-hr (9.47×10^6 W/m^2) in a pool of sodium at 1300°F (704°C). At higher temperatures, the heat flux for boiling in the sodium pool diminished sharply (Ref. 2.12).

The reason for the higher boiling limit in the heat pipe is that its wick is much thinner than the thermal boundary layer for natural convection in a pool. [The thermal boundary layer thickness in a pool of sodium is about 0.6 in. (0.0152 m) (Ref. 2.12)]. Consequently, the thermal conductance per unit area of the wick is considerably higher than that for the thermal boundary layer of a bottom-heated pool. Thus, at a specified superheat for boiling, the heat flux will be larger in the higher-conductance wick. [See Eq. (2.12).]

2.3.3. Entrainment Limit

In a heat pipe, the vapor and liquid flow past each other in opposite directions. The higher-velocity vapor exerts a drag force on the slower-moving liquid, which is opposed by the liquid surface tension. This situation is illustrated in Figure 2.11.

As the heat transport rate through the heat pipe increases, so do the fluid velocities and the drag force. If the drag becomes so large that it can no longer be balanced by surface tension, liquid droplets will be entrained in the vapor and carried to the condenser end of the heat pipe. Less liquid will then return to the evaporator than is being lost by evaporation, eventually leading to dryout of the evaporator region, a temperature excursion, and possible meltdown. Entrainment is initiated at the evaporator exit, where the fluid velocities are the highest.

The heat transport rate through the heat pipe at which entrainment occurs is called the entrainment limit. It is also expressed as an axial heat flux, the heat transport rate per unit of vapor space cross-sectional area.

The drag force on the heat pipe liquid is proportional to the liquid surface area in the wick pores, whereas the resisting surface tension force is proportional to the pore width normal to the drag force. Consequently, the ratio of the drag force to the surface tension force is proportional to the pore size, and decreases as the pore size diminishes. A higher vapor velocity is then needed to increase the drag force and cause entrainment. The entrainment limit is therefore an inverse function of the wick pore size.

If the wick is partially filled with liquid, the liquid-vapor interface will lie within the wick structure. The vapor in contact with the liquid should then be relatively stagnant, and the viscous shear stress on the liquid surface will be quite small. The entrainment limit would then be expected to be considerably larger than when the liquid-vapor interface lies at the edge of the wick structure.

In general, this approach to raising the entrainment limit would only be considered

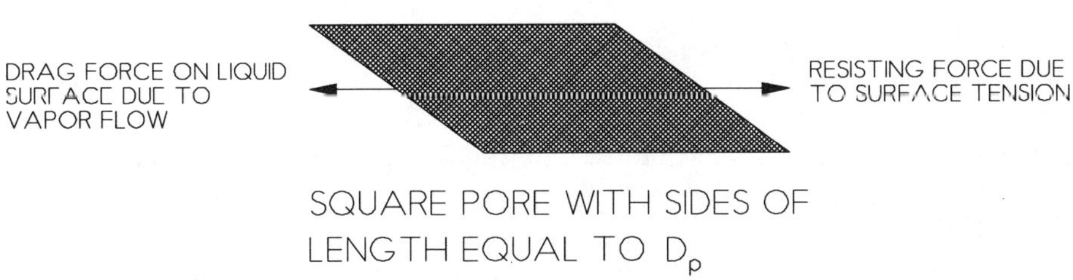

Figure 2.11. Force balance on the liquid-vapor interface in wick pores.

for a composite or two-layer wick, in which the flow channel layer remains full and liquid recession is confined to the capillary pumping layer. In single-layer wick structures consisting of open grooves cut into the heat pipe wall, even a slight degree of underfilling can severely reduce heat transport capability (Refs. 2.17, 2.18).

2.3.4. Sonic Limit

In a heat pipe, new vapor is continuously being added to the vapor space along the evaporator length. The vapor velocity then increases with evaporator length, reaching a maximum at the evaporator exit. As the heat transport rate, and hence the vapor generation rate, increases, the exit velocity becomes higher.

When the exit velocity reaches the sonic value, the vapor flow is said to be choked. At that point, no further increase in vapor flow or heat transport is possible without an increase in the vapor temperature. The heat transport rate at which the vapor velocity becomes sonic is called the sonic limit. Like the entrainment limit, the sonic limit may also be expressed as an axial heat flux (heat transport rate per unit of vapor space cross-sectional area).

The sonic limit is a rapidly increasing function of temperature. If the heat transport rate should exceed the sonic limit at some heat pipe temperature, the temperature will rise to the point at which the sonic limit is once more equal to the heat transport rate. As long as the new temperature is within the upper temperature limit of the heat pipe structural material, the heat pipe will continue to function normally.

In this respect, the sonic limit differs from the other heat transport limits. If the capillary pumping, boiling, or entrainment limit should be exceeded, disruption of heat pipe operation, dryout of the evaporator region, and a temperature excursion may be expected. On the other hand, if the sonic limit should be exceeded, a modest temperature rise to a new operating plateau will ensue. Heat pipe operation will not be disrupted unless the new operating temperature is beyond the acceptable temperature limits of the heat pipe structural material.

When a heat pipe is operating at the sonic limit, there is a substantial increase in the momentum of the vapor over the evaporator length. This momentum increase is accompanied by a vapor pressure drop, which may be equal to 50% or more of the pressure at the beginning of the evaporator. Since the vapor is saturated, there is an associated drop in temperature. For liquid metal heat pipes, this temperature drop can be in excess of 100°F (56°C).

For this reason, a heat pipe is normally designed to operate well below the sonic limit in order to maintain reasonably isothermal heat transport. The permissible axial variation in heat pipe temperature will depend on the specific design situation. Under normal design conditions the axial heat flux should not exceed 50% of the sonic limit, and in most cases will be less than 20%.

Although the axial heat flux is usually well below the sonic limit during normal operation of a heat pipe, the sonic limit will be encountered during startup if the initial vapor pressure is low enough. This situation can always be expected during the startup of alkali liquid metal heat pipes, since at room temperature the liquid metals are frozen and their vapor pressures are negligible. The startup process is discussed in Section 2.4.

Temperature Distribution near Sonic Limit. The nonisothermal temperature distribution that develops at or near the sonic limit is illustrated in Figure 2.12 for a sodium heat pipe (Ref. 2.7). Heat was added to the evaporator at a constant rate of 6400 W, and rejected from the condenser by conduction through a gas gap to a surrounding water calorimeter. The thermal resistance of the gap was controlled by varying the proportion of helium in a helium-argon mixture in the gap.

For curves B, C, and D of Figure 2.12, the maximum vapor temperature of 695°C and the evaporator temperature distribution remain unchanged, despite the lower vapor temperatures and associated pressures that develop in the condenser as the thermal resistance of the gas gap is lowered.

This behavior is characteristic of heat pipe operation at the sonic limit. Then, pressure changes in the condenser, which travel at sonic velocity, cannot be propagated back into the evaporator. It may be concluded, therefore, that the sonic limit for a maximum vapor temperature of 695°C is equal to the heat input rate of 6400 W.

Curve B represents the incipient sonic limit condition. The sonic vapor flow leaving the evaporator immediately starts to condense and decelerate upon entering the condenser, with resultant temperature and pressure recovery as the vapor momentum is lost.

For curves C and D, the evaporator exit vapor flow continues to accelerate in the condenser, becoming supersonic during condensation while the temperature and pressure decrease. Temperature and pressure recovery then occurs via a shocklike process

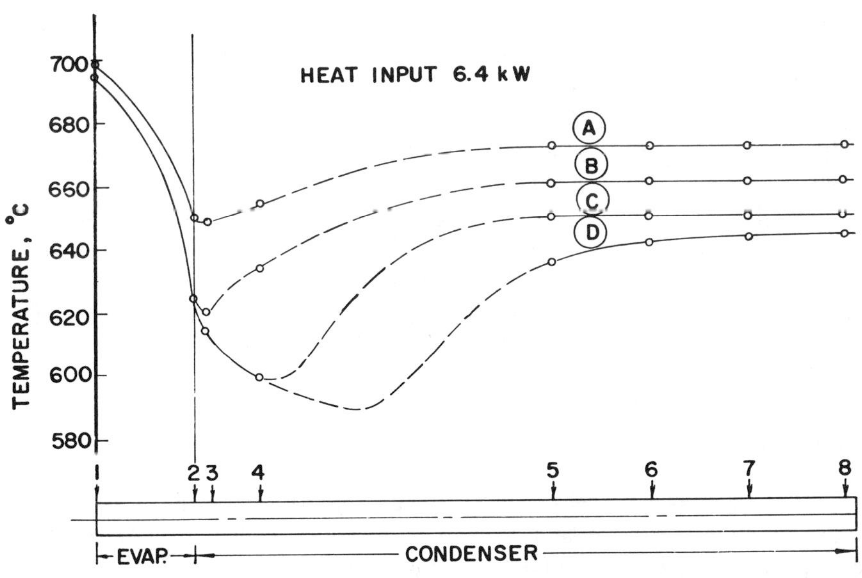

Figure 2.12. Nonisothermal temperature distribution in a sodium heat pipe at or near the sonic limit (Ref. 2.7).

that initially decelerates the vapor flow to the subsonic state, followed by further subsonic deceleration.

At 699°C, the maximum vapor temperature for curve A, the sonic limit for the heat pipe of Figure 2.12 is about 6700 W. Since the actual heat transport rate is 6400 W, curve A represents the heat pipe temperature distribution at about 95% of the sonic limit. The evaporator temperature drop is still substantial, but smaller than that corresponding to operation at the sonic limit (curves B, C, and D). As the subsonic vapor leaving the evaporator condenses, the vapor is decelerated, momentum is lost, and temperature and pressure recovery occurs.

At a maximum vapor temperature of 800°C, the sonic limit for the heat pipe of Figure 2.12 increases to 19,800 W. When the heat pipe was operated at 6400 W at temperatures close to 800°C, the axial temperature distribution in the vapor was almost isothermal. The heat transport rate was then about 30–32% of the sonic limit. In general, relatively isothermal operation can be expected when the heat transport rate through a heat pipe does not exceed 20–30% of the sonic limit.

2.4. HEAT PIPE STARTUP

In our discussion of heat pipe operational characteristics to this point, steady-state heat pipe operation has been assumed, with heat being added to and removed from the heat pipe at a constant rate. Under usual design conditions the heat pipe is relatively isothermal, and its temperature is determined from the requirement that the heat input and output rates are equal.

If this heat balance should be altered, as for example by an increase in the heat input rate or a change in thermal resistance between the heat pipe condenser and an adjacent heat sink, the operating temperature will change to a level at which the balance between heat input and output rates is restored.

We will now consider the process by which the heat pipe passes from a nonoperational condition at ambient temperature to the fully operational state at some higher temperature. By *fully operational state,* we mean that the heat pipe is relatively isothermal and that heat is being dissipated over the entire length of the condenser. If the heat input and output rates are then equal, the heat pipe will be functioning in the steady-state condition. If there is an imbalance between the heat input and output rates, the temperature of the fully operational heat pipe will continue to change with time.

2.4.1. The Startup Process (Refs. 2.7, 2.19)

The startup behavior of a heat pipe is significantly influenced by the vapor pressure and physical state of the heat pipe liquid at ambient temperature, as well as by the thermal resistance between the condenser and the adjacent heat sink.

Prior to startup, the temperature of a heat pipe is equal to the ambient temperature, and its internal pressure is equal to the vapor pressure of the heat pipe liquid at ambient temperature. Also, depending on its freezing point, the heat pipe liquid may be in the liquid or the solid state.

For example, at an ambient temperature of 70°F (21°C), the internal pressure in a water heat pipe is 0.36 psia (2480 N/m^2), and the water in the wick is in liquid form. The situation is quite different for a sodium heat pipe. Since the ambient temperature is well below the sodium melting point of 208°F (98°C), the wick structure is filled with frozen sodium, and the vapor pressure in the heat pipe is negligible.

Evaporator Conditions during Startup. The startup process is initiated by adding heat to the evaporator section. Heat may be added in several different ways, depending on the specific circumstances associated with startup. Heat may be added at a constant rate equal to the steady-state design value. The heat input rate may vary continuously with time in accordance with the requirements of a particular application. Or, the heat input rate may be increased in discrete steps to a final steady-state value, with the heat pipe allowed to reach steady-state operation between steps.

The general variation of the temperature distribution in a heat pipe with time during startup is illustrated in Figure 2.13. Here, startup is from a very low vapor pressure

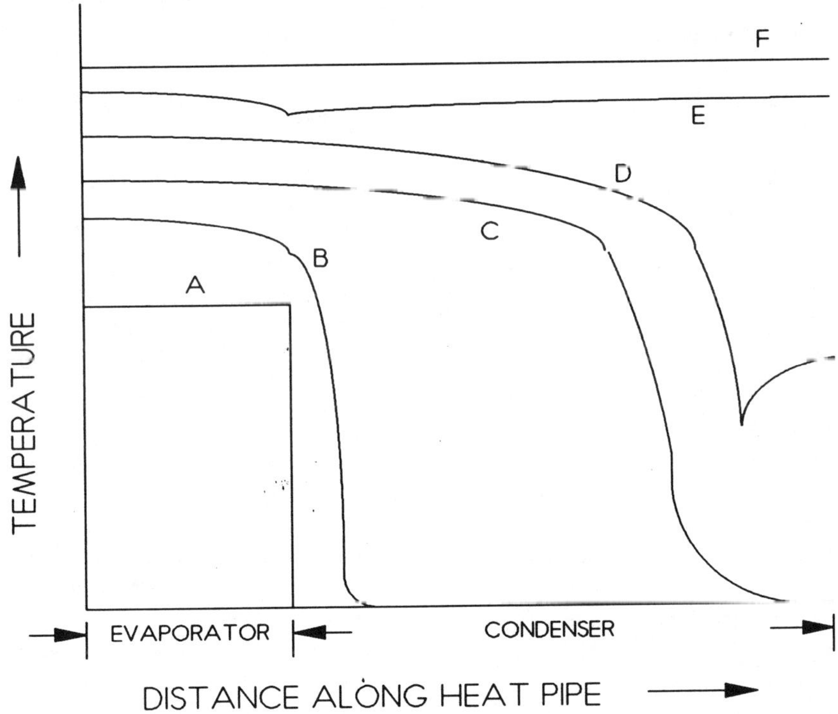

Figure 2.13. Temperature distribution in a liquid metal heat pipe at various times after startup. Time increases successively from A (shortest) to F (longest).

(typical of liquid metal heat pipes), and the heat input rate is constant. This figure is based on experimental data from Ref. 2.19, and will be referred to frequently in the course of describing the startup process.

Upon the addition of heat, liquid in the wick starts to evaporate at the liquid-vapor interface. If the liquid is frozen, vapor will be produced by sublimation. As vapor accumulates in the vapor space, its pressure, temperature, and density will increase above the values of comparable vapor properties in the condenser. Vapor will then flow out of the evaporator into the condenser at a rate that depends on the pressure differential along the evaporator and the vapor density.

The maximum rate at which vapor can leave the evaporator is achieved when the vapor velocity at the evaporator exit becomes sonic. The vapor flow is then said to be choked. The product of the maximum, or choked, vapor mass flow rate and the heat of vaporization represents the maximum rate at which heat can be transported out of the evaporator. This maximum heat transport rate is equal to the sonic limit.

If the vapor pressure is so low that the mean free path of the vapor molecules is comparable to or greater than the minimum vapor space dimension, molecular flow conditions will prevail. Vapor and heat transport out of the evaporator are then negligible, and the heat input is absorbed by an increase in the evaporator temperature. This is the situation that exists at time A in Figure 2.13. Eventually, the vapor in the evaporator will reach a temperature and pressure at which the mean free path is small compared to the minimum vapor space dimension. At that point, continuum flow will have been established.

For alkali liquid metal heat pipes, continuum flow may not be established until the evaporator has reached a temperature on the order of 1000°F (540°C). For water heat pipes, on the other hand, continuum flow conditions generally exist at room temperature.

Once continuum flow conditions exist, as at time B, the heat transport rate out of the evaporator will increase in proportion to the pressure differential across the evaporator. At some critical ratio of the vapor pressure at the evaporator exit to that at the start of the evaporator, the vapor flow will be choked and the sonic limit will have been reached. For primarily monatomic vapors such as the liquid metals, this critical pressure ratio is equal to 0.375. Since the pressure in the unheated portion of the condenser is generally negligible for liquid metal heat pipes, heat flow out of the evaporator is always at the sonic limit during startup.

As long as the heat input rate to the evaporator is greater than the rate at which heat can be transported out of the evaporator by the vapor, the evaporator temperature will continue to rise.

Vapor expands to supersonic velocity in the condenser before condensing. When the heated zone has reached the end of the condenser, as at time C, the temperature and pressure at the end of the condenser start to rise. There is further temperature and pressure recovery at time D as the supersonic vapor is decelerated to subsonic velocity through shocklike processes. The heat transport rate will drop below the sonic limit when the condenser pressure has risen sufficiently, as at time E. Finally, the heat transport rate approaches the heat input rate when the evaporator temperature stabilizes at time F.

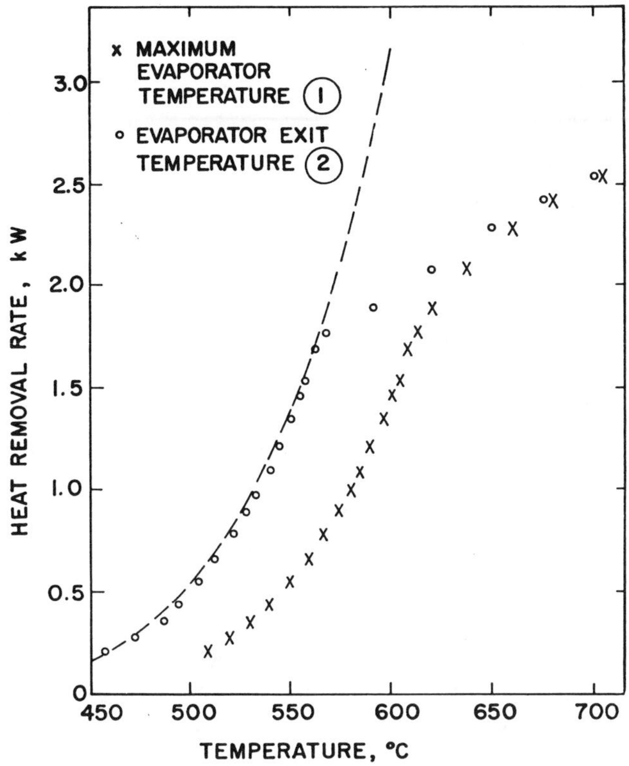

Figure 2.14. Temperature drop along the evaporator section of a sodium heat pipe at various heat transport rates (Ref. 2.7).

Since the vapor is at least nominally saturated, the pressure drop along the evaporator will be matched by an associated temperature drop. At the sonic limit, the evaporator temperature drop may be 100°F (56°C) or higher for liquid metal heat pipes.

The temperature drop along the evaporator section of the sodium heat pipe for which the data of Figure 2.12 were obtained is shown as a function of heat transport rate in Figure 2.14 (Ref. 2.7). The heat sink in this case was a water calorimeter that surrounded the condenser. Heat was rejected by conduction through an argon gas gap between the condenser and the heat sink. The heat transport rate was increased in small increments, with temperatures allowed to reach the steady state before adding the next increment.

The evaporator temperature drop remains at about 50°C until the heat transport rate has exceeded about 1500 W and the maximum vapor temperature has exceeded about 600°C. At that point the hot vapor continuum region has arrived at the end of the condenser. Further increases in the heat transport rate lead to larger increases in the maximum vapor temperature and a decrease in the evaporator temperature drop. The startup process is complete at 2500 W and 690°C, when the evaporator has become practically isothermal.

Condensation Front. When the hot vapor that leaves the evaporator enters the unheated condenser section, the vapor will cool and accelerate to supersonic velocity

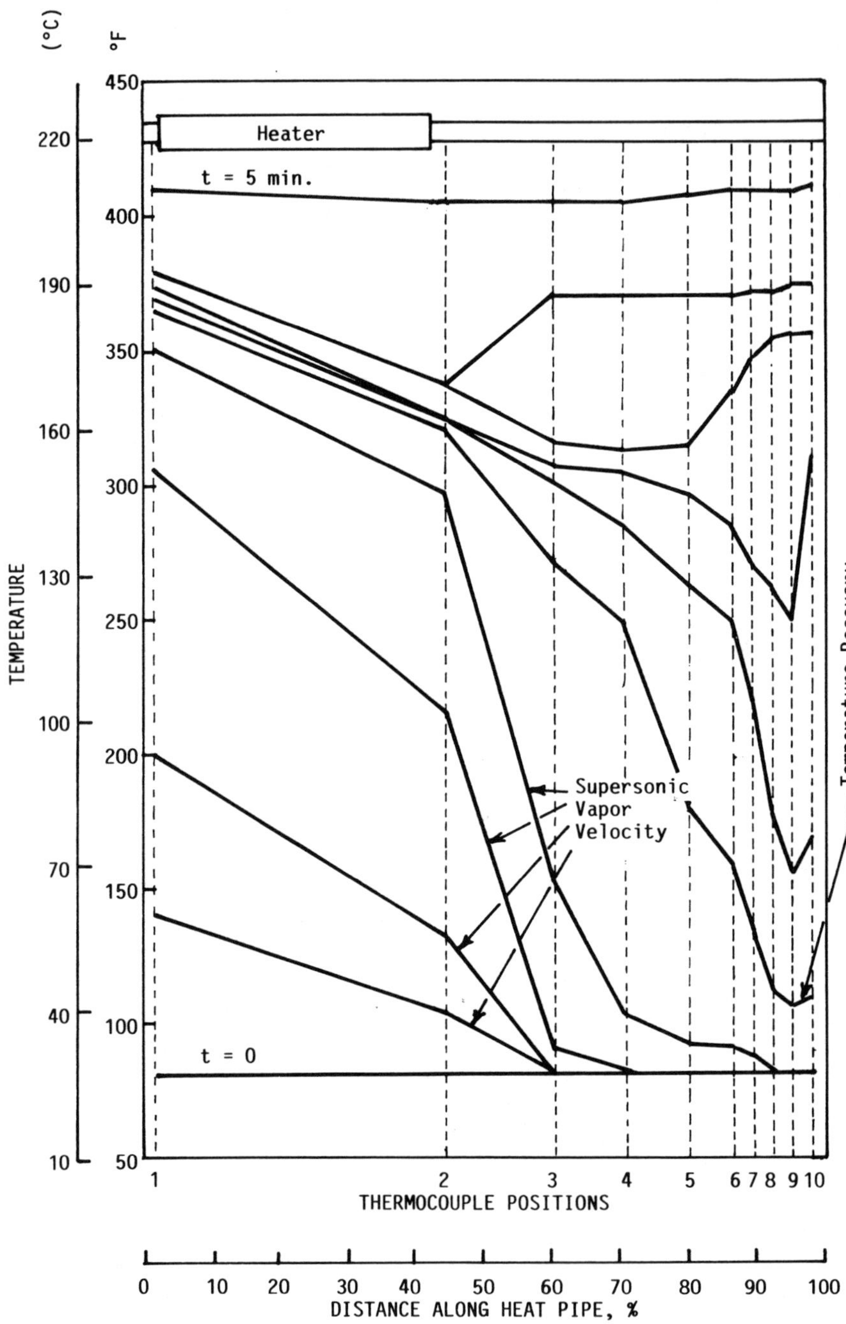

Figure 2.15. Temperature distribution in a mercury heat pipe during startup at a constant heat input rate (Refs. 2.10, 2.19).

while condensing on the adjacent wick. Since only the vapor at the outer boundary of the vapor space is in contact with the cooler wick, all of the vapor cannot condense simultaneously. A condensing zone of finite length develops, with the condensing region closest to the hot incident vapor being heated to the vapor temperature and succeeding regions being heated to a lesser extent. Thus, a temperature gradient forms along the condensing zone, with the temperature dropping to that of the unheated condenser when all the vapor has condensed. (See Figure 2.13.)

As vapor continues to enter the condenser, relatively little condenses on the already heated portion of the condenser. The bulk of the vapor condenses along the condensing zone, causing the zone to advance further into the condenser. Thus, the condensing zone acts like a wave or front, which moves through the condenser, condensing vapor and raising the temperature of the condenser as it passes by. Behind the front, condensation has mostly been completed, and the temperature is close to that of the entering vapor. Ahead of the front, condensation has not yet started, and there has been no heating of the condenser.

The length of the condensation front in a heat pipe depends on how rapidly vapor can move to the condensing surface of the wick in comparison to the axial vapor velocity. The condensation front will be referred to here as a *hot continuum front,* since it separates the hot continuum zone of the heat pipe from the unheated zone, where the vapor pressure may be low enough to support free molecular flow.

If the vapor were to condense simultaneously upon contacting an unheated section of the condenser, the condensation front would be vertical, with zero length. The vertical hot continuum front is a convenient concept that can be used in place of the actual condensation front to simplify analysis of the startup process.

When the condensation front reaches the end of the condenser, there is no longer any unheated section. As the front continues to advance, the temperature and pressure at the end of the condenser start to increase, and the forward end of the front may turn upward. These changes are a consequence of the fact that the supersonically flowing vapor must now be decelerated via a normal shock to the subsonic state in order to condense at the higher pressures at the end of the condenser.

The startup process is complete when the rear of the hot continuum front has reached the end of the condenser, and the temperature distribution no longer varies with time.

The time-dependent temperature distribution over the length of a mercury heat pipe that was subjected to a constant heat input rate of 300 W is shown in Figure 2.15. Steady state, isothermal operation was attained after 5 minutes.

The heat pipe was fabricated from 347 stainless steel. Its length was 19⅝ in. (0.499 m), and its outer diameter was 0.600 in. (0.0152 m). The wick consisted of three layers of 100 mesh, 304 stainless steel screen. Magnesium and titanium were added to the mercury to promote wetting of the stainless steel structure and wick. Heat rejection was to air by radiation and natural convection.

2.4.2. Failure Mechanisms during Startup (Ref. 2.19)

The heat transport process during startup can be disrupted under some circumstances, thereby preventing the heat pipe from achieving fully operational status. The likelihood of a startup failure is enhanced when the condenser is closely coupled to the heat sink, i.e., when the thermal resistance between the condenser and the heat sink is relatively small. Then, since the heat transport rate into the condenser is limited during the initial phase of startup, the condenser temperature, vapor pressure, and vapor density will tend to remain low. The possibility of a startup failure is enhanced under these conditions, as will be explained.

Failure of a heat pipe to start up may be the result of vapor freezeout, entrainment of the heat pipe liquid, or inadequate capillary pumping. Disruption of heat pipe startup due to boiling is not likely, because the boiling limit is quite high when the heat pipe temperature and vapor pressure are relatively low.

During startup, heat is transported into the condenser at or below the sonic limit. If the sonic limit at startup is well below the design heat transport rate, the temperature of a closely coupled condenser may not rise much above that of the heat sink. Then, if the heat pipe liquid was originally frozen, melting and vaporization may occur in the evaporator while condensing vapor continues to freeze out on the wick in the condenser. Since the liquid in the evaporator cannot be replenished as long as the condensate remains frozen, the evaporator wick may eventually dry out and overheat.

Heat pipe failure due to freezeout of vapor in the condenser is much less likely if the thermal coupling between the condenser and the heat sink is relatively poor during the initial phase of startup. This situation is encountered when heat rejection is by thermal radiation and/or natural convection, or when the condenser is fully insulated against heat loss. Then, heat rejection from the portion of the wick on which condensation is occurring is generally negligible over the lower range of startup temperatures, permitting the wick temperature to rise above the melting point of the heat pipe liquid. The liquid is then free to replenish liquid that vaporizes in the evaporator.

Since the vapor usually reaches sonic velocity during startup, the drag force at the liquid-vapor interface may be relatively high even though the vapor density is low. If entrainment is to be avoided, the entrainment limit should be greater than the sonic limit as long as the actual heat transport rate is equal to the sonic limit. Otherwise, the heat pipe liquid will be entrained by its vapor, reducing the rate of liquid return to the evaporator and leading to eventual dryout and overheating.

It has been shown that, if a small amount of noncondensable gas is added to the vapor space, an entrainment failure that would otherwise occur during startup can be avoided. At the low vapor pressure in a heat pipe prior to startup, the noncondensable gas occupies most of the vapor space in the condenser, limiting the effective condenser heat rejection area and raising the effective thermal resistance between the condenser and the heat sink. At the design operating temperature, the vapor pressure is high enough to confine the noncondensable gas to a negligibly small volume at the end of the condenser.

Failure during startup can also result from inadequate capillary pumping, again with the likelihood of failure enhanced if the heat sink is closely coupled. In that case, the temperature and vapor pressure will remain low in the condenser while rising in the evaporator due to continued heat addition. If the sum of the resulting relatively large vapor pressure drop and the pressure drop in the heat pipe liquid should exceed the maximum available capillary pressure in the wick pores, the liquid-vapor interface will be disrupted, leading to dryout and overheating of the evaporator.

2.4.3. Comparison of Heat Pipe Vapor Flow with Nozzle Flow (Ref. 2.19)

The flow of vapor through a heat pipe during startup bears a strong resemblance to the flow of a gas through a converging-diverging nozzle. We will now examine the flow process in a nozzle, and indicate the similarity of this process to the flow of vapor in a heat pipe during startup.

Nozzle Flow. In a nozzle, gas flowing at a constant rate is accelerated from an initially low velocity to some specified final velocity. The flow acceleration is accompanied by decreases in gas pressure, density, and temperature.

As the pressure at the nozzle exit decreases, the required exit flow area becomes smaller while the exit gas velocity increases. There is a critical ratio of exit pressure to inlet pressure at which the area reaches a minimum value and the velocity becomes sonic. A nozzle for which the exit-inlet pressure ratio is equal to or greater than the critical pressure ratio is called a converging nozzle.

For values of the exit-inlet pressure ratio that lie below the critical pressure ratio, the exit velocity becomes supersonic. The required nozzle area decreases to a minimum value at the point where sonic velocity is reached, and then increases continuously to the nozzle exit. The location where the flow area is a minimum and the velocity is sonic is called the nozzle throat. This type of nozzle is called a converging-diverging nozzle.

In a converging-diverging nozzle with specified inlet conditions, the gas mass flow rate will be constant, despite variations in the gas pressure just outside the nozzle exit, as long as the pressure ratio at the throat is equal to the critical value. This condition persists over a wide range of exit pressures, and is characterized by supersonic flow in the diverging portion of the nozzle. A similar situation exists in a heat pipe when operating at the sonic limit.

The effect of external gas pressure on the pressure distribution in a converging-diverging nozzle is indicated in Figure 2.16.

For exit pressures above the nozzle design value (of P_d), the supersonic flow is decelerated to subsonic flow via a normal shock as pressure recovery to the exit pressure takes place (curve C). As the exit pressure is increased, the shock weakens and moves toward the throat. Finally, at some exit pressure the shock has disappeared, and pressure recovery occurs via a smooth deceleration of the gas from sonic velocity at the throat (curve B).

If the exit pressure is increased further (curve A), the gas leaving the throat is decelerated in a similar manner. However, now the throat velocity is less than sonic, and the throat pressure ratio is greater than the critical value. The mass flow rate is then lower than when the critical pressure ratio exists at the throat.

For external pressures equal to or lower than the design exit pressure (curve D), the gas flow is supersonic throughout the nozzle. It may accelerate further outside the nozzle to a lower external pressure via expansion waves.

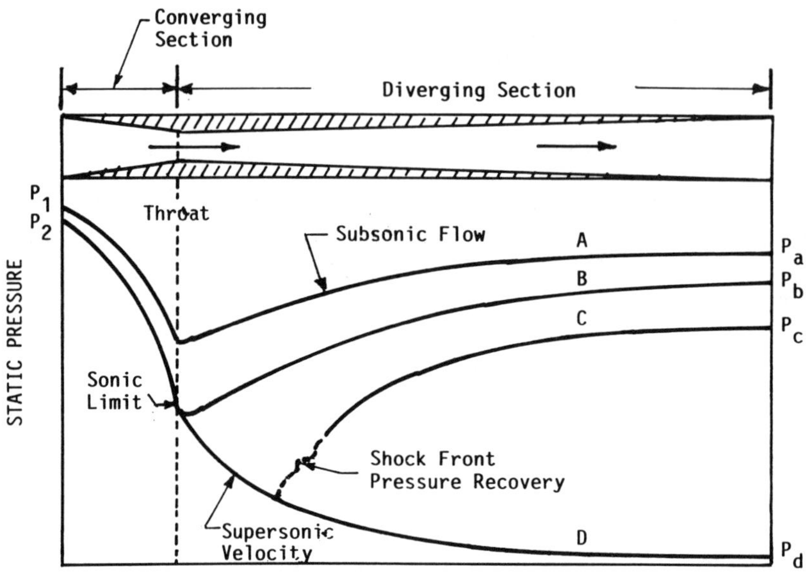

Figure 2.16. Effect of external gas pressure on pressure distribution in a converging-diverging nozzle (Ref. 2.19).

Vapor Flow in Heat Pipe. In a heat pipe, the evaporator is analogous to the converging section of a converging-diverging nozzle, and the heated condenser length is analogous to the diverging section. (See Figure 2.17.)

During steady-state operation, the vapor mass flow rate varies along the length of the constant-area vapor space. It reaches a maximum value at the evaporator exit and decreases to zero at the end of the condenser. In a nozzle, the mass flow rate remains constant while the flow area varies along the nozzle length. It reaches a minimum value at the nozzle throat, and then increases continuously to a maximum value at the nozzle exit. The mass flow rate per unit area is a maximum at both the evaporator exit of a heat pipe and the throat of a converging-diverging nozzle.

Despite these similarities, the nozzle and heat pipe differ in two important respects. First, the critical pressure ratio at which the flow is accelerated to sonic velocity is not the same for the two devices. Second, whereas the velocity at the nozzle exit is always finite, the velocity at the end of the active condenser length in a closed heat pipe is zero (i.e., all of the vapor is condensed).

The heat pipe startup process has been illustrated schematically in Figure 2.13, where the temperature distribution is presented as a function of time. The pressure distribution in the saturated vapor is qualitatively similar to the temperature distribution, but varies by a larger factor. It will now be shown that the behavior of the heat pipe vapor closely resembles that of the gas in a converging-diverging nozzle of varying length.

In the early phase of startup, when the vapor pressure in the unheated portion of the

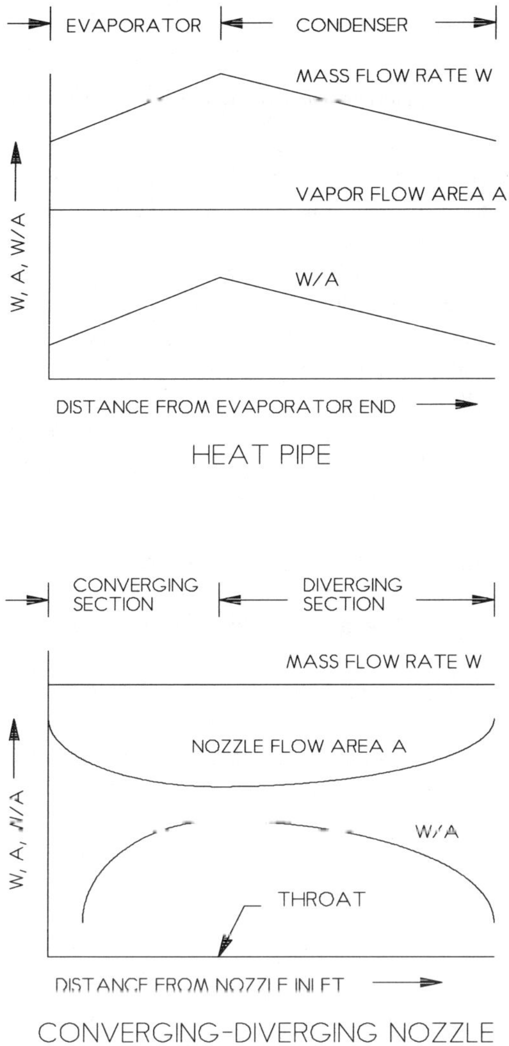

Figure 2.17. Comparison between mass flow rate and flow area in a heat pipe and a converging-diverging nozzle.

condenser is very low, hot vapor entering the condenser at the sonic limit will expand into the lower-pressure region and then rapidly condense. As the hot continuum front moves through the condenser, the supersonic flow behind the front will continue to expand into and condense in the low-pressure region ahead of the front, with essentially no pressure recovery.

This situation is analogous to expansion of supersonic vapor from the nozzle exit into a lower-pressure gas (curve D of Figure 2.16). In effect, the length of condenser behind the hot continuum front is equivalent to the diverging section of a nozzle whose length is continuously increasing with time. Since some condensation may also be occurring behind the front, the nozzle analogy may be more accurately portrayed as a diverging section in which there is some heat transfer out of the section.

The same type of behavior can occur under steady-state conditions if a heat pipe is operating at the sonic limit when the sonic limit is relatively low. Such was the case for the sodium heat pipe used to obtain the startup data of Figure 2.14, where the temperatures shown at a given heat transport rate are steady-state values. The heat transport rate was equal to the sonic limit for heat inputs up to 2500 W, at which level the hot continuum front finally reached the end of the condenser. For heat transport rates below 2500 W, the front was "stalled" somewhere in the condenser interior.

Once the hot continuum front has reached the end of the condenser, there is no longer any unheated condenser section. The temperature and pressure of the vapor at the end of the condenser will then rise, eventually becoming higher than at the evaporator exit. At that point the condenser section will start acting like the diverging section of a nozzle of fixed length, whose exit pressure is increasing with time. The supersonically flowing vapor will be decelerated to the subsonic state by a normal shock (as with curve C in Figure 2.16), and then to zero velocity as the vapor continues to be condensed and pressure recovery takes place.

This type of behavior can also occur under steady-state conditions if the heat pipe is operating at a sonic limit that is large enough to heat the entire heat pipe. See curves C and D of Figure 2.12 (where the saturation temperature is plotted rather than vapor pressure).

With further increases in condenser pressure and temperature during startup, the evaporator exit flow will become subsonic (as with curve A of Figure 2.16), and the actual heat transport rate will dip below the sonic limit. Eventually, the design operating temperature and pressure will be reached. In a properly designed heat pipe the actual heat transport rate will then be a relatively small fraction of the sonic limit, and the heat pipe will be virtually isothermal.

2.5. REFERENCES

2.1. "Surface Tension," *McGraw-Hill Encyclopedia of Science and Technology,* 6th Ed., Vol. 18, McGraw-Hill Book Co., New York, 1987, pp. 22–23.

2.2. "Interface of Phases," *McGraw-Hill Encyclopedia of Science and Technology,* 6th Ed., Vol. 9, McGraw-Hill Book Co., New York, 1987, pp. 278–279.

2.3. Davies, J. T., and Rideal, E. K., *Interface Phenomena,* Academic Press, New York, 1961, pp. 34–39.

2.4. Kunz, H. R., Langston, L. S., Hilton, B. H., Wyde, S. S., and Nashick, G. H., "Vapor Chamber Fin Studies: Transport Properties and Boiling Characteristics of Wicks," CR-812, NASA, Washington, D.C., June 1967.

2.5. "Structure and Function of Conductive Tissues: Plant Internal Transport," *The New Encyclopedia Brittanica,* 15th Ed., Vol. 28, 1989, pp. 690–693.

2.6. Hayward, A. T. J., "New Law for Liquids: Don't Snap, Stretch!" *New Scientist,* January 29, 1970, pp. 196–199.

2.7. Kemme, J. E., "Ultimate Heat Pipe Performance," *IEEE Transactions on Electron Devices,* Vol. ED-16, No. 8, August 1969, pp. 717–723.

2.8. Busse, C. A., and Kemme, J. E., "The Dry-Out Limits of Gravity-Assist Heat Pipes with Capillary Flow." Presented at the Third International Heat Pipe Conference, Palo Alto, California, May 1978.

2.9. Busse, C. A., "Theory of the Ultimate Heat Transfer Limit of Cylindrical Heat Pipes," *International Journal of Heat and Mass Transfer,* Vol. 16, 1973, pp. 169–186.

2.10. Deverall, J. E., "Mercury as a Heat-Pipe Fluid," ASME Paper 70-HT/SpT-8. Presented at the Space Technology and Heat Transfer Conference, Los Angeles, California, June 1970.

2.11. Chen, J. C., "Incipient Boiling Superheats in Liquid Metals," *Journal of Heat Transfer,* August 1968, pp. 303–312.

2.12. Kudryavtsev, A. P., Ovechkin, D. M., Sorokin, D. N., Subbotin, V. I., and Tsyganok, A. A., "Transfer of Heat from Sodium Boiling in a Large Vessel," *Liquid Metals,* TT F-522, NASA, Washington, D.C., May 1969, p. 265.

2.13. Lewis, J. P., Groesbeck, D. E., and Christenson, H. H., "Tests of Sodium Boiling in a Single Tube-in-Shell Heat Exchanger Over the Range 1720°F to 1980°F," TN D-5323, NASA, Washington, D.C., July 1969.

2.14. Silverstein, C. C., "Surface Heat Flux for Incipient Boiling in Liquid Metal Heat Pipes," *Nuclear Technology,* Vol. 12, September 1971, pp. 56–62.

2.15. Barry, R. E., and Springer, G. S., "Vapor Phase Resistance in Filmwise Condensation," Paper 69-WA/HT-26. Presented at the American Society of Mechanical Engineers Winter Annual Meeting, Los Angeles, California, November 1969.

2.16. Kemme, J. E., "High Performance Heat Pipes." Presented at the 1967 Thermionic Conversion Specialist Conference, Palo Alto, California, October 1967.

2.17. Kroliczek, E. J., and Brennan, P. J., "Axial Grooved Heat Pipes—Cryogenic Through Ambient," ASME Paper 73-ENAc-48. Presented at the Intersociety Conference on Environmental Systems, San Diego, California, July 1973.

2.18. Alario, J., Brown, R., and Kosson, R., "Monogroove Heat Pipe Development for the Space Constructible Radiator System," AIAA-83–1431. Presented at the AIAA 18th Thermophysics Conference, Montreal, Canada, June 1983.

2.19. Deverall, J. E., Kemme, J. E., and Florschuetz, L. W., "Sonic Limitations and Startup Problems of Heat Pipes," LA-4518, Los Alamos Scientific Laboratory, Los Alamos, New Mexico, November 1970.

CHAPTER

THREE

HEAT PIPE TECHNOLOGY

Various technologies associated with the design, fabrication, and operation of heat pipes are reviewed in this chapter. These include heat transport fluids, containment and wick materials, wick types and characteristics, heat pipe fabrication, environmental influences on heat pipe operation, and heat pipe systems. Methods for predicting the design and performance characteristics of heat pipes and heat pipe systems form the subject matter for the remainder of the book in Chapters 4 through 8.

3.1. HEAT TRANSPORT FLUIDS

Almost any fluid may be used in a conventional heat pipe, subject to its ability to wet the wick structure and the containment wall. Wettability of the wick structure is essential to the generation of capillary pressure in the wick pores of a conventional heat pipe. (See Section 2.1.2.) Effective wetting of the wall is desirable to minimize the size of nucleation sites and hence maximize the superheat required for the initiation of boiling. (See Section 2.3.2.)

Virtually all liquids meet the wetting requirement, with the metallic materials usually specified for the heat pipe wick and containment structures. In some instances, wetting can be enhanced by oxidation of the metallic surfaces. Such is the case for water on nickel or copper surfaces (Refs. 3.1, 3.2).

For an inverted heat pipe, the heat pipe liquid must be nonwetting to the wick structure in order to have a stable liquid-vapor interface at the wick pores where capillary pressure is generated. (See Section 2.2.3.) However, wetting of the internal wick structure and the containment wall is desirable to avoid premature boiling in the wick, just as is the case for conventional heat pipes.

Generally, a heat transfer application that calls for the use of heat pipes requires that the heat pipe operate at some particular temperature, or over a range of temperatures. Unless specified otherwise, the term *heat pipe temperature* is used in this book to denote the vapor temperature at the beginning of the evaporator section.

3.1.1. Operating Temperature Range

For a given heat pipe fluid and heat pipe design, there is a range of operating temperatures over which the heat pipe will be functional. This operating temperature range is determined by the heat transport limits of the heat pipe at a particular heat transport rate. (See Figure 3.1.)

At low heat transport rates, the operating temperature range is relatively large, with the upper temperature limit usually determined by the boiling limit and the lower temperature limit by the sonic limit. The operating temperature range narrows as the heat transport rate increases. At higher heat transport rates, the operating temperature range may be limited by the cessation of capillary pumping or the onset of entrainment, as well as by the onset of sonic velocity or boiling.

If the lower operational temperature limit is determined by the sonic heat transport limit, operation of the heat pipe in the regime below this limit will not be physically possible. The temperature can then be lowered only by reducing the heat transport rate.

A heat transport rate will ultimately be reached at which the operating temperature range is reduced to a single point, denoted as $T(Q_{max})$ in Figure 3.1. This is the maximum attainable heat transport rate, and is denoted as Q_{max} in Figure 3.1. Any attempt to operate the heat pipe at a heat transport rate above Q_{max}, or to operate at Q_{max}

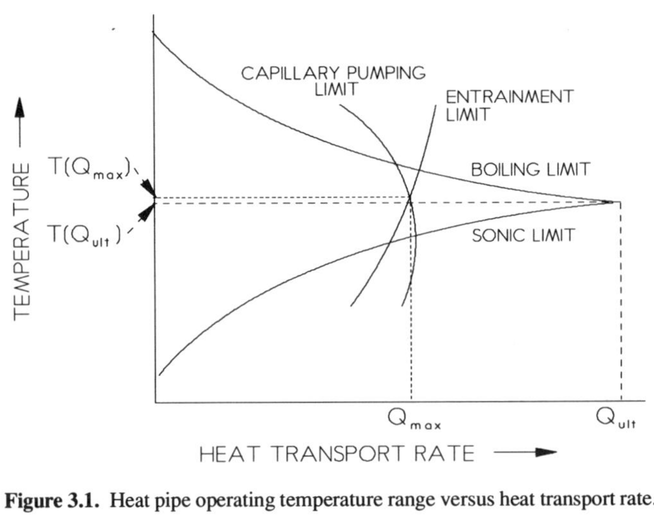

Figure 3.1. Heat pipe operating temperature range versus heat transport rate.

at a temperature other than $T(Q_{max})$, either will be impossible or will result in heat pipe operational failure.

It is evident from Figure 3.1 that determination of the operating temperature range for a particular heat pipe and heat transport rate requires that the temperature-dependent heat transport limits be known. Although the sonic limit, expressed as an axial heat flux, is dependent only on temperature, the other limits depend on the heat pipe geometry as well. Therefore, the operating temperature range varies with the specific heat pipe design as well as the heat transport rate.

An initial estimate of the operating temperature range can be found by noting that, for most heat pipes, the vapor pressure usually lies in the range between 1 and 100 psia (6895 and 689,500 N/m^2).

For vapor pressures below 1 psia (6895 N/m^2), heat transport capability is increasingly limited by the low density of the heat pipe vapor. The low vapor density in turn restricts the allowable axial heat flux for isothermal operation to relatively low values. For vapor pressures above 100 psia (689,500 N/m^2), heat transport capability is increasingly limited by the low superheat for boiling in the heat pipe liquid, as well as by the thicker and heavier heat pipe wall needed to contain the pressurized vapor.

The 1 to 100 psia (6895 to 689,500 N/m^2) vapor pressure range represents a convenient basis for initial estimates of operating temperature ranges for various heat pipe fluids, as well as for the selection of candidate fluids for preliminary heat pipe design studies. However, the design requirements of a particular application could require selection of a vapor pressure outside this range.

The operating temperature range for various heat pipe fluids is shown in Figure 3.2, based on the 1 to 100 psia (6895 to 689,500 N/m^2) vapor pressure range. It is evident that

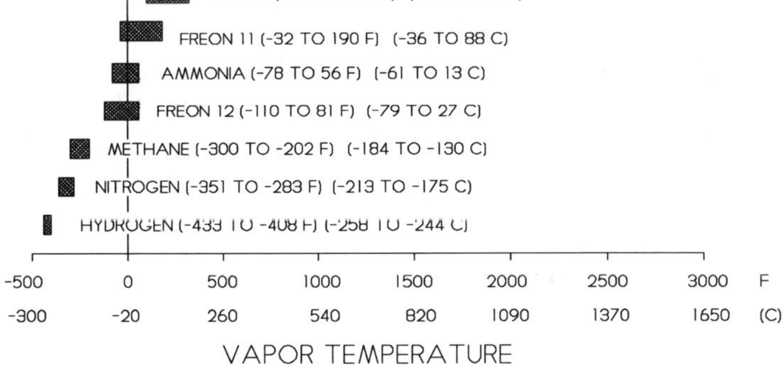

Figure 3.2. Operating temperature ranges of various heat pipe fluids.

heat pipes may be designed to operate at temperatures ranging from close to absolute zero to values in excess of 3000°F (1650°C). When gaps in temperature coverage occur, they can be minimized by extending the vapor pressure range of adjacent fluids to below 1 psia (6895 N/m^2) or above 100 psia (689,500 N/m^2).

For example, a water heat pipe could conceivably be operated at 500°F (260°C), at a vapor pressure of 681 psia (4.70×10^6 N/m^2). The wall of such a heat pipe would be relatively thick in order to withstand the internal pressure. Because of the low superheat for boiling at high vapor pressures, either the surface heat flux would have to be quite low to avoid boiling, or a design would have to be developed that could tolerate some level of boiling.

When more than one fluid is applicable to a particular temperature or temperature range, selection of the preferred fluid should be made on the basis of heat transfer, structural, and integrability criteria appropriate to a particular application.

3.1.2. Liquid Heat Transport Parameter

Although Figure 3.2 provides a basis for selection of heat pipe fluids to meet a given operating temperature requirement, no information is supplied about relative heat transport capabilities. A convenient measure of heat transport capability is the liquid heat transport parameter P_{ht}, defined as

$$P_{ht} = \rho_l \sigma \kappa / \mu_l \tag{3.1}$$

where ρ_l is the density of the heat pipe liquid, σ is the surface tension, κ is the heat of vaporization, and μ_l is the liquid viscosity.

In the final part of Section 5.1.2, it is shown that heat pipe heat transport capability is directly proportional to the liquid heat transport parameter, provided that the following conditions are satisfied:

1. The vapor pressure drop is negligible.
2. Liquid flow through the wick structure is laminar.
3. The liquid hydrostatic head is negligible.
4. The maximum heat transport rate is established by the capillary pumping limit.

These conditions are frequently satisfied for heat pipes operating at subambient and ambient temperatures, which have been designed for space applications. Even when one or more of these conditions are not satisfied, the liquid heat transport parameter is still a reasonable, but more approximate, indicator of relative heat transport capacity. The liquid heat transport parameter for a given liquid varies with temperature, and tends to reach a maximum value at temperatures corresponding roughly to vapor pressures in the 1 to 10 psia (6895 to 68,950 N/m^2) range.

In Figure 3.3, the value of the liquid heat transport parameter at the boiling point corresponding to a vapor pressure of 14.7 psia (10^5 N/m^2) is plotted as a function of the boiling point temperature for selected heat pipe liquids. The data fall along two distinct curves—one corresponding to the nonliquid metals and one corresponding to the liquid metals.

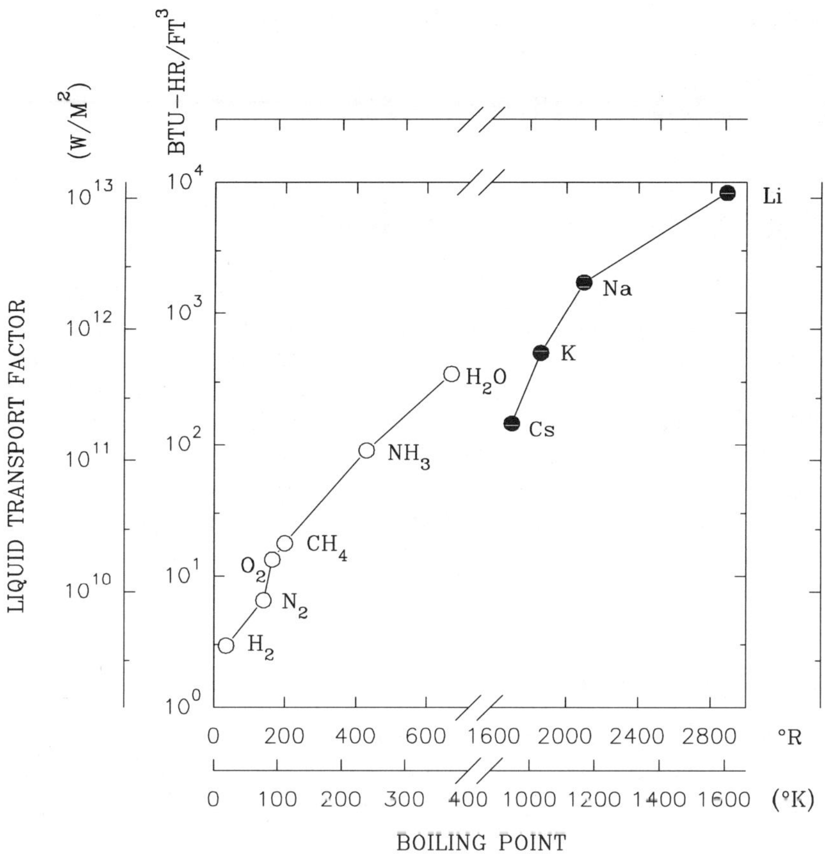

Figure 3.3. Liquid transport factor at boiling point versus boiling point.

For each curve, the liquid heat transport parameter is an increasing function of the boiling point. In general, the parameter is substantially higher for the liquid metals. However, for nonliquid metals with boiling points in excess of 500°R, the parameter is comparable to or greater than that of the lower-boiling-point liquid metals cesium and potassium.

Figure 3.3 can be used to estimate the relative heat transport capability of a given heat pipe design with various heat pipe fluids while operating at a vapor pressure of 14.7 psia (10^5 N/m^2). For example, the heat transport capacity of a water heat pipe is almost 4 times greater than that of an ammonia heat pipe, almost 20 times greater than that of a methane heat pipe, and almost 120 times greater than that of a hydrogen heat pipe. A water heat pipe also transports heat at a rate almost 2.4 times that of a cesium heat pipe, and almost 70% that of a potassium heat pipe.

The heat transport capability of a lithium heat pipe is almost 5 times greater than that of a sodium heat pipe, 17 times greater than that of a potassium heat pipe, 60 times greater than that of a cesium heat pipe, and 24 times greater than that of a water heat pipe.

Again, it must be emphasized that these comparisons are strictly valid only when the four conditions mentioned earlier are fulfilled.

3.2. CONTAINMENT AND WICK MATERIALS

The materials used to fabricate a heat pipe wick and container must be compatible with the selected heat pipe fluid over its anticipated operating temperature range. The containment material must also have adequate strength and resistance to deformation, and should be readily fabricable. Good thermal conductivity is another desirable attribute of the containment and wick materials, in order to minimize wall and wick temperature drops, thermal stress in the wall, and superheat in the wick.

Compatibility is required in at least two respects. First, the heat pipe liquid should not react chemically with the containment and wick materials. Second, the solubility of the containment and wick materials and associated impurities in the heat pipe liquid should be negligible.

3.2.1. Chemical Compatibility

The classic example of chemically incompatible heat pipe materials is the water-aluminum combination. Aluminum reacts with the oxygen in water to yield aluminum oxide and free hydrogen. The hydrogen formed can rapidly occupy the bulk of the vapor space, effectively inactivating the heat pipe.

Water will also react with chromium, iron, and to some extent nickel in stainless steel heat pipes to yield oxides of these metals and free hydrogen (Ref. 3.3). On the other hand, water is completely nonreactive with copper in a water-copper heat pipe. In fact, a water-copper heat pipe is believed to hold the world longevity record for heat pipes, having operated at 120°C for 135,000 hr (Ref. 3.4).

These results can be explained qualitatively by examination of the electromotive series of the metals, which lists the electrode potential of metals relative to hydrogen. Those metals above hydrogen in the series will act to displace hydrogen from compounds, while those metals below hydrogen show no such tendency. The electrode potential for a few key metals is given in Table 3.1.

The strong tendency of aluminum to reduce water to free hydrogen is evident from Table 3.1, as is the lesser tendency of chromium, iron, and nickel, the principal constituents of stainless steel. The chemical inertness of water and copper is also explained by the position of copper below hydrogen.

Lithium is compatible with the refractory metals niobium, molybdenum, tantalum, and tungsten and their alloys (Refs. 3.6, 3.7). These materials are also characterized by good high-temperature strength at typical lithium heat pipe operating temperatures, and thermal conductivities comparable to or greater than that of lithium. However, they

Table 3.1. Electromotive series of metals (Ref. 3.5)

Metal	Electrode potential, volts
Aluminum	−1.34
Chromium	−0.6
Iron	−0.44
Nickel	−0.22
Hydrogen	0.00
Copper	+0.34

require protective coatings for heat pipe operation in the atmosphere or in a combustion gas environment.

On the basis of compatibility, strength, fabricability, and thermal conductivity, molybdenum or its alloy TZM is a frequent choice for the containment and wick material in lithium heat pipes (Refs. 3.8, 3.9). A lithium-molybdenum heat pipe at Los Alamos National Laboratory operated for more than 25,000 hr at 2600°F (1427°C) (Ref. 3.9).

Sodium, potassium, and cesium are compatible with the nickel- and cobalt-based superalloys, such as the Hastelloys and Haynes alloys, as well as with the stainless steels (Refs. 3.10–3.13). These materials are also generally quite resistant to oxidation upon exposure to the atmosphere at operating temperatures consistent with these heat pipe fluids. Sodium is also compatible with molybdenum and tungsten (Ref. 3.14). Potassium is also compatible with nickel, titanium, and the refractory metals and their alloys (Refs. 3.9, 3.15, 3.16).

3.2.2. Mass Transfer (Refs. 3.7, 3.8, 3.11)

Virtually all solids and gases are soluble in liquids to some extent, the solubility generally increasing with temperature. In a heat pipe, this solubility is responsible for mass transfer from the condenser to the evaporator.

When the heat pipe vapor condenses, the condensate is highly purified, and hence has a maximum capability to dissolve low-solubility solids and gases. As the condensate flows through the wick toward the evaporator, small quantities of the wick and containment material will dissolve and be carried into the evaporator. Over a sufficiently long period of time, actual penetration of the condenser wall may occur. Small quantities of impurities in the wall and wick, such as oxygen, will also dissolve and be carried into the evaporator.

When the heat pipe liquid vaporizes, the solute remains in the evaporator, and its concentration gradually increases with time. If sufficient solid material should accumulate in the evaporator, blockage of the liquid flow path could reduce the capillary pumping limit below the actual heat transport rate, resulting in heat pipe failure. If sufficient impurities should accumulate in the evaporator, accelerated corrosion and/or the formation of low-melting-point eutectics could lead to damage of the wick structure and/or breaching of the heat pipe containment.

The rate of mass transfer in a heat pipe can be expected to increase with temperature (due to increased solubility) and heat transfer rate (due to the increased recirculation rate of the heat pipe fluid).

In general, mass transfer rates in all-liquid metal loops with sodium, potassium, and cesium have been observed to be much lower for cobalt-based alloys such as the Haynes alloys than for nickel-based alloys such as the Inconels and Hastelloys (Ref. 3.11). The same situation would be expected to hold for liquid metal heat pipes.

Appreciable temperature differences may exist across the thickness of a heat pipe wick at high incident heat fluxes and/or when the wick thermal resistance is relatively large. Since the heat pipe liquid has a radial as well as an axial velocity component in the evaporator and condenser regions, the liquid in the hotter section of the wick can dissolve the wick/container material, redepositing it as the liquid cools while flowing across the wick thickness.

This mass transfer mechanism can augment the accumulation of material in the evaporator due to axial mass transfer. In the condenser, it may result in some accumulation of material near the wall/wick interface.

Ordinarily, if the heat pipe liquid is chemically compatible with the wall/wick materials and precautions have been taken to limit the level of impurities present, the mass transfer rate will be so low that operation for many thousands of hours can be expected without deleterious effects on heat pipe operation. This is the situation for most of the nonliquid metal heat pipes.

Since liquid metal heat pipes characteristically operate at higher temperatures and power levels than nonliquid metal heat pipes, mass transfer can adversely influence operational lifetimes. This is particularly the case with respect to any oxygen impurities that may be present.

For example, lithium heat pipes fabricated from both niobium–1% zirconium and tantalum failed by breaching of the evaporator wall after 10 to 100 hr when operated at around 1500°C, with an oxygen content of the wall material as low as 100 ppm. When the oxygen content of the niobium–1% zirconium heat pipe was reduced to less than 1 ppm, the heat pipe operated for 1000 hr with no evidence of failure.

When a lithium heat pipe was fabricated from tantalum containing 10 ppm of oxygen and a small percentage of yttrium, the heat pipe operated at 1600°C for 1000 hr. No evidence of oxygen corrosion was evident during a postoperational examination of the heat pipe interior (Ref. 3.6).

In a lithium-molybdenum heat pipe operating at 1500°K (2240°F) and a power level of 14,000 W, the wick structure experienced significant damage after 514 hr. The damage was attributed to a high nitrogen level in the lithium in the evaporator region, said to result from the high ratio of condenser area to evaporator area (about 5). This damage mechanism was not observed in other lithium-molybdenum heat pipes with smaller condenser/evaporator area ratios, over operating periods as long as 30,000 hr (Ref. 3.8).

3.2.3. Different Wall and Wick Materials

The heat pipe wick is generally fabricated separately from the containment structure. An exception is the grooved wick, in which axial or circumferential grooves are cut directly into the heat pipe wall.

The preferred design choice is a heat pipe container and wick that are fabricated from the same material. However, this may not always be feasible because of material scarcity or fabrication difficulties. The use of different wall and wick materials that are individually compatible with the heat pipe fluid may then be an acceptable alternative. For example, a stainless steel screen wick might be a suitable choice in a sodium–Hastelloy X heat pipe if Hastelloy X screen with the desired mesh size were not available.

However, the selection of different wall and wick materials can lead to operational problems.

Satisfactory operation for 12,000 hr of an ammonia–stainless steel heat pipe with an aluminum wick has been reported. Yet operational difficulties were experienced with ammonia-aluminum heat pipes utilizing stainless steel wicks (Ref. 3.3).

Another interesting case has been reported in the literature in which material was incompatible with the heat pipe liquid when used in the wick, but was quite compatible when used only for the heat pipe containment. The material in question was 347 stainless steel. In a water heat pipe with a stainless steel wall and wick, hydrogen gas accumulated rapidly. When copper was used for the wick structure, retaining the 347 stainless steel wall, no gas generation was observed after an extended period of operation (Ref. 3.3).

3.2.4. Recommended Wall and Wick Materials

A summary of recommended wall and wick materials for various heat pipe liquids is given in Table 3.2. These recommendations are based on extensive life testing of heat pipes, in most cases covering periods of 10,000 or more hours. The lifetime tests were usually conducted at moderate temperatures and heat transport rates for the heat pipe fluid involved.

When a heat pipe is operated under extreme conditions of temperature and/or power, mass transfer effects will be accelerated and some adverse effect on heat pipe life is to be expected. For example, a lithium heat pipe of tungsten–26% rhenium was reported to have a lifetime of many years at 1600°C, around one year at 1700°C, and one month at 1800°C (Ref. 3.3). Unfortunately, the effects of operating temperature and heat transport rate on heat pipe lifetime have not been systematically evaluated. Therefore, extended testing at anticipated temperatures and power levels is necessary to establish the lifetime of a specific combination of heat pipe materials.

3.3. WICK CHARACTERISTICS

A number of processes that are vitally important to the transport of heat through a heat pipe take place in the wick structure. The heat pipe fluid evaporates from and condenses on the inner wick surface. Condensate returns to the evaporator section through the

Table 3.2. Recommended heat pipe wall and wick materials (Refs. 3.3, 3.4, 3.6–3.14)

Heat pipe liquid	Recommended material	Heat pipe liquid	Recommended material
Ammonia	Aluminum	Potassium	Nickel
	Carbon steel		Stainless steel
	Nickel		Inconel
	Stainless steel		Titanium
Acetone	Copper		Refractory metals and alloys
Methanol	Copper	Lithium	Niobium–1% zirconium
	Stainless steel		Molybdenum
Water	Copper		TZM molybdenum
	Monel		Molybdenum–13% rhenium
Sodium	Stainless steel		Tungsten
	Nickel		Tungsten–26% rhenium
	Inconel 800		SGS tantalum
	Hastelloy X	Mercury	Stainless steel
	Haynes 188	Silver	Tungsten–26% rhenium
	Molybdenum		
	Tungsten		

wick. Capillary pressure necessary to sustain circulation of the heat pipe fluid is generated at the inner wick surface. The surface tension barrier that prevents liquid in the wick from being entrained by adjacent high-velocity vapor also develops at the inner wick surface. Finally, heat enters and leaves the heat pipe by flowing across the wick thickness.

The liquid flow channel should be characterized by a relatively coarse pore structure in order to minimize resistance to liquid flow. On the other hand, relatively fine pores are desired to maximize the capillary pressure that balances the liquid-vapor pressure difference throughout the heat pipe. A relatively thick wick structure is also desirable to minimize the axial resistance to liquid flow, whereas resistance to heat flow across the wick increases with thickness.

The basic wick types that may be employed in a heat pipe are depicted in Figure 3.4. Selection of a particular type is dependent on the design requirements of the specific application. In general, the simplest wick structure that is appropriate to a particular design situation is preferable.

In situations in which a porous wick structure is used, the wick may be formed from a variety of small elements, including screens, particles, and fibers. As previously mentioned, the wick structure may also be formed from grooves cut into the heat pipe wall. A variant on the grooved wick is a roughened wall or a geometric pattern knurled or engraved in the wall (Refs. 3.15, 3.17). A slotted sheet or layers of screen may also be used to form multiple liquid flow channels without the necessity of cutting grooves in the heat pipe wall. (See Figure 3.5.)

3.3.1. Single-Layer Wick

The single-layer wick of Figure 3.4 is of simple construction, consisting of a relatively homogeneous assemblage of elements such as screens, fibers, or particles. The heat

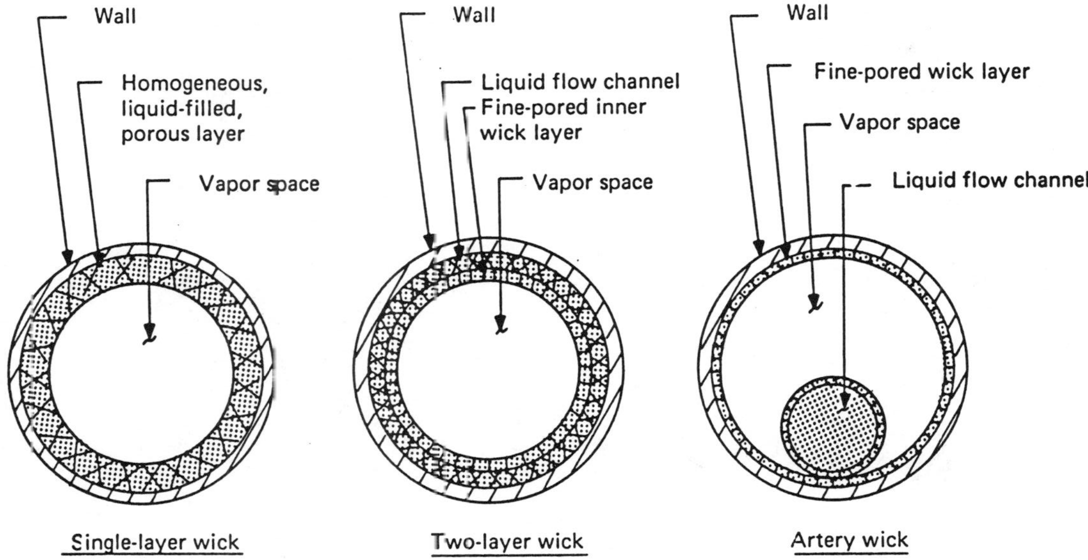

Figure 3.4. Basic wick types for heat pipes.

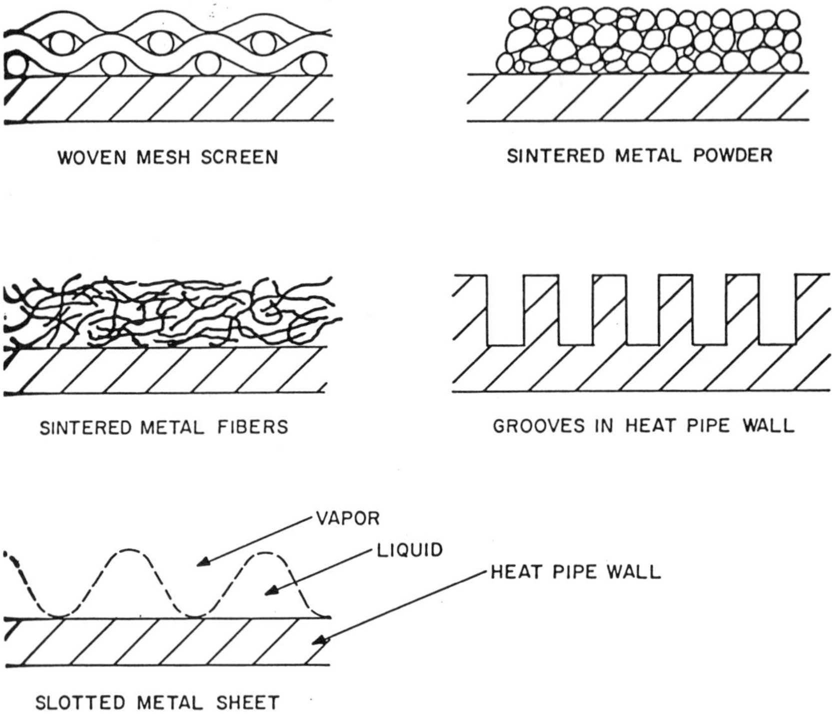

Figure 3.5. Heat pipe capillary wick structures.

transport capability of a heat pipe with a single-layer wick is generally limited by the need to select a single wick pore size that can accommodate both liquid flow channel and capillary pressure requirements. The conflicting requirements of large pore size to minimize frictional flow resistance and small pore size to maximize capillary pressure capability may be difficult to reconcile with a single-layer, homogeneous wick structure.

Consequently, the single-layer wick heat pipe may be unsuitable for an application involving one or more of the following conditions: a relatively small liquid heat transport parameter, a high heat transport rate, or a relatively long heat pipe length.

3.3.2. Two-Layer Wick

A two-layer or composite wick structure can be employed in which the distinct functions of liquid flow path and capillary pressure generation are assigned to separate wick layers, as indicated in Figure 3.4. The flow channel can then be characterized by coarse pores to minimize frictional resistance to flow of the heat pipe liquid, without regard for the effect of the pores on capillary pressure capability. A separate wick layer adjacent to the vapor flow passage provides the requisite fine pore structure needed to develop adequate capillary pressure.

A heat pipe with a two-layer wick has substantially greater heat transport capability than a single-layer wick, since the pore sizes of the liquid flow channel and capillary pumping layer are established independently of each other. In fact, the internal pore structure in the liquid flow channel of a two-layer wick can be completely eliminated to produce a completely open annular channel of minimum frictional resistance. Employment of an annular open liquid flow channel generally permits high heat transport rates with a relatively thin wick structure.

To maintain the annular dimension of the open liquid channel, axial spacers between the wall and the capillary pumping layer may be added at equal intervals around the heat pipe periphery. As the number of spacers is increased to the point at which the distance between spacers becomes comparable to the annular thickness, the open annular liquid flow channel is transformed into a series of axial grooves. (See Figure 3.5.)

The grooved flow channel must be thicker than the annular channel in order to retain the same flow area and heat transport capacity. However, the grooved channel is less sensitive to flow disruptions resulting from the presence of noncondensable gaseous impurities, and can generally be restarted more readily following such disruptions (Ref. 3.17).

If the groove width is small enough, significant capillary pressure can be developed in the grooves, and the inner capillary pumping layer may be dispensed with. However, the liquid-vapor interface in a groove can only develop curvature across the groove width, since the radius of curvature is essentially infinite along the groove length. In contrast, radii of curvature can develop along two normal axes of the pores of a homogeneous capillary structure. Therefore, the capillary pressure capability of a groove is only one-half that obtainable in a capillary structure with pores whose effective diameter is equal to the groove width.

3.3.3. Artery Wick

The conflicting requirements of a relatively thick wick to minimize frictional resistance in the liquid flow direction and a relatively thin wick to minimize thermal resistance across the wick thickness can be reconciled through use of an artery wick.

With an artery wick, the liquid flow channel is displaced from its usual position next to the heat pipe wall, so that the heat transferred into and out of the heat pipe is not required to traverse the thickness of the liquid flow passage. The thermal resistance to heat flow is then limited to a wick layer of minimal thickness along the wall in which capillary pressure is generated (i.e., the capillary pumping layer).

The liquid flow channel, usually located within the vapor flow channel and of cylindrical shape, is called an artery. The artery is bounded by a fine-pored wick layer to assure that the requisite capillary pressure is available at the liquid-vapor interface surrounding the artery. The artery may be formed as an integral part of the wall capillary pumping layer. It may be in contact with the capillary pumping layer, as shown in Figure 3.4, or connected via a stem of the same fine-pored material.

Condensate that forms on the capillary pumping layer along the wall of the condenser flows circumferentially into the artery, and axially through the artery to the evaporator. Liquid then flows from the artery circumferentially into the capillary pumping layer on the wall and is revaporized.

Although the capillary pumping layer is quite thin, its relatively wide liquid flow area and short circumferential flow path usually result in a tolerable liquid circumferential pressure drop. If necessary, multiple, equally spaced arteries can be used to shorten the flow path, and hence the resistance to liquid flow, in the circumferential direction.

The capillary pumping layer on the heat pipe wall serves the vital functions of circumferentially distributing liquid that leaves the artery (in the evaporator) and collecting liquid that enters the artery (in the condenser). Therefore, it is usually referred to as the *distribution wick.*

Since the arterial liquid flow path is no longer located in the heat flow path into and out of the heat pipe, its dimensions can be established independently of thermal resistance requirements. However, because capillary pressure developed in the artery itself is responsible for filling the artery with liquid, there is an upper limit on artery size. The larger the artery, the smaller will be the available capillary pressure for filling it. If the maximum diameter of a single artery based on liquid filling requirements should be insufficient to meet the flow needs of the heat pipe liquid, multiple arteries may be used.

The artery wick is generally employed when the temperature drop through the thickness of a single- or two-layer wick would cause either destructive boiling in the heat pipe liquid or excessively high wall temperatures.

This is frequently the case for non–liquid metal heat pipe fluids with low liquid heat transport parameters and low thermal conductivities. Consequently, non–liquid metal heat pipes with ambitious goals for heat transport rate and distance may be expected to employ an artery-type wick. The two-layer wick is more commonly used in high-performance liquid metal heat pipe applications.

Many variations on the basic artery design shown in Figure 3.4 have been considered, including open and porous flow channels, multiple arteries, and single and multiple supporting stems (Ref. 3.18).

3.3.4. Monogroove Heat Pipe

Figure 3.6 shows a variation on the artery wick concept in which the artery is now located completely outside the vapor space. This concept is known as the monogroove heat pipe (Refs. 3.19, 3.20). In the artery concept of Figure 3.4, the pressure difference between the vapor and the artery liquid is balanced by capillary pressure over the entire periphery of the artery. In the monogroove heat pipe, the vapor-liquid pressure difference is balanced by the capillary pressure that develops in a single fine groove.

Just as is the case with the artery wick, liquid is distributed over the inner circumference of the evaporator and condenser sections by a thin, fine-pored wick layer. This wick layer can be simply an array of fine circumferential grooves.

Since the monogroove liquid flow channel is bounded by a portion of the heat pipe periphery, heat flow must be restricted to some other segment of the periphery to avoid direct heating of the artery liquid. Thus, the monogroove concept is most applicable to

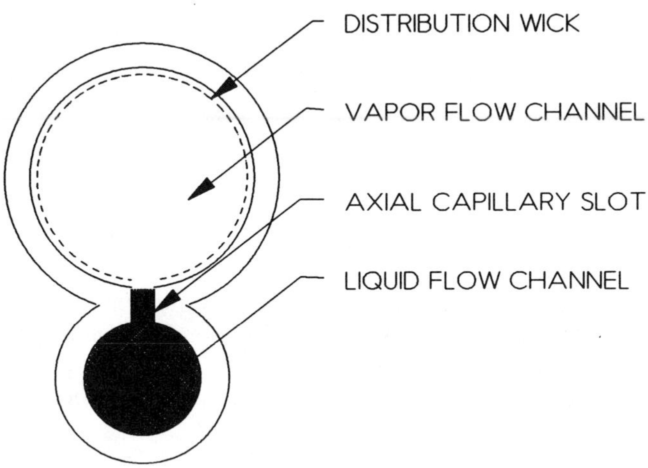

Figure 3.6. The monogroove heat pipe concept (Ref. 3.19).

situations in which heat transfer to and from the heat pipe occurs over only a limited portion of its periphery. Such is the case in space radiator heat rejection systems. Special precautions are necessary to limit heat leakage into the artery liquid to levels at which either the liquid can remain subcooled, or superheat will be insufficient to initiate boiling in the liquid flow channel (Ref. 3.20).

3.3.5. Variable-Thickness Wick

A simpler alternative to the artery wick for reconciling the conflicting wick thickness needs of the liquid and heat transfer flow paths involves the use of a variable-thickness

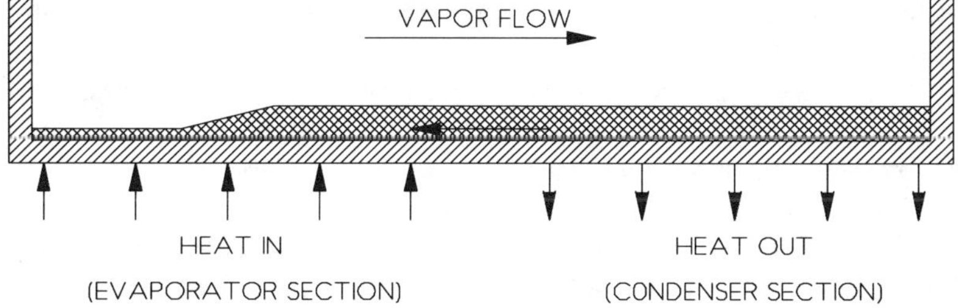

Figure 3.7. Variable-thickness wick.

wick. (See Figure 3.7.) In the region of the evaporator where the heat flux is highest, the liquid flow channel is made thin to minimize the temperature drop and superheat across the wick thickness. In other regions of the heat pipe, a thicker liquid flow channel is used to reduce the overall liquid pressure drop.

3.4. HEAT PIPE FABRICATION

A heat pipe is a closed, recirculating fluid heat transport system in which mass transfer phenomena may become significant over a period of time. Therefore, special precautions must be taken during fabrication to assure that impurities that can adversely affect heat pipe operational performance and longevity are reduced to an absolute minimum.

3.4.1. The Fabrication Process

Emphasis on the avoidance of extraneous material starts with the basic materials selection process. The heat pipe liquid should be of high purity, with little or no trace of dissolved gases or solids. Similarly, the heat pipe wick and containment materials should have a minimum content of gaseous and solid contaminants that can contribute to internal corrosion.

These considerations are especially important for high-temperature, high-power liquid metal heat pipes. In these systems, oxygen and other impurities in excess of a few parts per million in the liquid and/or materials of construction can limit lifetime through excess corrosion (Refs. 3.6–3.8).

If a grooved wick is required, the next step is to form the grooves in the heat pipe wall. The grooves can be formed by mechanical machining, although other techniques such as hot extrusion or chemical milling might be appropriate for specific design situations. One end of the containment structure is then sealed, generally by welding on an end cap.

Any porous wick structure required is then cleaned and inserted into the heat pipe containment structure. A porous wick might or might not be required as an inner layer for a grooved wick heat pipe. The elements of the porous structure may have been sintered together or mechanically drawn down prior to insertion, to assure adequate strength and stiffness. The porous wick structure may be mechanically secured to the heat pipe wall by means of a tight pressure fit, or may be bonded to the wall by sintering, brazing, or spot welding.

If a two-layer or artery-type wick is used, special precautions are necessary to assure that the heat pipe vapor does not contact the liquid flow channel. Otherwise, the smaller capillary pressure generated there may be insufficient to balance the vapor-liquid pressure difference that develops during heat pipe operation. Contact between the vapor and the liquid flow channel can be prevented by plugging either the vapor space or the liquid flow channel, as indicated in Figure 3.8 (Ref. 3.10). The plug must be effectively sealed by welding or brazing. During the fabrication of a liquid metal heat pipe, disks of suitable *getter* materials may be added at the ends of the heat pipe to aid

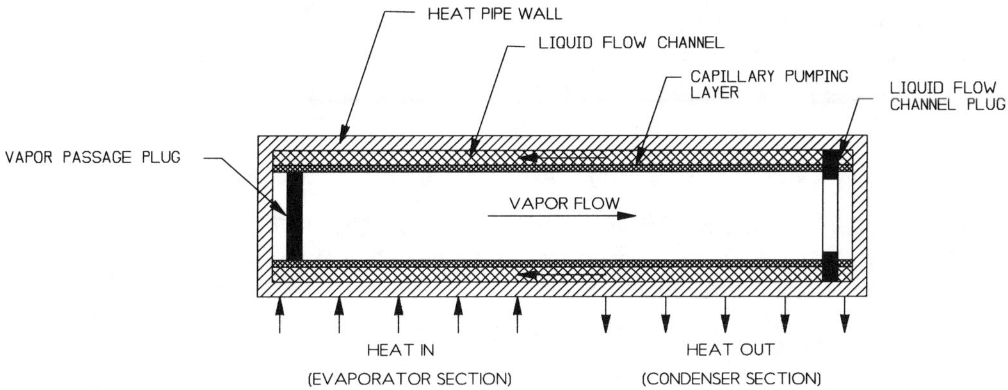

Figure 3.8. Vapor space/liquid flow channel plugs for two-layer wicks (Ref. 3.10).

in the removal of trace contaminants. In lithium-molybdenum heat pipes, hafnium and zirconium disks have been used for this purpose (Ref. 3.9).

Once the wick structure has been properly installed and secured in place, the heat pipe assembly must be thoroughly degreased and cleaned, and all cleaning agents and solvents completely removed. It may be necessary to repeat the cleaning step several times, including heating and *baking out* the assembly in vacuum to remove residual grease and solvents.

Despite all precautions, some residue may remain and generate gas during heat pipe operation. The noncondensable gas formed is generally confined to the condenser end of the heat pipe. If the noncondensable gas inactivates only a small fraction of the condenser, its presence may be tolerable.

One method of eliminating noncondensable gas arising from residual contaminants is to install a valve at the condenser end. During operation, the valve can then be opened to *burp* the heat pipe, permitting the noncondensable gas to be vented. Once the noncondensable gas has been clearly eliminated, the gas vent can be permanently sealed by welding.

After the cleaning and outgassing operation, the heat pipe liquid is added, usually by one of the following methods. In the first method, a measured quantity of the liquid is added directly to the heat pipe. If liquid metals are involved, the heat pipe liquid may be added to the heat pipe as solid pellets, since their melting point is generally above ambient temperature. In the second method, the liquid is vacuum distilled into a cooled heat pipe, so that the vapor introduced can condense on the interior heat pipe surfaces.

Once the liquid charge has been introduced, the heat pipe is sealed under high vacuum, usually by welding the fill tube shut. To avoid the loss of liquid in the first method while the heat pipe is being evacuated, the temperature should be maintained close to or below the melting point.

The sealed heat pipe is then maintained at an elevated temperature for several hours to assure that the liquid fills the wick pores and wets all internal surfaces. Arteries and open annular flow passages will fill completely if the capillary pressure associated with the flow channel dimension is greater than the hydrostatic pressure of the liquid in the filled channel and any associated stem (Ref. 3.2).

3.4.2. Effect of Liquid Density Change with Temperature

Enough liquid should be added to assure that the wick structure is completely full at the desired heat pipe operating temperature. If the operating temperature is significantly higher than ambient temperature, as is the case with liquid metal heat pipes, the liquid density can decrease by as much as 30% while the heat pipe temperature rises to the design operating level.

It follows that, if the wick has been completely filled with the heat pipe liquid at low temperature, a substantial fraction of the liquid will have spilled out into the vapor space by the time the operating temperature has been reached. Ordinarily, the excess liquid will be retained at the condenser end of the heat pipe by the heat pipe vapor, although some puddling may occur. If the heat pipe length is increased to accommodate the inactive length occupied by the excess liquid, heat pipe performance will not be affected.

However, the additional heat pipe length needed to accommodate the excess liquid might not be permissible, or an acceleration or gravity head large enough to move the excess liquid to the evaporator end may be present. In such instances, excess liquid in the vapor space should be avoided.

One way to handle this situation is to load the heat pipe with only enough liquid to just fill the wick structure at the operating temperature. Then, at lower temperatures when the liquid density is higher, the liquid will be partially retracted into the wick structure. As long as the liquid-vapor interface of the partially filled wick always lies within the capillary pumping layer, heat pipe performance should not be significantly affected.

Another approach is to provide for a reservoir next to the wick structure into which the heat pipe liquid may expand as the temperature rises.

The performance of heat pipes with uncovered grooved wicks, including axial grooves and circumferential grooves characteristic of artery wick heat pipes, is quite sensitive to underfilling. In a monogroove heat pipe, the heat transport capability was reduced by 50% when the wick was underfilled by just 3%, and dropped to zero at an underfill of 9% (Ref. 3.20).

For such cases, the initially underfilled heat pipe may be heated uniformly over its entire length to the operating temperature prior to the start of operation, to avoid heat transport through the heat pipe during the startup process. Alternatively, a supplemental liquid reservoir may be used to absorb excess liquid during startup of an initially filled wick.

3.5. ENVIRONMENTAL INFLUENCES ON HEAT PIPE OPERATION

Heat pipe performance and/or lifetime can be affected by the external environment in which the heat pipe operates. These external influences include acceleration, gravity, vibration, gaseous diffusion, and oxidation/corrosion. Their effect on heat pipe operation will now be considered.

3.5.1. Acceleration/Gravity/Vibration

The effect of acceleration and/or gravity on heat pipe operation is to increase or decrease the liquid pressure drop. This effect can ordinarily be allowed for in the design process. However, when the actual operating environment differs from the $1g$ environment of earth, problems may arise in experimental verification of heat pipe performance.

Heat pipes have operated successfully under a vibration environment typical of that encountered during a missile launch (Ref. 3.21). In one vibration test with a water heat pipe, the temperature drop along the heat pipe decreased from 2°C to 1°C, indicating an actual performance improvement. When heat pipe operation is required under vibrational conditions, operability should be verified experimentally for the expected vibrational environment.

3.5.2. Gaseous Diffusion

When a heat pipe is operated in air, water vapor in contact with the heat pipe surface can be reduced to free hydrogen by reacting with the constituents of the heat pipe wall. The hydrogen gas formed can diffuse through the wall into the heat pipe interior, with adverse effects on heat pipe operation.

Effect on Heat Pipe Operation. If the gas should accumulate in the vapor space at the condenser end, the effective heat pipe length could be shortened. The formation of discrete gas bubbles in the wick structure could impede liquid flow, disrupt the liquid-vapor interface, or facilitate the onset of boiling. Depending on the specifics of the heat pipe design and its operating conditions, these effects could reduce one or more heat transport limits, leading to premature dryout and overheating of the evaporator section.

Once a heat transport limit has been exceeded and dryout of the wick has occurred, hydrogen that has diffused into the heat pipe may occupy the dried-out zones. Refilling the wick with the heat pipe liquid may then be difficult, and subsequent heat pipe startup and operational capability may be greatly impaired (Ref. 3.17).

Susceptible Materials. Stainless steels and nickel- and cobalt-based superalloys contain significant quantities of elements that can reduce hydrogen compounds. (See Table 3.1.) Since these alloys are quite compatible with sodium, potassium, and cesium, they are frequently specified as containment and wick materials for heat pipes employing these fluids. Thus, sodium, potassium, and cesium heat pipes can be susceptible to the adverse effects of hydrogen diffusion if operated in a humid atmosphere.

This premise has been verified in an experiment in which a sodium–stainless steel heat pipe was first operated in argon saturated with water vapor. A rise in the evaporator temperature, indicating a disruption of normal heat pipe operation, occurred rapidly.

Normal operation was restored by surrounding the heat pipe with dry argon, thus allowing the accumulated hydrogen to diffuse out of the heat pipe (Ref. 3.11).

Similar results have been demonstrated with air. In dry air, no effects of hydrogen diffusion were observed, but with an increase in humidity, a significant accumulation of hydrogen occurred after a few hours (Ref. 3.4).

If the heat pipe surface is oxidized prior to operation in a moisture-bearing atmosphere, the water vapor will no longer be exposed to the reactive metal constituents of the heat pipe. Hydrogen generation will then be suppressed. A sodium–stainless steel heat pipe subjected to this type of treatment operated in the atmosphere at 800°C for 5200 hr without evidence of hydrogen diffusion (Ref. 3.4).

In other tests with 304 stainless steel heat pipes, extended life tests were carried out with potassium as the heat pipe fluid at 600°C and with sodium at 700°C. After 10,000 hr, there was no indication of adverse effects on heat pipe performance due to hydrogen diffusion. Some oxidation of these heat pipes may have occurred during short performance tests that were conducted prior to the lifetime tests (Ref. 3.12).

These observations suggest that, if a sodium, potassium, or cesium heat pipe is to be exposed to the atmosphere during operation, a short period of preoperation at relatively high temperature is desirable to oxidize the surface. Then a second preoperational period at elevated temperature should be included, in either vacuum or a dry atmosphere, to allow any initial accumulation of hydrogen to diffuse out of the heat pipe.

Influence of Wick Design on Diffusion Effects. The extent to which hydrogen diffusion adversely affects heat pipe operation, particularly with respect to restart capability, depends significantly on the wick design. A wick with an annular open liquid flow channel appears to be most susceptible to hydrogen diffusion, because a hydrogen bubble can block all or most of the channel. Two-layer wicks with grooved or porous liquid flow channels are much less susceptible, and effects on single-layer porous wicks may be negligible (Ref. 3.11).

In tests with two-layer-wick liquid metal heat pipes employing an annular open liquid flow channel, heat transport performance subsequent to dryout caused by hydrogen accumulation was poor. Normal performance capability was restored by heating the heat pipe uniformly at elevated temperature in a vacuum furnace to diffuse the accumulated hydrogen out of the heat pipe.

On the other hand, no significant performance impairment was detected following dryout of a two-layer wick structure with axial grooved liquid flow channels, although some hydrogen had apparently accumulated in a few channels (Ref. 3.17).

3.5.3. Oxidation/Corrosion

Liquid metal heat pipes operated in the atmosphere or in an environment of hot combustion gases must be capable of withstanding oxidation and/or corrosion.

Oxidation-Resistant Alloys. Extended testing of sodium heat pipes in air for periods in excess of 10,000 hr at peak temperatures ranging from 880°C to 1020°C indicated that

310S stainless steel, Hastelloy X, Hastelloy B, and Haynes 188 are suitable oxidation-resistant containment materials. These oxidation-resistant alloys develop a thin, smooth, hard, adherent coating of black oxide on their external surfaces (Ref. 3.13).

Excessive oxidation of 304L stainless steel heat pipes was observed after 2000 to 4000 hr of operation (Ref. 3.13). However, it should be noted that, as described previously, sodium–304 stainless steel heat pipes at 700°C apparently withstood atmospheric testing quite well for 10,000 hr.

More quantitative data are available from oxidation tests of the nickel-base superalloy René 41 and the cobalt-base superalloy Haynes 25. After 1000 hr of exposure to the atmosphere at 1600°F, the oxidation penetration was 0.0022 in. for René 41 and 0.0009 in. for Haynes 25 (Ref. 3.11).

The containment for a sodium heat pipe was also fabricated from Kanthal A-1, an oxidation-resistant alloy with 22 wt % chromium, 5.7 wt % aluminum, and 0.5 wt % cobalt. A 316 stainless steel wick was used. The heat pipe was initially operated in air to produce an oxide coating on the external surface, and then operated in a natural gas furnace with the evaporator section heated by hot combustion products and the condenser section convectively cooled. After 100 hr of testing at about 1200 to 1250°K, there was no deterioration of the heat pipe surface (Ref. 3.14). However, the heat pipe failed from internal corrosion after 200 hr (Ref. 3.7).

Protective Coating for Molybdenum and TZM. Refractory metal containment is generally required for lithium heat pipes. If such heat pipes are to operate for extended periods in the atmosphere, protective oxidation-resistant coatings are required.

Molybdenum disilicide ($MoSi_2$) has been shown to provide effective oxidation protection to TZM (an alloy of molybdenum, 0.5 wt % titanium, 0.08 wt % zirconium, and 0.03 wt % carbon maximum) (Ref. 3.22). This coating may be expected to protect unalloyed molybdenum as well. The silica layer that forms at the exposed surface of the coating protects the substrate from further oxidation.

Below approximately 1050°C (1922°F), coating lifetime tends to be limited by the formation of cracks during cooling that do not reseal upon reheating. At higher temperatures, cracks that have formed in the coating tend to reseal. Coating lifetime is then limited by the time required for $MoSi_2$ to be transformed into the nonprotective compound Mo_5Si_3 by interdiffusion of the molybdenum and silicon.

For coating thicknesses of 80 microns (0.003 in.), and exposure to still air in 17-hour cycles, coating lifetimes of 1000 to 2000 hr for temperatures up to 1205°C (2201°F) can be expected. At higher temperatures the transformation of $MoSi_2$ into Mo_5Si_3 occurs at an accelerated rate, leading to correspondingly shorter lifetimes. The coating lifetime is on the order of 150 hr at 1315°C (2399°F) and 50 hr at 1425°C (2597°F).

Ceramic Protective Coatings (Ref. 3.14). Operation of liquid metal heat pipes at higher temperatures, for longer lifetimes, and in combustion gas environments requires the use of ceramic protective coatings such as silicon carbide. The development of suitable ceramic protective coatings is still in the formative stage.

A sodium heat pipe fabricated from an outer layer of chemical vapor–deposited (CVD) silicon carbide and an inner layer of CVD tungsten (with a textured surface to provide wicking) was operated in air and combustion gases at a peak temperature of about 1200°K (1700°F) for about 100 hr. The heat pipe was restarted thirty times during this period. Some evidence of deterioration at the tungsten–silicon carbide interface was observed. Further tests using an extruded silicon carbide–carbon tube coated with silicon carbide on the outer surface and tungsten on the inner surface were planned.

A sodium-molybdenum heat pipe was coated with a 0.1-mm (0.004-in.) layer of CVD tungsten and an outer layer of about 0.25 mm (0.010 in.) of CVD silicon carbide. The evaporator section of the heat pipe was heated in a gas furnace and operated for about 150 hr at a maximum temperature of about 1300°K (1881°F). Other than oxidation of a small flaw that was originally present in the coating, the coating remained in good condition throughout the test.

Sodium-alumina heat pipes were also fabricated, but they failed during startup because of the low thermal shock resistance of the alumina. However, low-temperature toluene-alumina heat pipes were successfully fabricated and started up. One was operated intermittently for about a year in the 350 to 420°K (170 to 297°F) range with no apparent deterioration.

3.6. HEAT PIPE SYSTEMS

In practice, a heat pipe or an array of heat pipes functions as a component of a thermal system that in turn is integrated with some other system or device. Often, requirements of the overall system will determine heat pipe design requirements. Heat pipe thermal systems may be divided into three broad categories of application: heat dissipation, heat exchange, and temperature control. Some examples of how heat pipes can be used in such systems will now be presented.

3.6.1. Heat Dissipation

Heat dissipation is required for the following situations:

1. Heat incident on or generated by a device or structural component must be removed to avoid excessive temperatures.
2. Unutilized heat from an energy conversion device must be removed.
3. Heat is to be produced for process or thermal comfort purposes.

A common characteristic of heat dissipation systems is that heat produced at one location is transferred to another location for disposal, without subsequent recovery for productive use. Examples of three possible types of heat pipe heat dissipation systems are illustrated in Figures 3.9, 3.10, and 3.11.

Leading Edge Heat Pipe Cooling System. A heat pipe cooling system for the leading edge of a hypersonic aircraft wing is shown in Figure 3.9. In this case, heat pipes run

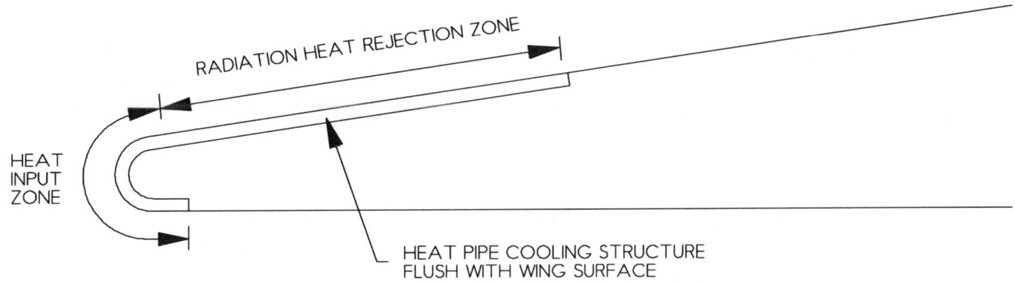

Figure 3.9. A leading edge heat pipe cooling system.

around the wing leading edge and upper surface. Aerodynamic heat incident on the leading edge is transported through the heat pipe and radiated to the environment from the upper wing surface, where the aerodynamic heating rate is relatively low.

The heat pipes operate at a temperature that will permit the heat load from the leading edge to be dissipated from the radiating section. However, the heat pipe temperature cannot be less than the radiation equilibrium temperature of the upper wing surface (the temperature that would be assumed by the surface if all the incident aerodynamic heat were radiated away). As the heat pipe temperature approaches the radiation equilibrium temperature, the area of the radiating section will approach infinity.

Space Power Plant Radiator. A second type of heat pipe heat dissipation system is shown in Figure 3.10. In this system, unutilized heat from a space energy conversion system is radiated to space.

Rejected power plant heat is absorbed by a heat transfer fluid, which then flows into

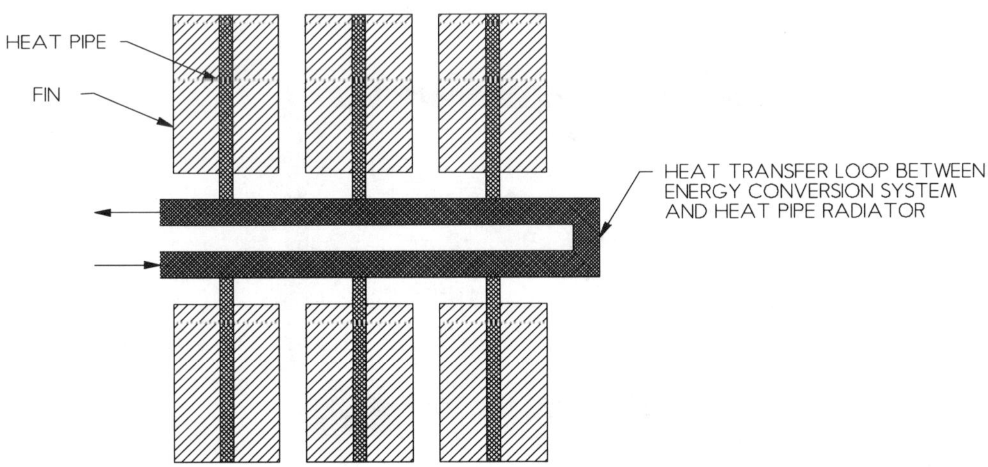

Figure 3.10. A heat pipe radiator for a space energy conversion system.

the space radiator. The radiator consists of an array of heat pipes oriented normal to and in contact with the heat transfer loop. The heat pipes in turn are attached to solid metal sheets or *fins*. As the heat transfer fluid flows through the loop, heat is transferred into the heat pipes. The heat flows from the heat pipes into the fins, and then from the fin surfaces into space by thermal radiation.

Since the heat pipes are isothermal, the mean fin temperature is higher than if the fins were connected directly to the heat transfer loop. Less fin area is then required to radiate waste heat from the energy conversion system to surrounding space. The radiating area could be further reduced by eliminating the solid fins and placing the heat pipes side by side in contact with each other, creating a virtually isothermal radiator. However, the heat pipes would then be more vulnerable to penetration and disablement by high-speed micrometeoroids present in the space environment.

Heat Distribution System for Thermal Comfort. A third example of a heat pipe heat dissipation system is shown in Figure 3.11. Here, heat pipes are used to transfer heat from a furnace to various rooms where heat is needed for the physical comfort of the inhabitants. In this instance, the dissipated heat is used productively before being discarded into the environment.

One or more heat pipes run successively through each room to be heated. The evaporator section is within the furnace. Condenser sections, finned for greater efficiency in transferring heat to room air, are located in each room. Insulated, adiabatic sections run between the furnace and the first room, and between adjacent rooms. In this example, the heat pipes are serving as components of a thermal distribution system, transferring heat from its point of generation to various heat dissipation locations.

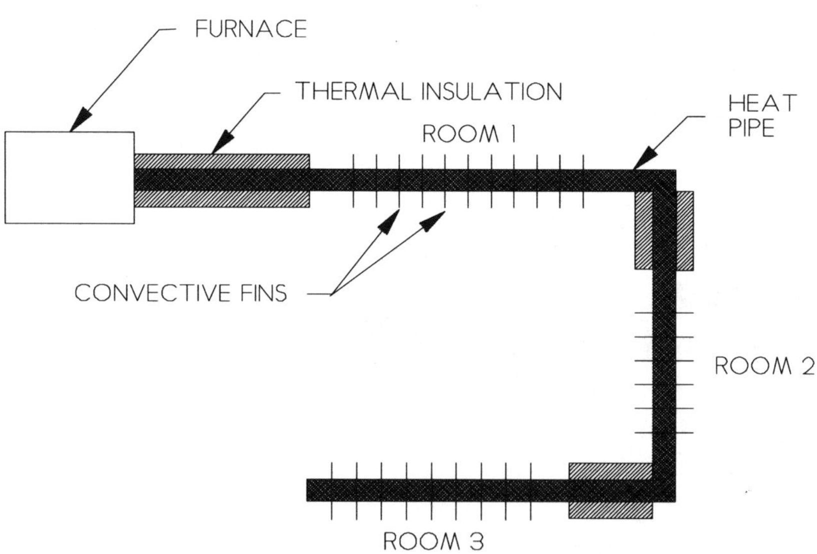

Figure 3.11. A heat pipe heat distribution system.

3.6.2. Heat Exchange

Heat exchange occurs when heat is transferred from one fluid stream to another. The purpose of the exchange is not to avoid excessive temperature in the hotter stream, but to increase the efficiency of thermal energy conversion or of a thermal process by recovering heat that would otherwise be wasted.

For example, a heat exchanger can be used in a gas turbine power plant to transfer heat from the hot turbine exhaust gas to cooler compressor discharge air. Less energy must then be added to the compressor discharge air by burning fuel in the combustor, contributing to increased overall cycle efficiency. This type of heat exchanger is referred to as a regenerator or recuperator.

A heat exchanger may also be used to improve the efficiency of process applications by recovery from a furnace exhaust stream of heat that would otherwise be dissipated into the atmosphere.

The configuration of a representative heat pipe heat exchanger is shown in Figure 3.12. An array of heat pipes extends through adjacent fluid passages, oriented normal to the flow direction. The heat pipes may be smooth cylinders, or they may be finned to increase the effective heat transfer area per unit length of heat pipe.

Heat is transferred from the hotter fluid to the heat pipes, through the heat pipes, and then from the heat pipes into the cooler fluid. With heat pipes, effective heat transport between the hot and cold fluids can be accomplished even if the flow passages are located at some distance from each other.

Additionally, the transfer of heat between the hot and cold fluids is accomplished with virtually no disruption of flow area or direction. This is in marked contrast to the situation in conventional heat exchangers, where the hot and cold streams must be broken down into multiple flow passages in intimate contact with each other.

The temperature of individual heat pipes is determined from the requirement that

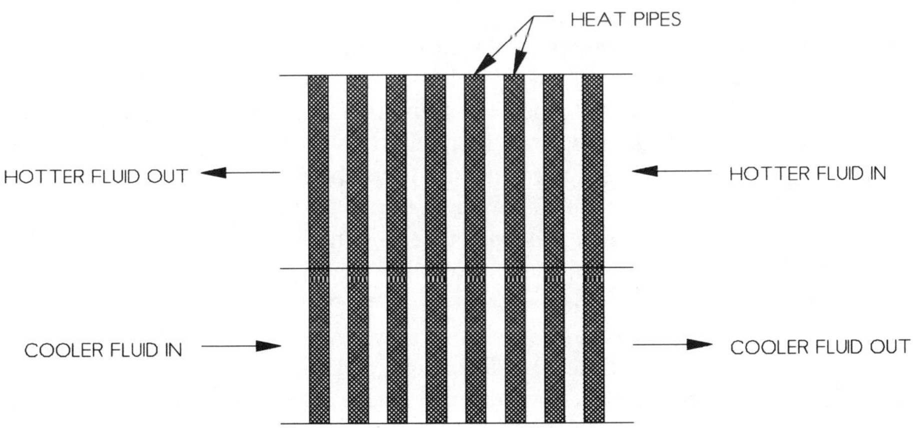

Figure 3.12. A heat pipe heat exchanger.

the heat transferred into a heat pipe from the hot fluid be equal to the heat transferred from that same heat pipe to the cold fluid.

Since the fluid temperatures vary with the axial distance along the heat exchanger, the temperature of the heat pipes will also vary along the axial length. If the axial temperature range is large, it may be necessary to use more than one heat pipe fluid in the heat exchanger in order to maintain the vapor pressure within the range (approximately 1 to 100 psia [6895 to 689,500 N/m^2]) necessary for efficient heat pipe heat transport.

3.6.3. Temperature Control

Ordinarily, the temperature of a heat pipe is established from the requirement that the heat input and output rates be equal. If conditions affecting heat transfer into or out of the heat pipe should change, the heat pipe temperature will change accordingly.

For example, the temperature of a leading edge heat pipe cooling system will increase if the aerodynamic heating rate increases. The temperature of a heat pipe heat rejection radiator in space will change if the effective heat sink temperature changes with orbital position, as will be the case if the radiator is oriented toward the sun at some orbital positions and away from the sun at others.

Variations of heat pipe temperature with time may be permissible in such situations, as long as some specified maximum temperature is not exceeded. In other situations, maintenance of a relatively constant heat pipe temperature, despite variations in heat input or output conditions, could be an essential design requirement.

For example, the heat input to a space radioisotope-powered thermoelectric generator (RTG) decreases with time in accordance with the decay of the radioisotope heat source. If the power output and generator efficiency are to remain invariant with time, the hot junction temperature must remain constant, and radioisotope heat generated in excess of the constant rate required at the hot junction must be dissipated to space. This can be accomplished by coupling the RTG to a variable conductance heat pipe. (See Figure 1.5.)

In this example, the radioisotope heat source extends around the heat pipe circumference along the constant-length *heat input* section, and the hot junction of the RTG is mounted on the left end of the heat pipe. Heat then flows through the heat pipe to the left into the hot junction at the constant rate required to power the RTG at a constant hot junction temperature. The remaining excess heat flows through the heat pipe to the right and is radiated to space from the variable-length *heat output* section.

As the excess heat rate decreases with time, the interface between the heat pipe vapor and the inert gas moves to the left, shortening the effective heat output length. If the inert gas reservoir volume is large compared to the volume of the heat pipe vapor space, the pressure of the inert gas will change only slightly with displacement of the interface. Then the pressure of the heat pipe vapor, and therefore its temperature, will remain almost constant as the heat output section self-adjusts to the length needed to radiate away excess heat at a lower rate.

With the variable conductance heat pipe, the heat pipe temperature can be maintained within a narrow range despite wide variations in conditions affecting the heat input and output rates. The variable conductance heat pipe has been described briefly in Section 1.4.1. It is not considered further in this book for the reasons cited in Section 1.6.

3.7. REFERENCES

3.1. Kunz, H. R., Langston, L. S., Hilton, B. H., Wyde, S. S., and Nashick, G. H., "Vapor Chamber Fin Studies: Transport Properties and Boiling Characteristics of Wicks," CR-812, NASA, Washington, D.C., June 1967, p. 91.

3.2. Johnson, G. D., and Saaski, E. W., "Arterial Wick Heat Pipes," 72-WA/HT-36. Presented at the ASME Winter Annual Meeting, New York, November 1972.

3.3. Basiulis, A., and Prager, R. C., "Compatibility and Reliability of Heat Pipe Materials," *Progress in Astronautics and Aeronautics,* Vol. 49, American Institute of Aeronautics and Astronautics, New York, 1976, pp. 515–529.

3.4. Ernst, D. M., and Eastman, G. Y., "High Temperature Heat Pipe Technology at Thermacore—An Overview," AIAA-85–0981. Presented at the AIAA 20th Thermophysics Conference, Williamsburg, Virginia, June 1985.

3.5. Eshbach, O. W., *Handbook of Engineering Fundamentals,* 1st Ed., Wiley, New York, 1936, pp. 10–37.

3.6. Busse, C. A., Geiger, F., Strub, H., Pötzschke, M., and Kraft, G., "High Temperature Lithium Heat Pipes." Presented at the 2nd International Conference on Thermionic Electrical Power Generation, Stresa, Italy, May 1968.

3.7. Lundberg, L. B., "Alkali Heat Pipe Corrosion Phenomena." Presented at the 6th International Heat Pipe Conference, Grenoble, France, May 1987.

3.8. Merrigan, M. A., and Feber, R. C., "Thermochemical Correlation of Material Transport in an Alkali Metal Heat Pipe," LA-UR-85–2648, Los Alamos National Laboratory, Los Alamos, New Mexico, 1985.

3.9. Merrigan, M. A., Keddy, E. S., and Sena, J. T., "Transient Performance Investigation of a Space Power System Heat Pipe," AIAA-86–1273. Presented at the AIAA/ASME 4th Joint Thermophysics and Heat Transfer Conference, Boston, Massachusetts, June 1986.

3.10. Kemme, J. E., "Ultimate Heat Pipe Performance," *IEEE Transactions on Electron Devices,* Vol. ED-16, No. 8, August 1969, pp. 717–723.

3.11. Silverstein, C.C., "A Feasibility Study of Heat-Pipe-Cooled Leading Edges for Hypersonic Cruise Aircraft," CR-1857, NASA, Washington, D.C., November 1971.

3.12. Ewell, G. J., Basiulis, A., and Lamp, T. R., "Reliability of Low-Cost Liquid Metal Heat Pipes." Presented at the 3rd International Heat Pipe Conference, Palo Alto, California, May 1978.

3.13. Kaufman, W. B., and Tower, L. K., "Compatibility of Sodium and Lithium in Superalloy Heat Pipes," AFWAL-TR-85–2006, Aero Propulsion Laboratory, Wright-Patterson Air Force Base, Ohio, April 1985.

3.14. Merrigan, M., Dunwoody, W., and Lundberg, L., "Heat Pipe Development for High Temperature Recuperator Application," *Advances in Heat Pipe Technology,* Reay, D. A. (Ed.), Pergamon Press, 1981, pp. 245–257.

3.15. Meier, K. L., Girrens, S. P., and Dickinson, J. M., "Titanium Heat Pipes for Space Power Systems." Presented at the 15th Intersociety Energy Conversion Conference, Seattle, Washington, August 1980.

3.16. Lundberg, L. B., "Titanium-Potassium Heat Pipe Corrosion Studies," *High Temperature Corrosion in Energy Systems,* Metallurgical Society of AIME, New York, 1985, pp. 697–708.

3.17. Kemme, J. E., Keddy, E. S., and Phillips, J. R., "Performance Investigations of Liquid Metal Heat Pipes for Space and Terrestrial Applications." Presented at the 3rd International Heat Pipe Conference, Palo Alto, California, May 1978.

3.18. Ponnappan, R., "Double Wall Artery Heat Pipe," AFWAL-TR-85–2041, Aero Propulsion Laboratory, Wright-Patterson Air Force Base, Ohio, June 1985.

3.19. Alario, J., Haslett, R., and Kosson, R., "The Monogroove High Performance Heat Pipe," AIAA-81-1156, American Institute of Aeronautics and Astronautics, New York, 1981.

3.20. Alario, J., Brown, R., and Kosson, R., "Monogroove Heat Pipe Development for the Space Constructible Radiator System," AIAA-83-1431. Presented at the AIAA 18th Thermophysics Conference, Montreal, Canada, June 1983.

3.21. Deverall, J. E., "Effect of Vibration on Heat Pipe Performance," LA-3798-MS, Los Alamos Scientific Laboratory, Los Alamos, New Mexico, 1967.

3.22. Tuominen, S. M., and Dahl, J. M., "Cyclic Oxidation Testing of Molybdenum Protected by Silicide Coatings," *Journal of Less Common Metals,* Vol. 81, 1981, pp. 249–260.

CHAPTER

FOUR

FLUID FLOW IN A HEAT PIPE

In this chapter, the fluid flow process in a heat pipe is examined. The two-dimensional nature of the flow process is emphasized, and one-dimensional approximations suitable for design purposes are discussed. Results are presented for the pressure drop in flow ducts with constant flow rates, and with flow rates that vary due to the injection or removal of fluid through the duct wall. These results are then used in Chapter 5 to formulate equations for the capillary pumping limit.

4.1. THE NATURE OF THE FLOW PROCESS

Flow in a heat pipe is a rather complex process in which the liquid and vapor phases move in counterflow to each other. Although the liquid and vapor phases are nominally saturated at their common interface, they may be superheated or subcooled at locations remote from the interface. If subcooling of the vapor is accompanied by the condensation of liquid droplets, a liquid-vapor mixture will be interspersed in the vapor space.

There is also a continuous interchange of mass between phases in the evaporator (liquid to vapor) and condenser (vapor to liquid) sections. In the evaporator, a net flux of molecules leaves the liquid surface at a given point in all directions, with a wide range of velocities. These molecules collide with adjacent vapor molecules, eventually assuming the direction and velocity of the bulk vapor flow. A similar situation exists in the condenser, where a net flux of molecules enters the liquid surface at a given point from all directions and with a wide distribution of velocities. These molecules then collide with adjacent liquid molecules, eventually assuming the direction and velocity of the bulk liquid flow.

Frictional resistance at flow boundaries produces a pressure drop in both the liquid and the vapor streams. The addition of mass to a flow stream increases the rate of momentum flow (mass flow rate times velocity), causing a pressure drop in addition to that resulting from friction. Similarly, the removal of mass from a flow stream reduces the rate of momentum flow, producing a pressure rise. Because flow velocities in a heat pipe are much higher in the vapor stream than in the liquid stream, pressure changes resulting from variations in the momentum flow are generally only significant for the vapor stream.

Additional pressure changes can be produced by gravity or by external acceleration applied to the heat pipe. These changes can produce a pressure drop or pressure rise, depending on whether the evaporator end of the heat pipe is above or below the condenser and on the direction of the external acceleration. The magnitude of the pressure change depends on the fluid density as well as gravity/acceleration and heat pipe orientation. Pressure changes due to gravity/acceleration are only significant in the liquid flow stream of a heat pipe, because the liquid density is much larger than that of the vapor.

Finally, as the mass flow rate increases beyond some critical value, the flow pattern can change from laminar to turbulent. *Laminar flow* is characterized by a parabolic distribution of velocity over the flow path cross section with the velocity a maximum at the center of the flow stream and zero at the wall. The flow stream at each point along the cross section remains parallel to the wall, with no intermingling of fluid. *Turbulent flow* is characterized by a velocity distribution that is relatively flat over most of the flow cross section, and decreases to zero through a thin fluid layer at the wall, which is called the boundary layer. Except for the boundary layer, there is a continuous interchange of fluid between adjacent flow elements.

Numerous investigators have addressed the problems of determining the liquid and vapor pressure drops in a heat pipe (Refs. 4.1–4.4). The liquid pressure drop is the more straightforward of the two problems.

The pressure drop in the liquid flow through the wick is treated as a one-dimensional incompressible flow problem in which momentum-induced pressure changes are negligible. The flow rate is variable in the evaporator and condenser sections, and constant in the adiabatic section. Laminar flow will prevail in porous liquid flow channels for virtually any conceivable heat pipe transport rate. However, in open annular or grooved liquid flow channels, the flow may become turbulent at high heat transport rates. The liquid frictional pressure drop is added to the pressure change due to acceleration/gravity to obtain the total liquid pressure drop (or rise).

In analyzing the vapor pressure drop problem, the vapor is also treated as a laminar, incompressible fluid, with a variable flow rate in the evaporator and condenser sections and a constant flow rate in the adiabatic section. In contrast to the situation for liquid flow, momentum-induced pressure changes are generally significant whereas gravity/acceleration pressure changes are negligible. As is the case for liquid flow in open or grooved channels, the possibility of turbulent flow in the vapor at high heat transport rates must be considered.

Accurate analysis of the vapor pressure drop requires a two-dimensional approach, in which the evaporation and condensation processes are approximated by the radial injection and suction (removal) of a single-phase fluid. Results of the two-dimensional analysis can then be applied as correction factors to an equivalent one-dimensional analysis.

4.1.1. Radial Reynolds Number

A key parameter in the two-dimensional analysis of heat pipe vapor flow is the radial Reynolds number Re_r, defined as

$$\mathrm{Re}_r = \frac{\rho_v V_r r_v}{\mu_v} \tag{4.1}$$

where ρ_v is the vapor density, μ_v is the vapor viscosity, r_v is the vapor space radius, and V_r is the average radial velocity at the vapor passage boundary. The radial Reynolds number can be expressed in terms of the average radial heat flux q_r at the vapor passage boundary by noting that

$$q_r = \frac{Q}{A_{lv}} \tag{4.2}$$

and

$$Q = W\kappa = \rho_v V_r A_{lv} \kappa \tag{4.3}$$

In Eqs. (4.2) and (4.3), Q is the heat transport rate, W is the maximum mass flow rate of the heat pipe fluid, κ is the heat of vaporization, and A_{lv} is the heat transfer area at the liquid-vapor interface (the vapor passage boundary) in the evaporator or condenser.

Using Eqs. (4.2) and (4.3), Eq. (4.1) can be written

$$\mathrm{Re}_r = \frac{q_r r_v}{\kappa \mu_v} \tag{4.4}$$

When $\mathrm{Re}_r \ll 1$, viscous forces dominate in the vapor flow, and the pressure drop is well represented by the frictional pressure drop for the laminar flow of an incompressible fluid through a circular tube, also known as Poiseuille flow (Ref. 4.5).

When $\mathrm{Re}_r \gg 1$, viscous forces are negligible compared to inertial forces. The pressure drop in the laminar, incompressible vapor is then equal to the one-dimensional change in the rate of momentum flow multiplied by a momentum factor that is derived from the two-dimensional flow solution.

The radial Reynolds number Re_r in the evaporator is given in Table 4.1 for a number of heat pipe fluids at representative operating temperatures. The data apply to a heat pipe with a vapor passage radius of 0.25 in. (0.0064 m) and an evaporator radial heat flux (based on the heat transfer area at the liquid-vapor interface) of 100 Btu/ft^2-sec (1.13×10^6 W/m^2). The evaporator length for each fluid has been adjusted so that the total heat transfer rate is 20% of the sonic limit for that fluid. (The significance of the last two columns in Table 4.1 will be discussed shortly.)

Table 4.1. Evaporator radial and axial Reynolds number and axial temperature drop for vapor of several heat pipe fluids

Fluid	Temperature, °F(°C)	Vapor pressure, psia (10^4 N/m^2)	Radial Reynolds number Re_r	Axial Reynolds number Re_a	$\Delta T_v/\%\Delta P_v$, °F(°C)
Lithium	2200 (1204)	5.5 (3.79)	20	2,180	2.4 (1.3)
Sodium	1500 (816)	7.2 (4.96)	83	5,050	2.0 (1.1)
Potassium	1300 (704)	8.8 (6.07)	134	6,940	1.9 (1.1)
Cesium	1100 (593)	6.6 (4.55)	557	10,880	1.7 (0.9)
Water	180 (82)	7.5 (5.17)	278	15,990	0.5 (0.3)
Ammonia	0 (−18)	30.4 (21.0)	683	104,300	0.4 (0.2)

It is evident from Table 4.1 that the radial Reynolds number is not within the range $Re_r << 1$ where viscous forces dominate. Since $Re_r >> 1$ for most fluids, inertial forces will be significant if not dominant. In most operating heat pipes, both viscous and inertial forces will contribute to the vapor pressure drop.

4.1.2. Vapor Pressure Drop in the Evaporator

Numerical solutions to the two-dimensional laminar incompressible vapor flow problem were obtained by Bankston and Smith (Ref. 4.4) for a radial Reynolds number range of 0.01 to 1000. It was found that the pressure drop in the evaporator can be expressed as the sum of frictional and momentum pressure drop terms, as would be expected.

A friction factor f and momentum factor Ω are associated with these terms. Both f and Ω are constants at a given Re_r, and are determinable from the vapor velocity distribution.*

When $Re_r \leq 0.1$, the velocity distribution is virtually indistinguishable from Poiseuille flow, and f and Ω take on the following values:

$$f_{po} = 64$$
$$\Omega_{po} = 4/3$$

where the subscript denotes Poiseuille flow.

As the radial Reynolds number increases beyond 1, the velocity distribution becomes flatter than the parabolic distribution characteristic of Poiseuille flow, and f and Ω become functions of Re_r. When $Re_r \geq 100$, f and Ω once again become constant.

The variation of f/f_{po} and Ω/Ω_{po} with Re_r is shown in Table 4.2.

It is evident that the friction factor f for vapor flow in the evaporator of a heat pipe is at most about 25% greater than that for Poiseuille flow, and that the momentum factor

* The value of the friction factor depends on how the frictional pressure drop term is formulated. Bankston and Smith (Ref. 4.4) use $f/4$ for the friction factor. Then, in the frictional pressure drop equation based on the vapor space diameter, the friction factor is multiplied by a factor of 4. In this book f is used for the friction factor, avoiding the need to include an extra factor of 4 in the frictional pressure drop equation.

Table 4.2. Variation of friction factor and momentum factor with radial Reynolds number

Radial Reynolds number Re_r	f/f_{po}	Ω/Ω_{po}
<< 0.1	1.000	1.000
0.1	1.010	0.998
1.0	1.068	0.984
10.0	1.202	0.947
100.0	1.245	0.929
>> 100.0	—	0.926

Ω is at the least about 8% less than for Poiseuille flow. Further, Poiseuille flow is dominant for $Re_r \leq 0.1$, and inertial flow dominates for $Re_r \geq 100$.

Although Table 4.2 could be used in conjunction with appropriate vapor frictional and momentum pressure drop terms to calculate the vapor pressure drop in the evaporator of a heat pipe, a simpler procedure is to use the friction factor $f_{po} = 64$ corresponding to $Re_r << 1$ and the momentum factor $\Omega = \pi^2/8 = 1.234$ corresponding to $Re_r >> 1$. Then the vapor pressure drop equation will give exact results at these extremes of Re_r, and acceptable accuracy for Re_r between these extremes.

The error associated with this simplification is shown in Figure 4.1. The maximum error of about –6% occurs at about $Re_r = 1$. For most cases of interest, Re_r will be on the

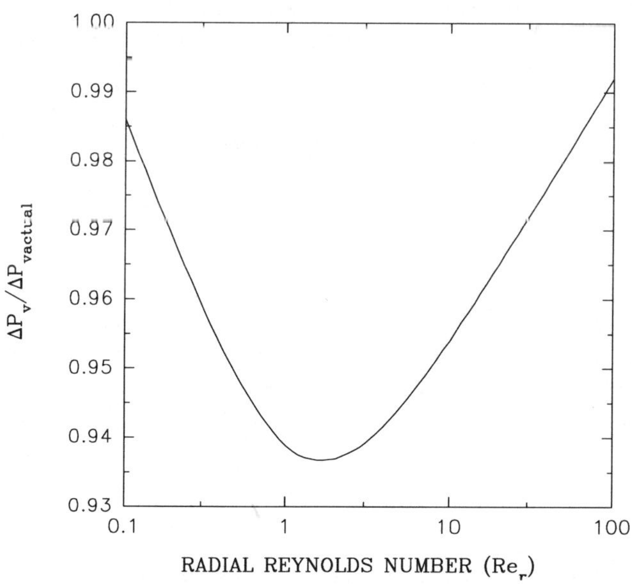

Figure 4.1. Ratio of simplified vapor pressure drop in a heat pipe evaporator to actual vapor pressure drop versus radial Reynolds number.

order of 30 or more, and the error will be less than 3%. An accuracy on the order of 10% is generally adequate for design calculations.

4.1.3. Vapor Pressure Drop in the Condenser

Although the vapor pressure drop in the evaporator is expressible as the sum of one-dimensional frictional and momentum terms modified by factors to correct for the two-dimensional nature of the flow, the situation is much more complex in the condenser section.

Generally, the condenser pressure drop depends on the evaporator radial Reynolds number, the condenser radial Reynolds number, and the condenser length (Ref. 4.4). Since the vapor mass flow rate decreases as vapor condenses along the condenser length, there is a decrease of vapor momentum, which is accompanied by an increase in pressure. If the radial Reynolds number in the condenser is greater than one, there can be some net recovery of vapor pressure in the condenser.

In the evaporator, the pressure variation in the radial direction is relatively small, and the radial velocity profile is relatively constant. On the other hand, in the condenser the pressure can vary sharply in the radial direction, and the radial velocity profile can change markedly with axial position.

If the radial Reynolds number is reasonably larger than 1, toward the end of the condenser the vapor pressure at the wall can be substantially lower than that at the heat pipe axis, largely negating pressure recovery at the wall. As condensation of vapor proceeds, the axial velocity at the wall can become negative, indicating that the vapor is flowing toward the evaporator. At the same time, the radial velocity gradient at the wall, upon which the viscous forces that determine the frictional pressure loss depend, may become zero or even negative.

In view of the complexities of vapor flow in the condenser, it would appear that the use of a simple one-dimensional equation to calculate the vapor pressure drop is not feasible.

4.1.4. Total Vapor Pressure Drop

In determining the capillary pumping limit, the total pressure drop in the heat pipe vapor and liquid are required rather than the individual drops in the evaporator and condenser. Cotter (Ref. 4.1) showed that, for $Re_r \ll 1$, the small deviations of the vapor from Poiseuille flow in the evaporator and condenser cancel out. Then the pressure drop for Poiseuille flow represents an exact solution for the total vapor pressure drop in a heat pipe.

Busse (Ref. 4.2) showed that the Poiseuille flow solution for the total vapor pressure drop is valid for any evaporator radial Reynolds number, and for condenser radial Reynolds numbers up to around 1. Bankston and Smith (Ref. 4.4) confirmed the general applicability of the Poiseuille solution for Re_r up to 1, and its applicability up to $Re_r = 10$ for heat pipe length-radius ratios of 40 or greater. Since $Re_r > 10$ for many heat pipe design situations (see Table 4.1), the Poiseuille pressure drop equation alone may be inadequate for predicting the total vapor pressure drop in most heat pipes.

4.1.5. Stability of Vapor Flow

In the two-dimensional analytical treatment of vapor flow in a heat pipe, it is assumed that the vapor flow remains laminar, even for radial Reynolds numbers much larger than 1. Inertial forces are then dominant and viscous forces are negligible. This condition usually leads to the development of turbulent flow when the flow rate is constant. The question of flow stability in the vapor of heat pipes will now be considered.

The stability of fluid flowing through a tube at a constant mass flow rate (i.e., without the radial injection or removal of fluid) is characterized by the *axial Reynolds number,* defined as

$$\mathrm{Re}_a = \frac{\rho V D}{\mu} \tag{4.5}$$

where ρ is the fluid density, μ is the fluid viscosity, V is the average fluid velocity, and D is the flow passage diameter (Ref. 4.6). Two distinct flow regimes can exist, depending on the value of Re_a.

When $\mathrm{Re}_a < 2000$, the flow is laminar, and all the fluid particles move parallel to the flow passage axis. The radial velocity distribution at any axial position is parabolic, falling from a maximum value at the axis to zero at the wall.

When $\mathrm{Re}_a > 4000$, the flow becomes turbulent, and its character changes radically. Extensive, random interchanging of adjacent flow particles takes place. The velocity distribution becomes relatively flat over most of the flow passage cross section, falling off sharply to zero in a thin boundary layer near the wall. The frictional resistance during turbulent flow is considerably larger than would be the case if the flow were to remain laminar.

Between an axial Reynolds number of 2000 and 4000, the flow is in transition and may exhibit both laminar and turbulent flow characteristics.

The stability of the vapor flow within a heat pipe is far less certain than for the constant flow rate situation in a tube. The flow rate increases continuously in the evaporator section, remains constant in the adiabatic section, and decreases continuously in the condenser section. There is evidence that the continuous suction of fluid from a flow stream maintains the flow in a laminar state for axial Reynolds numbers at which the flow would normally be expected to be turbulent (Ref. 4.7).

In a heat pipe, the abrupt transition from injection of vapor to removal of vapor at an evaporator-condenser junction could disturb the stability of what otherwise might be a laminar flow pattern, inducing turbulence. If the axial Reynolds number leaving the evaporator is greater than 4000, turbulence would be expected to develop in a following adiabatic section, persist well into the adjacent condenser section, and possibly propagate partially back into the evaporator section.

The axial Reynolds number Re_a of the vapor at the evaporator exit is shown in

Table 4.1 for the case in which the heat transfer rate is 20% of the sonic heat transport limit. Re_a is seen to be greater than 4000 for all of the heat pipe fluids except lithium. Thus, the potential for turbulent vapor flow exists for many heat pipe fluids during operation at high heat transport rates.

4.2. GENERAL PRESSURE DROP FORMULATION

A general one-dimensional pressure drop formulation applicable to any constant-density fluid will now be developed. The results will then be applied to obtain pressure drop relationships for the heat pipe vapor and liquid. The assumption of constant density for the vapor is reasonable as long as the vapor pressure drop is less than about 10%.

When the general pressure drop results are applied to heat pipes, frictional and momentum terms are included in the vapor pressure drop formulation, and frictional and gravity/acceleration terms are included in the liquid pressure drop formulation.

Reasonably isothermal heat transport is desired in most heat pipe design situations (i.e., the axial temperature drop in the heat pipe vapor should be no more than a few degrees). Table 4.1 shows that the vapor pressure drop will generally be less than a few percent if the isothermal heat transport requirement is to be met. Thus, the assumption of constant vapor density is valid for most heat pipe design situations.

When flow is laminar, the frictional pressure drop is assumed to be the same as that for laminar flow of an incompressible fluid flowing at a constant rate through a tube (i.e., for Poiseuille flow).

It is also assumed that turbulent flow will develop in the heat pipe fluid if the axial Reynolds number exceeds some critical value. Since the frictional resistance at a given flow rate is higher for turbulent flow than for laminar flow, this is a conservative design assumption. The situation for turbulent flow will be discussed shortly.

4.2.1. Pressure Drop for One-Dimensional Incompressible Flow

Consider the situation in Figure 4.2. An incompressible fluid with density ρ flows through a circular tube segment of diameter D and length dx. The tube is oriented at an angle θ with the vertical. The tube is also subjected to an external acceleration of n g's, directed at an angle of ϕ with the tube axis, where g is the acceleration of gravity.

Fluid enters the tube segment at an average velocity V, mass flow rate W, and pressure P. Fluid is also added radially through the wall of the tube segment at velocity V_r. The fluid leaves the tube segment at velocity $V + dV$, mass flow rate $W + dW$, and pressure $P + dP$. Within the segment, a shear stress τ_s acts at the wall in a direction opposite to that of the fluid flow.

The laws of conservation of mass and conservation of momentum will now be applied to the fluid flow.

Conservation of mass requires that the mass flow rate leaving the tube segment be equal to the sum of the mass flow rates that enter the segment axially and radially. The

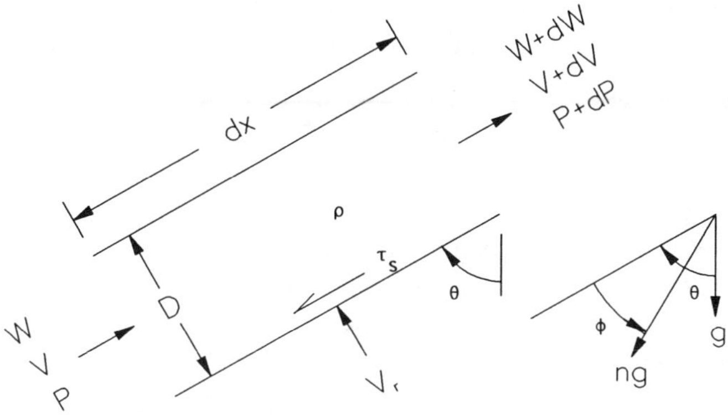

Figure 4.2. The one-dimensional incompressible flow model.

mass flow rate is equal to the product of the density, the mean fluid velocity, and the flow area normal to the velocity. Thus, we may write

$$\rho V(\pi/4)D^2 + \rho V_r \pi D dx = \rho(V + dV)(\pi/4)D^2$$

from which it follows that

$$dV = (4V_r/D)dx \tag{4.6}$$

Equation (4.6) relates the increment in axial velocity to the radial inlet velocity.

Conservation of momentum requires that the rate of axial momentum flow leaving the tube segment be equal to the sum of the axial momentum flow rate entering the segment and the net axial force acting on the segment. The net axial force acts in the direction of momentum flow rate increase. The rate of momentum flow is equal to the product of the mass flow rate, the velocity, and the momentum factor Ω.

The momentum factor Ω, which was introduced during the discussion of the two-dimensional vapor pressure drop problem in Section 4.1.2, is equal to the ratio of the total momentum flow rate over the flow tube flow area to the momentum flow rate based on the average velocity. As such, the momentum factor will be recognized as a correction factor to the momentum flow rate calculated from one-dimensional considerations.

We may then write the following momentum-force balance:

$$\Omega(W + dW)(V + dV) = \\ \Omega WV + (\pi/4)D^2P - (\pi/4)D^2(P - dP) - \tau_s \pi D dx + (\pi/4)D^2(-\rho g \cos\theta\, dx + \rho n g \cos\phi dx) \tag{4.7}$$

The component of gravity along the tube axis produces a body force in the fluid that acts opposite to the fluid flow direction. The component of external acceleration along the tube axis, on the other hand, acts to accelerate the flow tube, in effect producing a body force in the fluid in the same direction as the fluid flow.

Assuming that the product of the differentials $dWdV = 0$, Eq. (4.7) can be manipulated to yield

$$dP = (4/D)\tau_s dx + 2\Omega\rho V dV + (\cos\theta - n\cos\phi)\rho g dx \tag{4.8}$$

Upon integrating Eq. (4.8) over the tube length from $x = 0$ to $x = L$ and assuming that Ω is constant, the general pressure drop equation is obtained:

$$\Delta P = \int_0^L \frac{4}{D}\tau_s\, dx + \Omega\rho(V_o^2 - V_i^2) + (\cos\theta - n\cos\phi)\rho g L \tag{4.9}$$

Equation (4.9) shows that the pressure drop in the fluid is the sum of a frictional shear force term (the integral), a momentum term, and a gravity/acceleration term.

It is customary to express the frictional shear force acting on the tube wall as equivalent to a differential frictional pressure drop dP_f acting over the tube cross-sectional area. Then,

$$\pi D\tau_s dx = (\pi/4)D^2 dP_f \tag{4.10}$$

The differential frictional pressure drop is defined as proportional to the product of the tube length-diameter ratio dx/D and the dynamic pressure $\rho V^2/2$, with a proportionality factor f called the friction factor. Thus,

$$dP_f = f\,\frac{dx}{D}\,\rho\,\frac{V^2}{2} \tag{4.11}$$

Upon substituting Eq. (4.11) into (4.10), it follows that

$$\tau_s = \frac{f}{4}\left(\rho\,\frac{V^2}{2}\right) \tag{4.12}$$

Therefore, the frictional shear force term in Eq. (4.9) can be expressed in terms of the frictional pressure drop ΔP_f as

$$\Delta P_f = \int_0^L \frac{4}{D}\tau_s\, dx = \int_0^L f\rho\,\frac{V^2}{2}\,\frac{dx}{D} \tag{4.13}$$

where, from Eq. (4.12),

$$f = \frac{8\tau_s}{\rho V^2} \tag{4.14}$$

In order to perform the required integration of Eq. (4.13), the friction factor f and the distribution of the average axial velocity V along the flow path must be known. Suitable expressions for f and V will now be developed.

4.2.2. Friction Factor for Laminar Flow

For laminar flow, the friction factor can be found as an explicit function of the axial Reynolds number, as will be shown below for Poiseuille flow (laminar flow in a tube).

The shear stress at the wall is defined by the relation

$$\tau_s = \mu\left(\frac{dV_a}{dy}\right)_{y=0} \tag{4.15}$$

where y is the radial distance from the wall, V_a is the axial fluid velocity at y and μ is the fluid viscosity.

For Poiseuille flow, the radial distribution of the axial velocity at any point along the tube is parabolic (Ref. 4.5), and may be expressed in the form

$$V_a = V_{max}\left[1-\left(\frac{r_o-y}{r_o}\right)^2\right] \tag{4.16}$$

where V_{max} is the maximum velocity (on the tube axis) and r_o is the tube radius.

The average flow velocity V over the flow tube cross-section is given by

$$V = \frac{\int_0^{r_o} V_a\, 2\pi r dr}{\pi r_o^2} \tag{4.17}$$

If Eq. (4.16) is substituted for V_a in Eq. (4.17), it follows that

$$V = V_{max}/2 \tag{4.18}$$

That is, the average velocity is half the maximum velocity.

Now if Eqs. (4.16) and (4.18) are substituted into Eq. (4.15), and noting that $D = 2r_o$, the wall shear stress τ_s is expressible as

$$\tau_s = 8\mu V/D \tag{4.19}$$

On substituting τ_s from Eq. (4.19) into Eq. (4.14) and using Eq. (4.5), it follows that

$$f = \frac{64\mu}{\rho V D} = \frac{64}{\mathrm{Re}_a} \tag{4.20}$$

Thus, for laminar flow the friction factor f varies inversely as the axial Reynolds number.

4.2.3. Friction Factor for Turbulent Flow

For turbulent flow, f is a slowly varying inverse function of the axial Reynolds number and the relative roughness of the wall (ratio of average height of surface irregularities to the tube diameter). The friction factor varies from a minimum of 0.018 at $\mathrm{Re}_a = 100{,}000$ and a relative roughness of 0.00001 (a very smooth tube) to a maximum of 0.050 at $\mathrm{Re}_a = 4000$ and a relative roughness of 0.01 (a very rough tube) (Ref. 4.6).

It is difficult to characterize the relative roughness in a heat pipe. Although the relative roughness of the materials used inside a heat pipe may be on the order of 0.0001 or less, the wick structure itself may produce a considerably higher relative roughness.

For example, the wires of 200 mesh screen, a representative wick material, are 0.0023 in. (5.84×10^{-5} m) in diameter. If the average height of wire that protrudes above the liquid in the adjacent pores is one-quarter of the wire diameter and the vapor space diameter is 0.5 in. (0.0127 m), the relative roughness of the vapor passage would be (0.0023/4)/0.5 = 0.001. The turbulent friction factor would then be as follows:

Re_a	f
5,000	0.039
10,000	0.037
100,000	0.022

Since the curvature of the liquid-vapor interface between wires varies from a maximum at the evaporator end to essentially zero at the condenser end, the relative roughness of the vapor passage can be expected to vary in a similar manner. Even if the relative roughness, measured by the effective extension of the wick structure beyond the liquid interface, could be determined precisely over the heat pipe length, the author is not aware of any experimental correlations between the turbulent friction factor and the relative roughness due to an internal wick structure. Further, the relative roughness can be expected to increase with the heat transport rate, as the diameter of curvature in the wick pores becomes smaller.

In view of the considerable uncertainty associated with the magnitude of the turbulent friction factor in a heat pipe, a conservative approach is warranted. A value of $f = 0.04$—which is judged likely to be equal to or greater than the actual turbulent friction factor for most heat pipe design situations—is suggested for design purposes.

4.2.4. Dependence of Mean Axial Velocity on Axial Distance

In order to obtain the frictional pressure drop by integrating Eq. (4.13), the dependence of the average axial velocity V on the axial distance x must be known. From Eq. (4.6), it is evident that the x dependence of V is dependent on the x dependence of the radial velocity V_r at the flow duct wall.

In developing the pressure drop equations, it is more convenient to use the surface mass flux w, or the mass flow rate entering at the tube wall per unit wall area, instead of V_r. The surface mass flux is defined as

$$w = \rho V_r \tag{4.21}$$

Upon substituting Eq. (4.21) into Eq. (4.6) and integrating over the axial distance x,

$$V = V_i + \frac{4}{\rho D}\int_0^x w\,dx \tag{4.22}$$

where V_i is the mean axial flow velocity at $x = 0$.

If the surface mass flux distribution over the tube length L is given, the axial velocity distribution can be found from Eq. (4.22).

4.3. FRICTIONAL PRESSURE DROP FOR CONSTANT SURFACE MASS FLUX

The frictional pressure drop is found by substituting appropriate expressions for V and f into Eq. (4.13).

If the surface mass flux w is constant, then evaluation of Eq. (4.22) gives

$$V - V_i = \frac{4wx}{\rho D} \tag{4.23}$$

At the end of the tube, where $x = L$, the mean axial velocity is V_o. Therefore, for $x = L$ the above equation becomes

$$V_o - V_i = \frac{4wL}{\rho D} \tag{4.24}$$

Substituting $4w/(\rho D)$ from Eq. (4.24) into Eq. (4.23) gives the desired result for V:

$$V = V_i + (V_o - V_i)\left(\frac{x}{L}\right) \tag{4.25}$$

The average axial velocity V_{aa} along the flow length, obtained by integrating V over the flow length L and then dividing by L, is

$$V_{aa} = \frac{(V_i + V_o)}{2} \tag{4.26}$$

We now consider the frictional pressure drop ΔP_{fl} for laminar flow.

Upon substitution of Eq. (4.20) for f into Eq. (4.13), we obtain

$$\Delta P_{fl} = \frac{32\mu}{D}\int_0^L V\,dx \tag{4.27}$$

Thus, the laminar frictional pressure drop is directly proportional to the integral of the axial velocity V over the flow length. This is a general result for laminar flow, valid for any surface mass flux distribution.

On substituting Eq. (4.25) for V into Eq. (4.27) and using Eq. (4.26), the laminar frictional pressure drop ΔP_{fl} for a constant-surface heat flux is found to be

$$\Delta P_{fl} = 16\mu L\,\frac{(V_i + V_o)}{D^2} \tag{4.28}$$

For turbulent flow, f = constant = 0.04, according to the previous discussion. Letting f be constant in Eq. (4.13) and again substituting Eq. (4.25) for V, the pressure drop ΔP_{ft} for turbulent flow is obtained:

$$\Delta P_{ft} = \frac{f\rho L V_o^2}{6D}\left[1 + \frac{V_i}{V_o} + \left(\frac{V_i}{V_o}\right)^2\right] \tag{4.29}$$

In an adiabatic region, the surface mass flux is zero, and the axial velocity and mass flow rate are constant over the flow length. Then $V_o = V_i$, and Eq. (4.28) for laminar flow becomes

$$\Delta P_{fl} = 32\mu L\,\frac{V_i}{D^2} \tag{4.30}$$

Similarly, Eq. (4.29) for turbulent flow becomes

$$\Delta P_{ft} = f\rho L \frac{V_i^2}{2D} \tag{4.31}$$

If the surface mass flux is negative, fluid is being removed from the flow passage. This is the situation that occurs during the condensation of vapor in a heat pipe. The mean axial velocity V then decreases with distance along the flow tube, and the exit velocity V_o is smaller than the inlet velocity V_i. Equations (4.28) and (4.29) are valid for injection or suction, as long as the surface mass flux w remains constant.

Further, it should be noted that the velocity terms in Eqs. (4.28) and (4.29) are symmetrical with respect to the inlet and exit velocities. This means that, if the inlet and exit velocities of an injection region are equal, respectively, to the exit and inlet velocities of a suction region, the velocity terms for the two cases will be equal. This fact is utilized when the above results are applied to the pressure drop in a heat pipe.

4.4. FRICTIONAL PRESSURE DROP FOR NONCONSTANT SURFACE MASS FLUX

The laminar frictional pressure drop can be found for other surface mass flux distributions by following the same procedure as was used in the case of constant w. That is, the surface mass distribution is used in Eq. (4.22) to obtain the velocity distribution, which is then introduced into Eq. (4.13) to obtain the pressure drop.

Laminar flow results for a number of distributions relative to the constant-w case are summarized in Table 4.3. In Table 4.3, the entering velocity V_i is taken to be zero for simplicity. The flow tube geometry, the mean surface mass flux, and the total mass injection rate are the same for each case. The magnitude of the maximum surface mass flux w_{max} differs for each case.

For the symmetrical distributions of Cases 4 and 5, the pressure drop is the same as for a constant w (Case 1). The pressure drop is greater than that in the constant-w case when the w distribution decreases with x (Cases 2a and 3a) and less than that in the constant-w case when the w distribution increases with x (Cases 2b and 3b).

Unlike the situation for laminar flow, the frictional pressure drop for turbulent flow

Table 4.3. Relative laminar frictional pressure drop for various surface mass flux distributions

Case	Mass flux distribution	Relative pressure drop
1	$w = \text{constant}$	1
2a	$w = w_{max}(1 - x/L)$	4/3
2b	$w = w_{max}(x/L)$	2/3
3a	$w = w_{max} \cos[(\pi/2)(x/L)]$	$4/\pi$
3b	$w = w_{max} \cos[(\pi/2)(1 - x/L)]$	$2(1 - 2/\pi)$
4	$w = 2w_{max}(x/L); \quad 0 \le x \le L/2$ $w = 2w_{max}(1 - x/L); \quad L/2 \le x \le L$	1
5	$w = w_{max} \cos[(\pi/2)(1 - 2x/L)]$	1

Table 4.4. Hydraulic diameter for various flow passage cross sections

Shape	Hydraulic diameter
Circle (diameter D)	D
Rectangle (width t_w, height t_h)	$2t_w t_h/(t_w + t_h)$
Thin annulus (inner diameter D, thickness t_1, $t_1/D << 1$)	$2t_1$
Circle with enclosed arteries (diameter D, outer artery diameter d_o, N arteries)	$D[1 - N(d_o/D)^2]/[1 + N(d_o/D)]$

with a symmetrical surface mass flux distribution will be greater than that for the constant-w case.

4.5. HYDRAULIC DIAMETER

The development of the pressure drop equations presented thus far is based on the assumption of a circular flow passage cross section. The results may be generalized to any other cross-section by using the hydraulic diameter instead of the diameter of a circle in the pressure drop and Reynolds number expressions.

The *hydraulic diameter* D_h is defined as

$$D_h = 4A_{cs}/B_{cs} \tag{4.32}$$

where A_{cs} is the flow passage cross-sectional area and B_{cs} is the boundary of the cross section. B_{cs} is also referred to as the *wetted perimeter.* The ratio A_{cs}/B_{cs} is also known as the *hydraulic radius* (Ref. 4.6).

For a circular cross section, $A_{cs} = \pi D^2/4$ and $B_{cs} = \pi D$. Therefore, from Eq. (4.32), $D_h = D$. Thus, the hydraulic diameter and actual diameter are the same for a circular cross section. Expressions for the hydraulic diameter of circles, rectangles, and annuli are given in Table 4.4.

4.6. FRICTIONAL PRESSURE DROP IN A POROUS FLOW PASSAGE

A special situation arises when the flow passage is porous, being formed from an assembly of particles, fibers, or screens. Although flow through a porous medium can be characterized as flow through individual pores, modified by a tortuosity factor that accounts for flow path irregularities, it is more convenient to consider the flow passage as a whole.

According to Darcy's law (Ref. 4.8), the pressure gradient through a porous medium can be expressed as

$$\frac{dP_{fp}}{dx} = K\left(\frac{\mu}{\rho}\right)\left(\frac{W}{A_{cs}}\right) \tag{4.33}$$

where μ is the fluid viscosity, ρ is the fluid density, W is the mass flow rate, A_{cs} is the total cross-sectional area of the flow passage, K is the friction factor for the porous medium, and dP_{fp} is the differential frictional pressure drop over the differential length dx. The reciprocal of the friction factor, $1/K$, is called the *permeability*.

Darcy's equation applies only for laminar flow and negligible inertial effects. These conditions are generally met in the porous wick structure of a heat pipe.

From Eqs. (4.11) and (4.20), and noting that $W = \rho V A_{cs}$, the frictional pressure gradient for laminar flow in an open channel is

$$\frac{dP_{fl}}{dx} = \left(\frac{32}{D_h^2}\right)\left(\frac{\mu}{\rho}\right)\left(\frac{W}{A_{cs}}\right) \tag{4.34}$$

If we let ε_{po} be the ratio of the laminar pressure drop in a porous channel to the laminar pressure drop of the same fluid at the same flow rate in an open channel with the same overall cross section, then

$$\varepsilon_{po} = dP_{fp}/dP_{fl} = KD_h^2/32 \tag{4.35}$$

Thus, the pressure drop of a fluid in a porous channel can be found by multiplying its laminar pressure drop through an open channel of the same dimensions by the correction factor ε_{po}.

To illustrate the process, the correction factor will now be calculated for a thin annular liquid flow channel in a heat pipe that has been fabricated from four layers of 100 mesh screen. From Table C.8 in Appendix C, the wire diameter is 0.0045 in. (1.14×10^{-4} m) and the wick friction factor is 4.21×10^6 in.$^{-2}$ (6.53×10^9 m^{-2}). The thickness of each screen layer is twice the wire diameter, and that of four layers is eight times the wire diameter, or 0.036 in. (8.89×10^{-4} m). This is the thickness t_1 of the annular flow channel. From Table 4.4, the hydraulic diameter D_h of a thin annular flow channel is $2t_1 = 0.072$ in. (1.13×10^{-3} m).

The porous channel correction factor is then

$$\varepsilon_{po} = KD_h^2/32 = (4.21 \times 10^6 \text{ in.}^{-2})(0.072 \text{ in.})^2/32 = 682$$

Thus, the pressure drop in the porous channel is almost 700 times larger than that expected if the channel were open.

It is also instructive to compare the thickness of annular porous and open flow channels for the same pressure drop. This can be done by equating Eqs. (4.33) and (4.34), and solving for the thickness of the open channel. The open channel thickness is found to be 0.00408 in. (1.04×10^{-4} m). Thus, for the same pressure drop the thickness of the open annular channel is about one-ninth that of the porous channel.

It is apparent that the open channel offers decisive advantages over the porous channel when high flow rates and heat transport rates are required. However, potential problems with restarting a heat pipe with an open liquid flow channel could develop if the heat pipe is subject to gaseous diffusion. (See Section 3.5.2.) When high reliability during heat pipe restart is needed, a porous or grooved liquid flow channel may be preferable.

4.7. FRICTIONAL PRESSURE DROP FOR CONVECTIVE COOLING

A common technique for removing heat from a heat pipe is to flow a coolant through an annular jacket surrounding the condenser section. Since the pressure drop in the heat

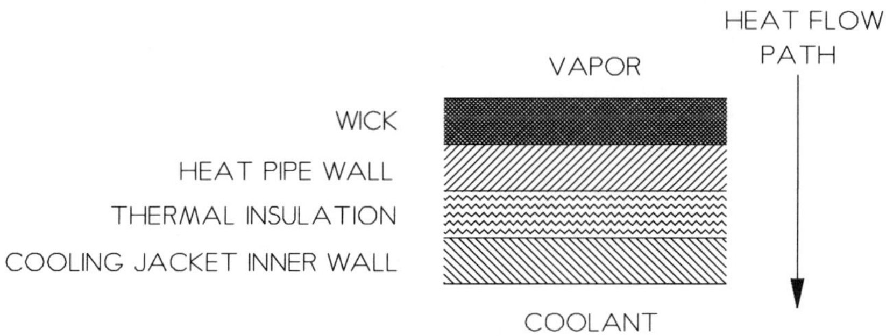

Figure 4.3. Heat flow path from heat pipe condenser to coolant.

pipe fluid during convective cooling of a heat pipe is of considerable importance, it is treated in some detail here.

If the heat pipe operating temperature is appreciably greater than that of the coolant, the addition of an insulating layer between the heat pipe and the coolant may be necessary. The heat rejection path is then as depicted in Figure 4.3. Heat from condensing vapor flows through the wick, the heat pipe wall, the insulating layer, and the cooling jacket inner wall, and finally into the coolant.

The surface mass flux distribution in the condenser during convective cooling (i.e., the vapor condensation rate per unit surface area) is dependent on the coolant flow rate and the thermal resistance between the condensing vapor and the coolant. Therefore, the heat transfer process during convective cooling must be analyzed before the pressure drop can be determined.

4.7.1. Temperature Distribution of Coolant

The initial step is to determine the temperature distribution of the coolant over the heat pipe condenser length. As indicated in Figure 4.4, the heat pipe vapor temperature T_v is assumed to be constant, whereas the coolant enters the cooling jacket at temperature T_{ci} and leaves at T_{co}. At the distance x from the point where vapor enters the condenser, the coolant temperature is T.

The differential rate dQ at which heat is transferred from the condensing heat pipe vapor to the coolant at x is given by

$$dQ = U_c B_c (T_v - T)dx = WcdT \tag{4.36}$$

where W is the coolant mass flow rate, c is the coolant specific heat, B_c is the inner periphery of the cooling jacket, and U_c is the overall thermal conductance per unit area between the heat pipe vapor and the coolant. The overall thermal conductance can be found from the relation

$$\frac{1}{U_c B_c} = \frac{1}{h B_c} + \sum_{i=1}^{m} \frac{t_{1i}}{k_i B_{mci}} \tag{4.37}$$

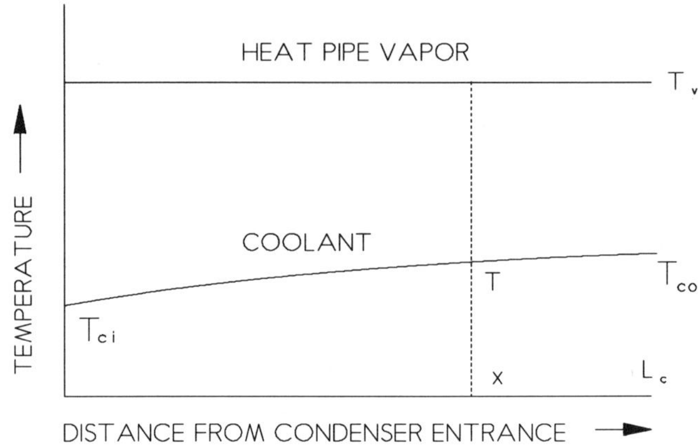

Figure 4.4. Temperature distribution over a heat pipe condenser with convective cooling.

The first term on the right of the equation is the convective resistance to heat flow at the cooling jacket inner periphery. The second term is the sum of the conductive resistances between the heat pipe vapor and the coolant.

The additional variables are defined as follows: h is the heat transfer coefficient of the coolant, t_{1i} is the thickness of the ith conductive layer, B_{mci} is the mean periphery in the ith conductive layer, and k_i is the thermal conductivity of the ith conductive layer. The convective resistance at the condensing surface of the heat pipe wick is assumed to be negligible, and has not been included in Eq. (4.37).

Integration of Eq. (4.36) over T gives the total heat transfer rate Q, or

$$Q = Wc(T_{co} - T_{ci}) \tag{4.38}$$

The maximum possible heat transfer rate Q_{max} would occur if T_{co} were equal to T_v. Thus,

$$Q_{max} = Wc(T_v - T_i) \tag{4.39}$$

Equation (4.36) can also be written in the form

$$\frac{dT}{T_v - T} = \frac{U_c B_c}{Wc} dx \tag{4.40}$$

Now let

$$\mathrm{NTU} = \frac{U_c B_c L_c}{Wc} \tag{4.41}$$

The dimensionless variable NTU is called the *number of heat transfer units,* and is a measure of the needed heat transfer area (Ref. 4.9). L_c is the condenser length. Equation (4.40) can then be written

$$\frac{dT}{T_v - T} = \frac{\mathrm{NTU}}{L_c} dx \tag{4.42}$$

Assuming that NTU is constant, Eq. (4.42) can be integrated over x to yield the *coolant temperature distribution:*

$$T = T_v - (T_v - T_{ci})\, e^{-NTU(x/L_c)} \tag{4.43}$$

When $x = L_c$, $T = T_{co}$. Therefore, from Eq. (4.43),

$$T_{co} = T_v - (T_v - T_{ci})\, e^{-NTU} \tag{4.44}$$

4.7.2. Heat Rate Distribution

Now let

$$\varepsilon = \frac{Q}{Q_{max}} = \frac{T_{co} - T_{ci}}{T_v - T_{ci}} \tag{4.45}$$

ε is called the *heat exchanger effectiveness,* and is equal to the ratio of the actual heat transfer rate Q to the maximum possible heat transfer rate Q_{max}. Upon substituting Eq. (4.44) into Eq. (4.45),

$$\varepsilon = 1 - e^{-NTU} \tag{4.46}$$

Thus, the heat exchanger effectiveness is dependent only on the number of heat transfer units. Equation (4.46) is characteristic of any parallel or counterflow heat exchanger with one condensing or evaporating fluid (Ref. 4.9). The subject of heat exchangers is treated in more detail in Section 6.6.

The heat rate distribution will now be found. It is needed to determine the velocity distribution in the heat pipe vapor, which in turn is used to find the frictional pressure drop. First, Eq. (4.43) is substituted in Eq. (4.36) to obtain

$$dQ = U_c\, B_c\, (T_v - T_{ci})\, e^{-NTU(x/L_c)} dx \tag{4.47}$$

Then, substituting for $T_v - T_{ci}$ from Eq. (4.39) and using Eq. (4.41), we obtain

$$dQ = NTU \left(\frac{Q_{max}}{L_c} \right) e^{-NTU(x/L_c)} dx \tag{4.48}$$

Equation (4.48) is the desired form for the heat removal rate distribution over the condenser length with convective cooling.

The total heat removal rate at the distance x is obtained by integrating Eq. (4.48). Then,

$$\frac{Q}{Q_{max}} = 1 - e^{-NTU(x/L_c)} \tag{4.49}$$

It should be noted that when $x/L_c = 1$, $Q/Q_{max} = \varepsilon$.

4.7.3. Velocity Distribution

The velocity distribution of the heat pipe vapor over the condenser length will now be obtained.

The heat transport rate at x is equal to the rate at which heat is being transported into the condenser at $x = 0$, minus the rate at which heat has been transferred out of the condenser to the coolant. Thus, we may write

$$\rho_v V_v A_v \kappa = \rho_v V_{vi} A_v \kappa - Q \tag{4.50}$$

where V_{vi} is the mean vapor velocity at the condenser entrance, V_v is the mean velocity at x, and A_v is the vapor passage flow area.

Substituting Eq. (4.49) for Q and solving for V_v,

$$V_v = V_{vi} - \frac{Q_{\max}}{\rho_v A_v \kappa} [1 - e^{-\mathrm{NTU}(x/L_c)}] \tag{4.51}$$

At the end of the condenser, $x = L_c$ and $V_v = 0$. Using this information along with Eq. (4.46), we obtain from Eq. (4.51)

$$V_{vi} = \frac{Q_{\max}}{\rho_v A_v \kappa} (1 - e^{-\mathrm{NTU}}) = \frac{Q_{\max}}{\rho_v A_v \kappa} \varepsilon \tag{4.52}$$

Equation (4.51) can then be written

$$\frac{V_v}{V_{vi}} = 1 - \frac{1}{\varepsilon} [1 - e^{-\mathrm{NTU}(x/L_c)}] \tag{4.53}$$

Equation (4.53) is the desired form of the velocity distribution for the heat pipe vapor in the condenser.

4.7.4. Pressure Drop Correction Factor for Convective Cooling

From Eq. (4.27), the frictional vapor pressure drop for laminar flow is proportional to the integral of V over x. Carrying out this integration with V_v from Eq. (4.53), and using Eq. (4.46),

$$\int_0^{L_c} V_v \, dx = V_{vi} L_c \left(1 + \frac{1}{\mathrm{NTU}} - \frac{1}{\varepsilon}\right) \tag{4.54}$$

For a constant heat removal rate, the vapor velocity decreases linearly with x and may be written as

$$V_v = V_{vi} \left(1 - \frac{x}{L_c}\right) \tag{4.55}$$

Integrating this velocity distribution over the condenser length, we obtain for the constant heat removal rate

$$\int_0^{L_c} V_v \, dx = \frac{V_{vi} L_c}{2} \tag{4.56}$$

The ratio CF of the laminar frictional vapor pressure drop in the condenser for convective heat removal to that for heat removal at a constant rate is equal to Eq. (4.54) divided by Eq. (4.56), or

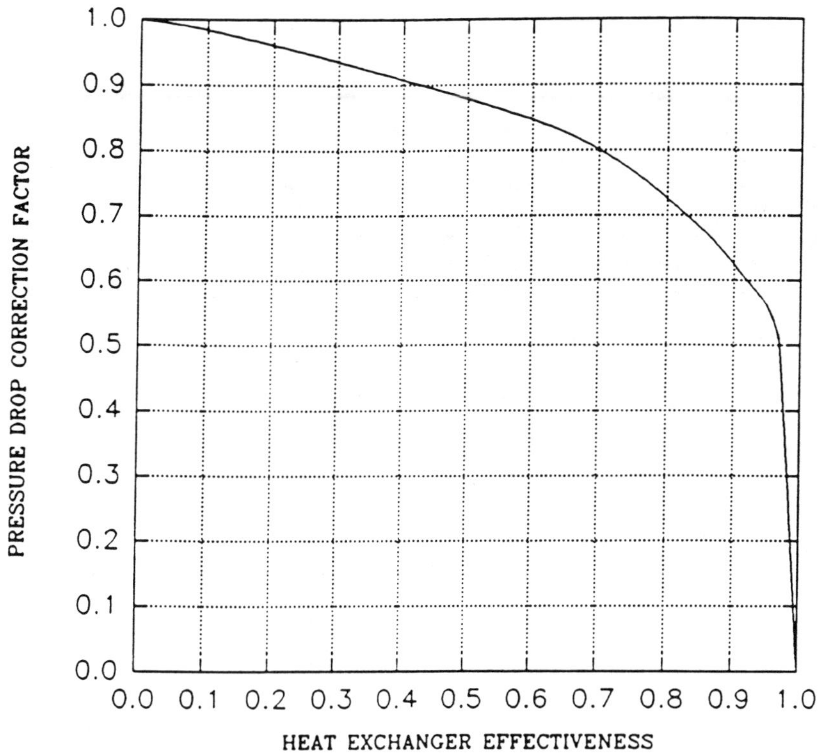

Figure 4.5. Pressure drop correction factor for convective heat removal from a heat pipe condenser (laminar flow).

$$CF = 2\left(1 + \frac{1}{NTU} - \frac{1}{\varepsilon}\right) \tag{4.57}$$

Since NTU and ε are related by Eq. (4.46), Eq. (4.57) shows that the pressure drop correction factor for convective cooling can be expressed solely as a function of the heat exchanger effectiveness ε (or, alternatively, of the number of heat transfer units NTU).

The correction factor is plotted against ε in Figure 4.5. It can be seen that the correction factor is always less than unity, and is greater than 0.9 if the effectiveness does not exceed 0.4. The pressure drop equation for a constant heat removal rate may then be used conservatively for this situation, with a relatively small error.

If the effectiveness is 0.8 or higher, the correction factor is 0.72 or less. For this situation, use of the pressure drop equation for a constant heat removal rate, although still conservative, will significantly overestimate the pressure drop for convective heat removal.

Although the pressure drop correction factor was developed for the condensing heat pipe vapor, it applies equally well to the accumulating liquid condensate.

4.8. REFERENCES

4.1. Cotter, T. P., "Theory of Heat Pipes," LA-3246-MS, Los Alamos Scientific Laboratory, Los Alamos, New Mexico, March 1965.

4.2. Busse, C. A., "Pressure Drop in the Vapor Phase of Long Heat Pipes." Presented at the 1967 Thermionic Conversion Specialist Conference, Palo Alto, California, October 1967.

4.3. Kemme, J. E., "Heat Pipe Design Considerations," LA-4221-MS, Los Alamos Scientific Laboratory, Los Alamos, New Mexico, August 1969.

4.4. Bankston, C. A., and Smith, H. J., "Incompressible Laminar Vapor Flow in Cylindrical Heat Pipes," 71-WA/HT-15. Presented at the ASME Winter Annual Meeting, Washington, D.C., November 1971.

4.5. Eckert, E. R. G., and Drake, R. M., Jr., *Heat and Mass Transfer,* 2nd Ed., McGraw-Hill, New York, 1959, pp. 153–154.

4.6. Baumeister, T. (Ed.), *Standard Handbook for Mechanical Engineers,* 7th Ed., McGraw-Hill, New York, 1967, pp. 3-58–3-60.

4.7. Wageman, W. E., and Guevara, F. A., "Fluid Flow Through a Porous Channel," *Physics of Fluids,* Vol. 3, No. 6, Nov.-Dec. 1960, pp. 878–881.

4.8. Kunz, H. R., Langston, L. S., Hilton, B. H., Wyde, S. S., and Nashick, G. H., "Vapor Chamber Fin Studies: Transport Properties and Boiling Characteristics of Wicks," CR-812, NASA, Washington, D.C., June 1967, p. 17.

4.9. Kays, W. M., and London, A. L., *Compact Heat Exchangers,* McGraw-Hill, New York, 1964, pp. 13–18.

CHAPTER

FIVE

HEAT TRANSPORT LIMITS

In the design of heat pipes, it is important to establish that heat transport will take place at the required rate without encountering limits on heat transport capability. These limits can arise from the cessation of capillary pumping, the attainment of sonic velocity in the heat pipe vapor, entrainment of the heat pipe liquid by its vapor, or the onset of boiling in the heat pipe liquid.

Loss of capillary pumping, entrainment of the heat pipe liquid, or boiling of the liquid can lead to dryout of the heat pipe wick, overheating, and possibly destruction of the heat pipe if the heat input is not rapidly terminated. Although the attainment of sonic velocity also represents a limiting heat transport rate, this limit is not inherently destructive in character.

During heat pipe operation, pressure drops in the heat pipe vapor and liquid produce a net pressure difference between the vapor and the liquid along the entire heat pipe length. This pressure difference is balanced by capillary pressure that develops in the wick pores at the curved liquid-vapor interface that separates the two phases. The pressure difference increases with the heat transport rate. Loss of capillary pumping occurs when the heat transport rate exceeds the level at which maximum capillary pressure is developed in the wick pores.

When the heat pipe vapor reaches sonic velocity, the vapor flow is said to be choked. Heat transport is then occurring at the maximum rate possible at a given heat pipe temperature. A further increase in the heat transport rate can only be accommodated by raising the sonic heat transport rate through an increase in the evaporator temperature. If the higher temperature can be tolerated structurally, heat pipe operation above the sonic limit at a particular temperature has no adverse consequences.

Operation at the sonic limit is generally accompanied by a significant temperature drop along the evaporator length, which can be on the order of 100 to 200°F (56 to 112°C). Where relatively isothermal heat transport is important, as is the case in most design situations, the heat transport rate generally should not exceed 20–30% of the sonic limit. A more precise criterion for isothermal heat transport is that the pressure drop in the heat pipe vapor along the heat pipe length not exceed 1–3% of the vapor pressure at the evaporator end.

When the vapor velocity is high, drag exerted on the adjacent heat pipe liquid may be sufficient to entrain liquid droplets in the vapor. The droplets will then be carried to the end of the condenser, impinging on the condenser end closure. The consequence of entrainment is to reduce the flow rate of liquid into the evaporator, causing dryout and overheating in a manner similar to that following a loss of capillary pumping.

The onset of boiling in liquid metal heat pipes can be explosive in nature, producing large pressure and temperature fluctuations, which may or not be destructive in character. Boiling in non–liquid metal heat pipes may be tolerable until the onset of film boiling and the associated large increase in the evaporator wall temperature. This may occur at heat transfer rates perhaps 20–30% above that at which boiling is initiated.

In this chapter, analytical expressions are developed for predicting capillary pumping, sonic, entrainment, and boiling heat transport limits. A criterion for isothermal heat transport, based on a specified percentage of allowable drop in vapor pressure, is also presented. These expressions are based on simplified representations of actual heat transport phenomena in a heat pipe, and are suitable for design purposes.

One method of establishing the adequacy of a heat pipe design is to compare the heat transport limits for a particular heat pipe to the actual heat transport rate. An alternate approach, which can often be preferable for design purposes, is to find the value of a key heat pipe design parameter at the actual heat transport rate, and then compare this value with that corresponding to a heat transport limit.

For example, the minimum diameter of curvature at the liquid-vapor interface can be calculated for the actual heat transport rate and compared to the effective wick pore diameter. Heat is being transported at the capillary pumping limit when the diameter of curvature is equal to the effective pore diameter. If the effective pore diameter is smaller than the calculated diameter of curvature, the actual heat transport rate will be less than the capillary pumping limit. Often, this is all the information needed about the capillary pumping limit for design purposes.

In a similar manner, the wick pore diameter above which entrainment will occur can be calculated for the actual heat transport rate. If the effective wick pore diameter is less than the calculated value, the design heat transport rate will be less than the entrainment limit.

Finally, the diameter of curvature at the liquid-vapor interface below which boiling will occur at the actual heat transport rate can be calculated, and then compared with the diameter of curvature required for capillary pumping at the same rate. If the diameter of curvature for boiling inception is smaller than that for capillary pumping, the actual heat transport rate will be less than the boiling limit.

There are several reasons why calculating the value of a key heat pipe design parameter for the actual heat transport rate can be preferable to the calculation of heat transport limits.

First, design characteristics such as pressure drop, extent of isothermality, and superheat in the wick liquid will then be determined for the actual heat pipe design condition rather than some hypothetical limiting condition.

Second, the validity of simplifying assumptions used to derive the heat pipe heat transport limit equations may be questionable if heat transport rates are high at these limiting conditions.

Finally, the solution of the capillary pumping equation is simpler if the diameter of curvature is calculated for the actual heat transport rate, rather than calculating the maximum heat transport rate (the capillary pumping limit) corresponding to the minimum possible diameter of curvature (the effective pore size).

5.1. CAPILLARY PUMPING LIMIT

The capillary pumping limit is the highest heat transport rate that can be sustained by the capillary pressure in a heat pipe wick. The basic relationship from which the capillary pumping limit can be derived is Eq. (2.11), which is repeated below:

$$\Delta P_{vl} - |\Delta P_v + \Delta P_l| \tag{2.11}$$

Equation (2.11) states that the maximum pressure difference ΔP_{vl} between the heat pipe vapor and its liquid is equal to the absolute value of the algebraic sum of the pressure drops in the heat pipe vapor (ΔP_v) and the heat pipe liquid (ΔP_l). This pressure difference is sustained by and equal to the capillary pressure ΔP_c that develops in the wick pores. ΔP_c in turn is defined by Eq. (2.4), repeated below, as

$$\Delta P_c = 4\sigma/D_c \tag{2.4}$$

where σ is the surface tension of the heat pipe liquid and D_c is the diameter of curvature in the wick pores at the interface between the heat pipe liquid and its vapor.

The capillary pumping limit is reached when D_c assumes its minimum possible value, which is the effective pore diameter D_p. Ideally, D_p is just the minimum dimension of the wick pores. In actuality, the effects of wick pore geometry and imperfect wetting (i.e., a wetting angle greater than zero) can result in a D_p that is appreciably larger than the minimum pore size.

The capillary pumping limit for a given heat pipe is found from Eqs. (2.11) and (2.4) by expressing the liquid and vapor pressure drops in terms of the heat transport rate Q and then solving for Q when $D_c = D_p$. To do so requires that heat pipe dimensions, wick capillary and frictional characteristics, and relevant properties of the heat pipe fluid be known. Suitable expressions for the liquid and vapor pressure drops must also be available. We will now proceed to consider the development of such expressions.

5.1.1. Fluid Pressure Drop in Heat Pipes

The pressure drop equations obtained in Chapter 4 will now be used to develop expressions for the vapor and liquid pressure drops in a heat pipe. Pressure drops are determined for the evaporator, adiabatic, and condenser regions, and then added together to obtain the total fluid pressure drop in the heat pipe. The procedure is applied separately for the heat pipe vapor and liquid.

In order to calculate the capillary pumping limit, the vapor and liquid pressure drops must be expressed in terms of the heat transport rate. This is readily done with the aid of Eq. (2.10), which is repeated here:

$$Q = \kappa W \tag{2.10}$$

where Q is the heat pipe heat transport rate, W is the total mass rate at which fluid evaporates and condenses in the heat pipe, and κ is the heat of vaporization of the heat pipe fluid.

W is also equal to the maximum axial mass flow rate in the heat pipe. Therefore, W can be expressed as the product of the fluid density ρ, the maximum mean axial velocity V_{max}, and the flow passage area A_{cs}, for both the heat pipe vapor and liquid. Thus, using Eq. (2.10),

$$W = Q/\kappa = \rho_v V_{v\max} A_{vcs} = \rho_l V_{l\max} A_{lcs} \tag{5.1}$$

where the subscript v denotes the heat pipe vapor and the subscript l denotes the liquid.

The radial mass flow rate due to vaporization or condensation per unit surface area will be designated as the surface mass flux w, and the associated surface heat flux as q. Depending on circumstances, the surface area that defines w and q may be the inner wick surface, the outer wall surface, or any surface in between.

In a heat pipe, the surface heat flux distribution q is usually prescribed rather than the surface mass flux distribution w. The two are related by the simple expression

$$q = \kappa w \tag{5.2}$$

Axial Velocity Distribution. Figure 5.1 shows the mean axial velocity distribution over the heat pipe length for the vapor and the liquid, for the case of a constant surface heat flux in the evaporator and the condenser. Note that the vapor and liquid flow in opposite directions.

The vapor velocity increases linearly from zero at the start of the evaporator to a maximum value at the evaporator exit, as vapor is introduced into the vapor passage at a constant rate. Since the liquid in the evaporator is being depleted at the same rate, the liquid velocity decreases linearly from its maximum value as it enters the evaporator to zero at the evaporator end.

The situation is reversed in the condenser, where heat is being removed and the vapor condenses. Here, the vapor velocity decreases linearly from its maximum value as

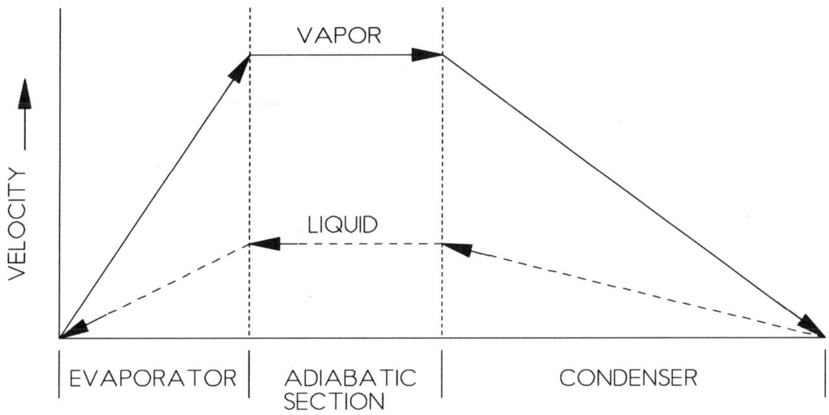

Figure 5.1. Axial velocity distribution along a heat pipe length for constant surface heat flux.

it enters the condenser to zero at the end of the condenser, while the liquid velocity increases linearly from zero at the condenser end to its maximum value as it exits the condenser.

Since heat is being neither added nor removed in the adiabatic section of the heat pipe, the vapor and liquid velocities there remain constant.

From Figure 5.1, it is evident that the vapor and liquid velocity distributions along the heat pipe length are identical in form, although differing in absolute magnitude. This situation exists even when the heat addition and removal rates are not constant.

Since the frictional pressure drop is proportional to the integral of the velocity distribution or of the velocity distribution squared [see Eqs. (4.27) and (4.13)], it follows that the same equations may be used to calculate the vapor and liquid frictional pressure drops. Of course, specific variables such as density take on values characteristic of the fluid phase for which pressure drop is being calculated, and the nature of the flow (laminar or turbulent) is independently established for each phase.

Criterion for Turbulent Flow. The flow pattern in a fluid stream changes from laminar to turbulent when the axial Reynolds number Re_a has exceeded some critical value. There remains considerable uncertainty over whether and under what conditions the flow in a heat pipe initially becomes turbulent or becomes reestablished in a laminar pattern from the turbulent state.

In ordinary pipe flow, in which the flow rate is constant along the pipe length, the flow is generally laminar if Re_a is less than about 2000, and turbulent if Re_a is greater than about 4000. The status of the flow is uncertain and in transition in the range from roughly 2000 to 4000.

Since the laminar friction factor is substantially smaller than the turbulent friction factor at a given Re_a, and the actual status of the fluid flow pattern in a heat pipe is highly uncertain, it is prudent to make the conservative design assumption that the flow is turbulent if Re_a exceeds roughly 2000.

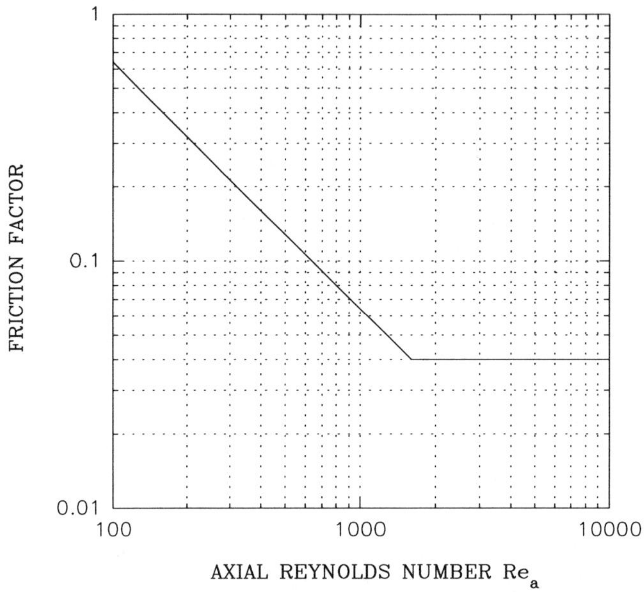

Figure 5.2. Recommended variation of friction factor with axial Reynolds number for a heat pipe.

In the laminar range, the friction factor f is an inverse function of Re_a, as defined by Eq. (4.20). In the turbulent range, the friction factor is dependent on the relative roughness of the flow passage wall as well as an inverse fractional power of Re_a. (Relative roughness is the ratio of the mean height of asperities in the wall to the flow passage cross-sectional dimension.) As was indicated in Section 4.2.3, the turbulent friction factor varies from around 0.05 for very rough flow passages at $Re_a = 4000$ to around 0.02 for very smooth passages at $Re_a = 100{,}000$.

In a heat pipe, relative roughness is probably related to the size of the elements comprising the wick structure and, for vapor flow, the diameter of curvature of the liquid-vapor interface in the wick pores. Currently available information provides little or no basis for characterization of the relative roughness in a heat pipe. For these reasons, an appropriate value of f based on engineering judgment is suggested when calculating the turbulent pressure drop in a heat pipe. A value of 0.04 is suggested for the turbulent friction factor as a reasonably conservative approach for conditions likely to be encountered in a heat pipe.

It is further suggested that flow be assumed turbulent when Re_a exceeds the value at which the laminar and turbulent friction factors are equal. Then, using Eq. (4.20) for the laminar friction factor,

$$64/Re_a = 0.04$$

or

$$Re_a = 1600$$

The recommended friction factor–Reynolds number relation is shown in Figure 5.2.

When a heat pipe liquid flows through a porous channel, the Reynolds number based on the pore size will be within the laminar range for virtually all cases of interest. Therefore, the flow pattern through a porous liquid flow channel can always be assumed laminar. The liquid frictional pressure drop is then calculated from the laminar pressure drop equation, using the correction factor of Eq. (4.35) to account for the increased friction of the porous flow channel.

Aside from the special case of liquid flow through a porous channel, it is necessary to establish whether the flow is laminar over the entire heat pipe length, or whether it becomes turbulent over some portion of that length.

If the vapor flow becomes turbulent, it will do so at some point along the evaporator section, remain turbulent in the adiabatic section, and then become reestablished as laminar flow at some point along the condenser section. The situation is reversed for the liquid flow in a nonporous channel. The liquid flow would become turbulent at some point in the condenser, remain turbulent in the adiabatic section, and become reestablished as laminar flow at some point along the evaporator.

The maximum mean axial velocity V_l at which the flow is laminar is found by equating the axial Reynolds number to the critical value of 1600 that has already been established. Thus,

$$\rho V_l D_h/\mu = 1600$$

Noting that $Q = \kappa W = \kappa \rho V A_{cs}$, *we may write*

$$Q_l = 1600 \kappa \mu A_{cs}/D_h \tag{5.3}$$

where A_{cs} is the cross-sectional area of the flow passage and Q_l is the maximum heat transport rate at which the heat pipe fluid flow remains laminar.

For the heat pipe vapor, Eq. (5.3) takes the form

$$Q_{vl} = 1600 \kappa \mu_v A_v/D_{hv} \tag{5.4}$$

and for the liquid,

$$Q_{ll} = 1600 \kappa \mu_l A_l/D_{hl} \tag{5.5}$$

where the subscripted A represents the flow passage area, the subscript v denotes the vapor and the subscript l denotes the liquid (as well as the limiting condition at which the flow is laminar).

It is convenient to define the variables

$$\gamma_v = Q_{vl}/Q \tag{5.6}$$

$$\gamma_l = Q_{ll}/Q \tag{5.7}$$

$$\alpha = A_v/A_l \tag{5.8}$$

$$\xi = Q/A_v \tag{5.9}$$

With these definitions, Eqs (5.4) and (5.5) may be written

$$\gamma_v = \frac{1600\ \kappa\ \mu_v}{D_{hv}\ \xi} \tag{5.10}$$

$$\gamma_l = \frac{1600\ \kappa\ \mu_l}{D_{hl}\ \alpha\ \xi} \tag{5.11}$$

where γ_v is the ratio of the maximum heat transport rate for laminar flow in the heat pipe vapor to the total heat transport rate, γ_l is the ratio of the maximum heat transport rate for laminar flow in the heat pipe liquid to the total heat transport rate, α is the ratio of the vapor and liquid flow passage areas, and ξ is the axial heat flux (ratio of total heat transport rate to vapor passage flow area).

The following heat pipe flow criteria can now be written:

If $\gamma_v \geq 1$, the *vapor* flow is entirely laminar.
If $\gamma_v < 1$, the *vapor* flow is partly laminar and partly turbulent.
If $\gamma_l \geq 1$, the *liquid* flow is entirely laminar.
If $\gamma_l < 1$, the *liquid* flow is partly laminar and partly turbulent.

These criteria are used in formulating expressions for the frictional pressure drop in the heat pipe vapor and liquid, which are presented subsequently. The expressions are derived for the case of constant (but not necessarily equal) surface heat flux in the evaporator and condenser. They are also generally adequate for situations in which the surface heat flux is not constant, unless the surface heat flux distribution is highly asymmetrical with respect to the evaporator and/or condenser length.

In that event, the velocity distribution in the asymmetrical sections will have to be determined from the surface mass flux distribution (proportional to the surface heat flux distribution), using the procedure described in Section 4.4, and then integrated to obtain the pressure drop. The solution of Section 4.7 will be found useful when the surface heat flux distribution is characteristic of that for convective cooling.

The nomenclature adopted for heat pipe distances and velocities in the pressure drop calculations is shown in Figure 5.3.

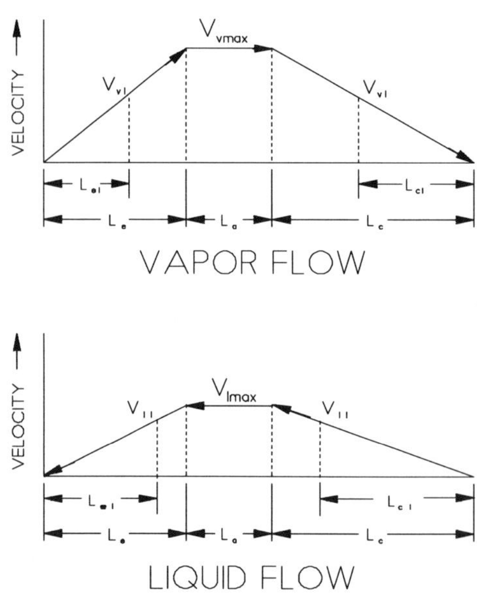

Figure 5.3. Nomenclature for heat pipe lengths and fluid velocities.

The distance from the evaporator end of the heat pipe over which the flow is laminar is called L_{el}. Similarly, the distance from the condenser end over which the flow is laminar is called L_{cl}. In general, L_{el} and L_{cl} are different for the heat pipe vapor and liquid. The maximum mean axial velocity at which the flow is laminar is called V_{vl} for the vapor and V_{ll} for the liquid. The maximum mean axial velocity in the heat pipe is called $V_{v\max}$ for the vapor and $V_{l\max}$ for the liquid.

Frictional Pressure Drop in Vapor. *When* $\gamma_v < 1$, the frictional pressure drop in the heat pipe vapor is found by summing the individual pressure drops over the following heat pipe segments: laminar evaporator length, turbulent evaporator length, adiabatic length (all turbulent), turbulent condenser length, and laminar condenser length. Equations (4.28) and (4.29) are used for the laminar and turbulent sections, respectively, of the evaporator and condenser. Within the adiabatic section, Eq. (4.31) is used.

In these equations, the inlet and outlet velocities V_i and V_o are replaced with zero, V_{vl}, or $V_{v\max}$, as appropriate for a particular heat pipe segment. Also, D is replaced by the hydraulic diameter D_h.

When these steps are performed and similar terms are consolidated, the following equation results:

$$\Delta P_{vf} = 16\mu_v V_{vl}\frac{(L_{el}+L_{cl})}{D_{hv}^2} + \left[\frac{f_t\rho_v V_{v\max}^2}{6D_{hv}}\right]\left[1+\frac{V_{vl}}{V_{v\max}}+\left(\frac{V_{vl}}{V_{v\max}}\right)^2\right][(L_e-L_{el})+(L_c-L_{cl})] + \frac{f_t\rho_v V_{v\max}^2 L_a}{2D_{hv}} \tag{5.12}$$

In Eq. (5.12), the first term is the pressure drop in the laminar regions of the evaporator and condenser, the second term is the pressure drop in the turbulent regions of the evaporator and condenser, and the last term is the turbulent pressure drop in the adiabatic region. The turbulent friction factor is denoted by f_t.

The form of this equation can be simplified, and the pressure drop expressed in terms of the axial heat flux ξ, by using the following relationships:

$$Q_{vl} = \kappa\rho_v A_v V_{vl} = q_{el}L_{el}B_p = q_{cl}L_{cl}B_p$$

$$Q = \kappa\rho_v A_v V_{v\max} = q_e L_e B_p = q_c L_c B_p$$

where q_{el} and q_{cl} are the average surface heat fluxes over the laminar lengths L_{el} and L_{cl}, q_e and q_c are the average surface heat fluxes over the evaporator and condenser lengths L_e and L_c, and B_p is the heat pipe periphery corresponding to the surface heat fluxes.

Also, the parameter $g(\gamma)$ will be defined as

$$g(\gamma) = (1-\gamma)(1+\gamma+\gamma^2) \tag{5.13}$$

It follows that γ_v, which has been defined by Eq. (5.6), can be expressed as

$$\gamma_v = V_{vl}/V_{v\max} = (q_{el}/q_e)(L_{el}/L_e) = (q_{cl}/q_c)(L_{cl}/L_c) \tag{5.14}$$

If the surface heat flux q is constant over the evaporator and condenser (but not necessarily the same), the above expression for γ_v reduces to

$$\gamma_v = V_{vl}/V_{v\max} = L_{el}/L_e = L_{cl}/L_c \tag{5.15}$$

Also, the heat pipe length is expressible as the sum of the evaporator, condenser, and adiabatic section lengths, or

$$L_p = L_e + L_c + L_a \tag{5.16}$$

Upon introducing Eqs. (5.13), (5.15), and (5.16) into Eq. (5.12), the *final equation for the frictional vapor pressure drop when* $\gamma_v < 1$ is obtained as

$$\Delta P_{vf} = \frac{16\,\mu_v\,(L_p - L_a)\,\gamma_v^2\,\xi}{\kappa\,\rho_v\,D_{hv}^2} + \frac{f_t\,(L_p - L_a)\,\xi^2}{6\,\kappa^2\,\rho_v\,D_{hv}}\left[g\,(\gamma_v) + \frac{3\,L_a}{L_p - L_a}\right] \tag{5.17}$$

In this equation, D_{hv} is the hydraulic diameter of the vapor flow channel.

When $\gamma_v \geq 1$, laminar flow exists throughout the heat pipe vapor. The frictional pressure drop is then the sum of the laminar pressure drop terms for the evaporator, condenser, and adiabatic section. Using Eqs. (4.28), (4.30), and (5.16), letting the maximum mean vapor velocity be V_{vl}, and noting that Q may then be expressed as

$$Q = \kappa\rho_v A_v V_{vl}$$

the *final equation for the frictional vapor pressure drop with all-laminar flow* is obtained in the form

$$\Delta P_{vf} = \frac{16\,\mu_v\,(L_p - L_a)\,\xi}{\kappa\,\rho_v\,D_{hv}^2}\left(1 + \frac{2L_a}{L_p - L_a}\right) \tag{5.18}$$

Frictional Pressure Drop in Liquid. For the liquid flow channel, the frictional pressure drop equations have essentially the same form as those for the heat pipe vapor. Of course, liquid rather than vapor properties are used, the correction factor ε_{po} for a porous flow channel must be included, and the hydraulic diameter is calculated for the liquid flow channel.

Also, it is desirable for the axial heat flux term to be based on the vapor passage flow area rather than the liquid flow area, so that a single heat flux term appears in both the vapor and liquid pressure drop terms in the capillary pumping limit equation. To achieve this, the axial heat flux term that would normally appear in the liquid pressure drop equation is modified as follows:

$$\frac{Q}{A_l} = \left(\frac{Q}{A_l}\right)\left(\frac{A_v}{A_v}\right) = \left(\frac{A_v}{A_l}\right)\left(\frac{Q}{A_v}\right) = \alpha\,\xi \tag{5.19}$$

Thus, when converting the vapor pressure drop equations to liquid form, the variable ξ is replaced by $\alpha\xi$.

We may then write, for the case of both laminar and turbulent liquid flow ($\gamma_l < 1$),

$$\Delta P_{lf} = \frac{16\,\mu_l\,(L_p - L_a)\,\gamma_l^2\,\alpha\,\xi}{\kappa\,\rho_l\,D_{hl}^2} + \frac{f_t\,(L_p - L_a)\,\alpha^2\,\xi^2}{6\,\kappa^2\,\rho_l\,D_{hl}}\left[g\,(\gamma_l) + \frac{3\,L_a}{L_p - L_a}\right] \tag{5.20}$$

and for laminar liquid flow only ($\gamma_l \geq 1$),

$$\Delta P_{lf} = \frac{16\,\mu_l\,(L_p - L_a)\,\varepsilon_{po}\,\alpha\,\xi}{\kappa\,\rho_l\,D_{hl}^2}\left(1 + \frac{2\,L_a}{L_p - L_a}\right) \tag{5.21}$$

In these equations, D_{hl} is the hydraulic diameter of the liquid flow channel.

Equation (5.20) is applicable to an open or grooved liquid flow channel only, since turbulent flow has not been considered possible in a porous channel. Equation (5.21) is applicable to an open, grooved, or porous liquid flow channel, as long as the flow is laminar.

Distribution Wick. When the heat pipe wick structure is of the artery type, an additional liquid frictional pressure drop ΔP_{lfd} is associated with the circumferential flow of liquid through the distribution wick lining the heat pipe walls in the evaporator and condenser. We will now proceed to derive an expression for ΔP_{fld}, and establish its magnitude relative to the axial pressure drop through the arteries.

To facilitate the discussion, the cross section of a two-artery wick structure is shown schematically in Figure 5.4.

Two parallel transverse liquid flow paths exist for each artery. The flow path length is half the peripheral distance between adjacent arteries. If there are N equally spaced arteries, there will be $2N$ parallel flow paths. The flow path length L_d and heat transport rate per flow path Q_d are then

$$L_d = B_v/(2N) \tag{5.22}$$

$$Q_d = Q/(2N) = \kappa W/(2N) \tag{5.23}$$

where B_v is the distance around the heat pipe vapor passage periphery (πD_v for a circular cross section).

The cross-sectional area of the distribution wick for liquid flow in the circumferential direction is A_{lde} in the evaporator and A_{ldc} in the condenser. Assuming that the distribution wick thickness t_d is the same in the evaporator and condenser, it follows that

$$A_{lde}/A_{ldc} = L_e t_d/(L_c t_d) = L_e/L_c \tag{5.24}$$

We can also define the distribution wick axial heat flux ξ_d as follows:

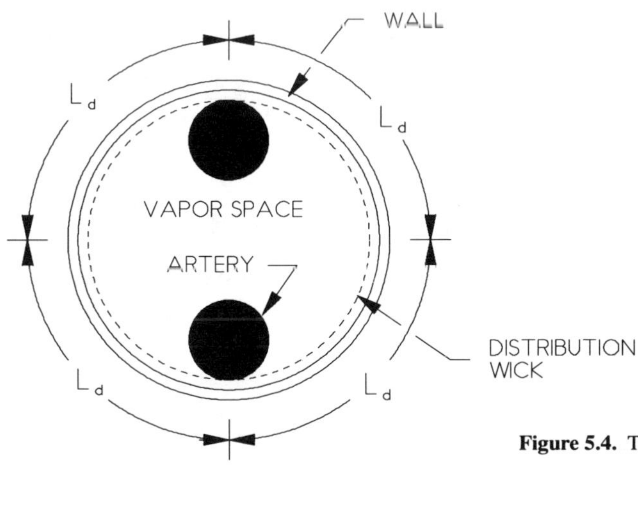

Figure 5.4. Two-artery heat pipe wick structure.

$$\xi_d = Q_d/A_v = Q/(2NA_v) = \xi/(2N) \tag{5.25}$$

where ξ_d is the heat flow rate per parallel flow path through the distribution wick per unit vapor space flow area.

Also, the distribution wick area ratios α_{de} for the evaporator and α_{dc} for the condenser are defined as

$$\alpha_{de} = A_v/A_{lde} \quad \text{and} \quad \alpha_{dc} = A_v/A_{ldc} \tag{5.26}$$

The distribution wick may consist of one or more layers of screen, or of a series of circumferential grooves in the heat pipe wall, which extend around the periphery of the vapor flow passage. In either case, the combination of low velocity and parallel flow paths, plus a very small hydraulic diameter for the liquid flow passage, will result in laminar flow for almost all design situations. Therefore, all-laminar flow through the distribution wick is assumed.

The frictional pressure drop through the distribution wick can be found from Eq. (5.21) for the axial laminar pressure drop by introducing the following modifications. The adiabatic length is taken to be zero. The equation is separated into the evaporator and condenser pressure drop terms, and distribution wick variables are used. We can then write for ΔP_{lfd},

$$\Delta P_{lfd} = \frac{16\,\mu_l\,L_d\,\varepsilon_{po}\,\alpha_{de}\,\xi_d}{\kappa\,\rho_l\,D_{hld}^2} + \frac{16\,\mu_l\,L_d\,\varepsilon_{po}\,\alpha_{dc}\,\xi_d}{\kappa\,\rho_l\,D_{hld}^2} \tag{5.27}$$

where the first term is the distribution wick pressure drop in the evaporator, the second term is the distribution wick pressure drop in the condenser, and D_{hld} is the hydraulic diameter for the distribution wick flow area.

Using Eqs. (5.22), 5.24), (5.25), and (5.26), Eq. (5.27) can be consolidated into the *final form for the distribution wick pressure drop:*

$$\Delta P_{lfd} = \frac{16\,\mu_l\,B_v\,\varepsilon_{po}\,\alpha_{de}\,\xi}{4\,N^2\,\kappa\,\rho_l\,D_{hld}^2}\left(1 + \frac{L_e}{L_c}\right) \tag{5.28}$$

The *total liquid frictional pressure drop in a heat pipe with an artery-type wick* ΔP_{lfa} is then

$$\Delta P_{lfa} = \Delta P_{lf} + \Delta P_{lfd} \tag{5.29}$$

ΔP_{lfd} will now be compared to the laminar frictional pressure drop ΔP_{lf} through the heat pipe arteries. For simplicity, the adiabatic length L_a will be taken as zero. Then, from Eq. (5.21),

$$\Delta P_{lf} = \frac{16\,\mu_l\,(L_e + L_c)\,\alpha\,\xi}{\kappa\,\rho_l\,D_{hl}^2} \tag{5.30}$$

Dividing Eq. (5.28) by Eq. (5.30) yields the *ratio of pressure drops through the distribution and arterial wicks:*

$$\frac{\Delta P_{lfd}}{\Delta P_{lf}} = \left(\frac{B_v\,\varepsilon_{po}}{4\,N^2\,L_c}\right)\left(\frac{\alpha_{de}}{\alpha}\right)\left(\frac{D_{hl}}{D_{hld}}\right)^2 \tag{5.31}$$

For a circular heat pipe with vapor space diameter D_v, a screen distribution wick of thickness t_d and friction factor K, and an artery inner diameter of d_i, and with the use of Eq. (4.35), it may be shown that Eq. (5.31) takes the form

$$\frac{\Delta P_{lfd}}{\Delta P_{lf}} = \frac{\pi^2 K D_v d_i^4}{512 N L_e L_c t_d} \tag{5.32}$$

Eq. (5.32) will now be evaluated for a distribution wick consisting of one layer of 100 mesh screen and the following design conditions:

$t_d = 0.009$ in. $(2.29 \times 10^{-4}$ m)
$K = 4.21 \times 10^6$ in^{-2} $(6.53 \times 10^9$ m$^{-2})$
$D_v = 1$ in. (0.0254 m)
$d_i = 0.050$ in. $(1.27 \times 10^{-3}$ m)
$N = 4$
$L_e = L_c = 12$ in. (0.305 m)

Then, from Eq. (5.32), $\Delta P_{lfd}/\Delta P_{lf} = 0.0978$.

In this particular example, the liquid frictional pressure drop through the distribution wick is almost 10% of that through the arteries.

Momentum Pressure Drop in Vapor. The momentum pressure drop is only significant in the heat pipe vapor. From the general pressure drop formulation of Eq. (4.9), the momentum pressure drop term can be expressed as

$$\Delta P_{vm} = \Omega \rho_v (V_o^2 - V_i^2) \tag{5.33}$$

where V_i is the mean axial velocity at the inlet to a vapor injection or suction section and V_o is the mean axial velocity at the section outlet.

The momentum factor Ω is the ratio of the integral of the momentum flow rate over the flow passage cross section to the momentum flow rate based on the average velocity.

Two terms comprise the momentum pressure drop in a heat pipe. The first term is a pressure drop associated with the vapor velocity increase in the evaporator. The second is a pressure rise (negative pressure drop) associated with the decrease of vapor velocity in the condenser.

In the evaporator of a heat pipe, $V_i = 0$, $V_o = V_{vmax}$, and $\Delta P_{vm} = \Omega_e \rho_v V_{vmax}^2$, where Ω_e is the momentum factor in the evaporator.

In the condenser, $V_i = V_{vmax}$, $V_o = 0$, and $\Delta P_{vm} = -\Omega_c \rho_v V_{vmax}^2$, where Ω_c is the momentum factor in the condenser.

Using results from Cotter (Ref. 5.1), the evaporator and condenser momentum factors in a heat pipe with very large radial Reynolds numbers can be shown to have the following values:

$$\Omega_e = \pi^2/8 \quad \text{and} \quad \Omega_c = 1/2$$

As was pointed out in Section 4.1.2, the evaporator momentum factor Ω_e can be taken as the value corresponding to large radial Reynolds numbers with relatively little error in the overall evaporator pressure drop, provided that the friction factor f corresponding to small radial Reynolds numbers is used.

The net momentum pressure drop for an entire heat pipe is expressible as the algebraic sum of the evaporator and condenser terms. Using the momentum factors corresponding to large Re_r,

$$\Delta P_{vm} = \Omega_e \rho_v V_{v\max}^2 - \Omega_c \rho_v V_{v\max}^2 = (\Omega_e - \Omega_c)\rho_v V_{v\max}^2$$
$$= 0.734 \rho_v V_{v\max}^2 \tag{5.34}$$

Although the momentum pressure drop in the evaporator is represented reasonably well in Eq. (5.34), considerable uncertainty exists about momentum-based pressure recovery in the condenser. (See Section 4.1.3.) Therefore, the net momentum pressure drop as given by Eq. (5.34) must be regarded as quite approximate. Despite its limitations, Eq. (5.34) is recommended for use in estimating the heat pipe momentum pressure drop at any radial Reynolds number, for reasons to be given shortly.

An expression will now be developed for the *ratio of the momentum pressure drop to the total pressure drop* in the vapor of a heat pipe with laminar flow and no adiabatic section. This expression will be useful in assessing conditions for which uncertainties in the momentum pressure drop can significantly impact the accuracy of capillary pumping limit calculations.

Noting that $Q = \kappa \rho_v V_{v\max} A_v$ and $\xi = Q/A_v$, ΔP_{vm} from Eq. (5.34) may be represented in the form

$$\Delta P_{vm} = \frac{0.734\, \xi^2}{\rho_v\, \kappa^2} \tag{5.35}$$

Dividing Eq. (5.35) by Eq. (5.18),

$$\frac{\Delta P_{vm}}{\Delta P_{vf}} = \frac{0.734\, D_{hv}^2\, \xi^2}{16\, \kappa\, \mu_v\, L_p} \tag{5.36}$$

Equation (5.36) will be expressed in terms of the radial Reynolds number in the evaporator, defined as

$$\mathrm{Re}_{re} = \frac{\rho_v\, V_{vr}\, D_{hv}}{2\, \mu_v} \tag{5.37}$$

where V_{vr} is the radial vapor velocity at the vapor passage periphery and D_{hv} is the hydraulic diameter of the vapor passage.

For a constant value of V_{vr}, the heat transport rate Q can be expressed as

$$Q = \kappa \rho_v V_{vr} B_v L_e \tag{5.38}$$

where B_v is the periphery of the vapor space and L_e is the evaporator length. $B_v L_e$ in turn can be expressed as

$$B_v L_e = 4 A_v L_e / D_{hv} \tag{5.39}$$

where A_v is the vapor flow passage area. The indicated substitution for B_v in the equation is based on the definition of the hydraulic diameter D_h. [See Eq. (4.32).]

Solving Eq. (5.38) for V_{vr}, using Eq. (5.39) for $B_v L_e$, and noting that $\xi = Q/A_v$, V_{vr} can be written

$$V_{vr} = \frac{D_{hv}\, \xi}{4\, \kappa\, \rho_v\, L_e} \tag{5.40}$$

Upon substituting this expression for V_{vr} into Eq. (5.37),

$$L_e \, \mathrm{Re}_{re} = \frac{D_{hv}^2 \, \xi}{8 \, \kappa \, \mu_v} \tag{5.41}$$

Upon substituting Eq. (5.41) into Eq. (5.36),

$$\frac{\Delta P_{vm}}{\Delta P_{vf}} = \frac{0.367 \, L_e \, \mathrm{Re}_{re}}{L_p} \tag{5.42}$$

Equation (5.42) gives the *ratio of the momentum and frictional pressure drops in the heat pipe vapor.*

Ratio of Momentum and Total Vapor Pressure Drops. The total vapor pressure drop ΔP_v is equal to the sum of the momentum and frictional pressure drops, or

$$\Delta P_v = \Delta P_{vm} + \Delta P_{vf} = \Delta P_{vm}(1 + \Delta P_{vf}/\Delta P_{vm}) \tag{5.43}$$

Solving for $\Delta P_{vm}/\Delta P_v$, and using Eq. (5.42) in Eq. (5.43), gives the result

$$\frac{\Delta P_{vm}}{\Delta P_v} = \frac{\mathrm{Re}_{re} \, (L_e/L_p)}{\mathrm{Re}_{re} \, (L_e/L_p) + 2.73} \tag{5.44}$$

Equation (5.44) is the *desired expression for the ratio of momentum and total pressure drops in the heat pipe vapor,* which is seen to be a function of the product of Re_{re} and L_e/L_p. This expression is plotted in Figure 5.5.

L_e/L_p is less than 0.5 for most cases of interest. For Re_{re} up to 1, the momentum contribution to the vapor pressure drop is small, being 16% or less for $L_e/L_p \leq 0.5$. For

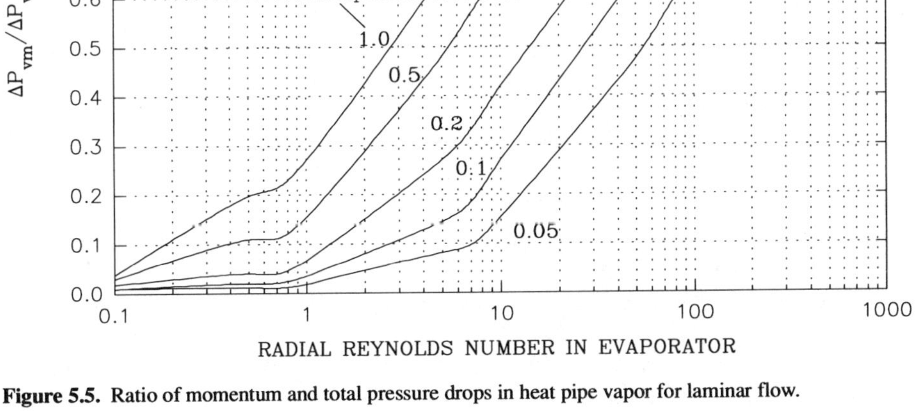

Figure 5.5. Ratio of momentum and total pressure drops in heat pipe vapor for laminar flow.

$Re_{re} > 1$, a range at which many heat pipes may operate (see Table 4.1), the momentum contribution can be quite significant.

As has been pointed out, the extent of momentum-produced pressure recovery in the condenser is uncertain, particularly along the vapor passage periphery. Therefore, the total vapor pressure drop calculation can be expected to be imprecise when the calculated momentum contribution is significant.

However, the uncertainty in the magnitude of the vapor pressure drop is of little consequence in establishing capillary pumping capability, as long as the vapor pressure drop is small compared to the liquid pressure drop. This condition prevails in many heat pipe design situations.

When a design situation is encountered in which the momentum pressure drop is a significant fraction of the total vapor pressure drop, and at the same time the total vapor pressure drop is comparable to or greater than the liquid pressure drop, additional design conservatism is advised in the capillary pumping limit calculation. For example, a smaller-than-calculated wick pore size at the liquid-vapor interface could then be specified.

Gravity/Acceleration Pressure Drop in Liquid. The pressure drop ΔP_{la} due to gravity and/or acceleration is only significant in the heat pipe liquid, because its density is much greater than that of the heat pipe vapor. The expression for ΔP_{la} that was developed in the general pressure drop formulation is directly applicable to heat pipes. [See Eq. (4.9).] Thus, in slightly modified form,

$$\Delta P_{la} = (\cos \alpha_b - n \cos \beta_b)\rho_l g L_b \tag{5.45}$$

where g is the acceleration of gravity, n is the external acceleration in number of gs, ρ_l is the liquid density, L_b is the straight-line distance between the lowest and highest points of the heat pipe, α_b is the angle between the acceleration of gravity and L_b, and β_b is the angle between the external acceleration and L_b.

The heat pipe geometry is shown in Figure 5.6. Notice that the liquid column along the straight-line distance L_b is used in calculating the gravity and acceleration liquid head components, rather than the liquid column along the heat pipe axis. Using L_b rather than the heat pipe axis to define the liquid column length simplifies calculation of the gravity/acceleration pressure drop when the heat pipe is curved along its length. The procedure is conservative in that it overestimates the liquid head and hence the static pressure drop,. However, the overestimate is small when $L_b/D_{hp} >> 1$.

The evaporator end E is elevated above the condenser end C, and the external acceleration vector points toward the condenser. With this orientation, the condensate undergoes a pressure drop (considered to be positive here) as it returns to the evaporator against gravity. The effect of the external acceleration is to facilitate the return of the condensate to the evaporator, producing a pressure rise (considered negative here).

If the condenser is at a higher elevation than the evaporator, the angle α_b becomes greater than 90 deg, gravity will facilitate the return of the condensate, and the pressure will rise. If the angle β_b is greater than 90 deg, acceleration will act to impede the return of condensate, and the pressure will drop. The effect of deceleration is exactly the opposite of that of acceleration.

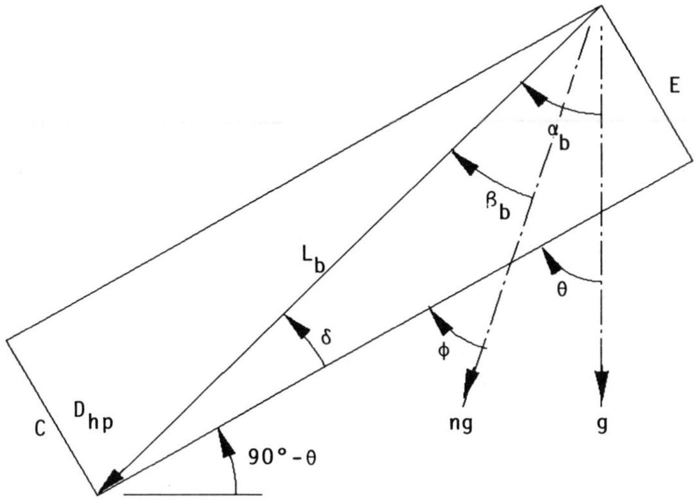

Figure 5.6. Geometry of a heat pipe subjected to gravity/acceleration.

These pressure changes due to gravity and acceleration are automatically accounted for by Eq. (5.45), provided that the angles α_b and β_b are measured with respect to the positive axis L_b that extends from the evaporator end to the condenser end of the heat pipe.

In Figure 5.6, the angles between the heat pipe axis and the gravity and acceleration vectors are θ and ϕ, respectively. They are related to the angles α_b and β_b by the following equations:

$$\alpha_b = \theta - \delta \tag{5.46}$$

$$\beta_b = \phi - \delta \tag{5.47}$$

where

$$\delta = \tan^{-1}(D_{hp}/L_p) \tag{5.48}$$

D_{hp} is the heat pipe diameter or an equivalent outside heat pipe dimension, and L_p is the heat pipe axial length.

5.1.2. Calculation of Capillary Pumping Limit

In this section, all the information required to calculate the capillary pumping limit is summarized and included in a *comprehensive calculational procedure*. The frictional pressure drop equations to be used are dependent on whether the vapor and liquid flows are laminar over the entire heat pipe length, or become turbulent over a portion of that length. The four possible flow condition options are listed in Table 5.1.

Fluid Coefficients. The various terms comprising the capillary pumping limit equations contain groups of temperature-dependent heat pipe fluid properties. To simplify

Table 5.1. Flow condition options for frictional pressure drop calculations

Vapor state	Liquid state
Laminar	Laminar
Laminar and turbulent	Laminar
Laminar	Laminar and turbulent
Laminar and turbulent	Laminar and turbulent

the calculational procedure, these groups are designated by single variables called fluid coefficients. The fluid coefficients are defined as follows:

$$a_f = 4\sigma \tag{5.49}$$

$$b_f = 16\mu_v/(\kappa\rho_v) \tag{5.50}$$

$$c_f = 0.734/(\kappa^2\rho_v) \tag{5.51}$$

$$d_f = 16\mu_l/(\kappa\rho_l) \tag{5.52}$$

$$e_f = f_t/(6\kappa^2\rho_v) \tag{5.53}$$

$$f_f = f_t/(6\kappa^2\rho_l) \tag{5.54}$$

$$i_f = 1600\kappa\mu_v \tag{5.55}$$

$$j_f = 1600\kappa\mu_l \tag{5.56}$$

The fluid coefficients have been incorporated into all subsequent forms of the capillary pumping limit equations. Fluid coefficient data for a number of heat pipe fluids are given in Table C.2 of Appendix C.

Calculational Procedure. The calculational procedure now follows. Equation numbers are indicated for specified expressions when appropriate. All terms should be checked for dimensional consistency when variables with dimensional units are introduced. As was pointed out in Section 1.7, g, the acceleration of gravity, may have to be included in the numerator or denominator of a term with dimensions in English or engineering units.

1. *Input data:*

Heat pipe fluid, heat pipe geometry, wick structure, wick friction factor K (if liquid flow channel is porous), heat pipe orientation (α_b), external acceleration (n and β_b), vapor temperature T_v, vapor pressure P_v, and heat transport rate Q.

Fluid coefficients a_f through f_f, i_f, j_f (5.49)–(5.56)

2. *Design parameters:*

$$D_{hv} = 4A_v/B_v \quad \text{and} \quad D_{hl} = 4A_l/B_l$$

$$\alpha = A_v/A_l$$

$$\xi = Q/A_v$$

$$\gamma_v = i_f/(\xi D_{hv}) \quad \text{and} \quad \gamma_l = j_f/(\alpha \xi D_{hl})$$

$$\text{If } \gamma_v < 1, \quad g(\gamma_v) = (1 - \gamma_v)(1 + \gamma_v + \gamma_v^2)$$

$$\text{If } \gamma_l < 1, \quad g(\gamma_l) = (1 - \gamma_l)(1 + \gamma_l + \gamma_l^2)$$

3. *Vapor momentum pressure drop:*

$$\Delta P_{vm} = c_f \xi^2 \tag{5.35}$$

4. *Liquid gravity/acceleration pressure drop:*

$$\Delta P_{la} = (\cos \alpha_b - n \cos \beta_b)\rho_l g L_b \tag{5.45}$$

5. *Vapor frictional pressure drop:*
If $\gamma_v \geq 1$, vapor flow is laminar, and

$$\Delta P_{vf} = \frac{b_f\,(L_p - L_a)\xi}{D_{hv}^2}\left(1 + \frac{2L_a}{L_p - L_a}\right) \tag{5.18}$$

If $\gamma_v < 1$, vapor flow is both laminar and turbulent, and

$$\Delta P_{vf} = \frac{b_f\,(L_p - L_a)\,\gamma_v^2\,\xi}{D_{hv}^2} + \frac{e_f\,(L_p - L_a)\,\xi^2}{D_{hv}}\left[g(\gamma_v) + \frac{3L_a}{L_p - L_a}\right] \tag{5.17}$$

6. *Liquid frictional pressure drop:*
If $\gamma_l \geq 1$ or the liquid flow channel is porous, liquid flow is laminar, and

$$\Delta P_{lf} = \frac{d_f\,(L_p - L_a)\,\varepsilon_{po}\,\alpha\,\xi}{D_{hl}^2}\left(1 + \frac{2L_a}{L_p - L_a}\right) \tag{5.21}$$

If the liquid flow channel is open,

$$\varepsilon_{po} = 1$$

If the liquid flow channel is porous,

$$\varepsilon_{po} = K D_{hl}^2/32 \tag{4.35}$$

If $\gamma_l < 1$ and the liquid flow channel is open, liquid flow is both laminar and turbulent, and

$$\Delta P_{lf} = \frac{d_f\,(L_p - L_a)\,\gamma_l^2\,\alpha\,\xi}{D_{hl}^2} + \frac{f_f\,(L_p - L_a)\,\alpha^2\,\xi^2}{D_{hl}}\left[g(\gamma_l) + \frac{3L_a}{L_p - L_a}\right] \tag{5.20}$$

If the wick is an artery type, add the term

$$\Delta P_{lfd} = \frac{d_f\,(L_p - L_a)\,\varepsilon_{po}\,\alpha_{de}\,\xi}{D_{hld}^2}\left(\frac{B_v}{4L_c N^2}\right) \tag{5.28}$$

7. *Total vapor and liquid pressure drops:*

$$\Delta P_v = \Delta P_{vf} + \Delta P_{vm}$$

$$\Delta P_l = \Delta P_{lf} + \Delta P_{la} \qquad \text{(no artery wick)}$$

$$\Delta P_l = \Delta P_{lf} + \Delta P_{lfd} + \Delta P_{la} \qquad \text{(with artery wick)}$$

8. *Capillary pressure:*

$$\Delta P_c = |\Delta P_v + \Delta P_l| \tag{2.10}$$

9. *Diameter of curvature:*

$$D_c = a_f / \Delta P_c \tag{2.4}$$

The symbols used in the above calculational procedure are defined in Table 5.2 for convenient reference.

When the diameter of curvature D_c is equal to the effective wick pore diameter D_p at the liquid-vapor interface, the heat transport rate Q is equal to the capillary pumping limit Q_c. If $D_c > D_p$, the actual heat transport rate Q is smaller than the capillary pumping limit. If $D_c < D_p$, Q exceeds the capillary pumping limit.

If $D_c \neq D_p$, the actual magnitude of the capillary pumping limit for a specified design can be found by systematically varying Q until

$$(1 - \delta_1)D_p \leq D_c \leq (1 + \delta_1)D_p \tag{5.57}$$

where δ_1 is some arbitrarily specified small quantity (0.01, for example). The Q at which Eq. (5.57) is satisfied is then taken to be equal to the capillary pumping limit Q_c.

It should be noted that the axial power density ξ can be used in place of the heat transport rate Q to find the diameter of curvature at the capillary pumping limit. In that case, the value of ξ for which Eq. (5.57) is satisfied is taken to be ξ_c, the axial heat flux at the capillary pumping limit.

Vapor Pressure Drop and Minimum Liquid Pressure. In the process of determining the capillary pumping limit, it is also advisable to calculate the percentage drop in vapor pressure and the minimum liquid pressure.

The vapor pressure drop cannot be greater than 100% (i.e., the vapor pressure cannot be less than zero). From a practical standpoint, the vapor pressure drop should not exceed a few percent if heat is to be transported through the heat pipe relatively isothermally. (See Table 4.1.)

If the vapor pressure drop is greater than about 10%, the calculational methods described previously for the capillary pumping limit will be inaccurate. The calculational accuracy can be improved by using the procedure to be suggested subsequently for a large vapor pressure drop.

The percentage vapor pressure drop is readily found, since ΔP_v has already been determined as part of the capillary pumping limit calculations, and the vapor pressure P_v at the start of the evaporator is a known function of the specified vapor temperature T_v.

In contrast to the situation for the heat pipe vapor, the minimum pressure for the liquid can be negative (i.e., the liquid can be under tension). [See Section 2.1.4.] This can happen when the liquid pressure drop due to gravity is very large, as for example in a long vertical heat pipe with the evaporator section on top. Heat pipe operation under negative liquid pressure may be acceptable. However, where there is a choice, negative liquid pressure is best avoided to avoid the uncertainty presently associated with this operating mode.

Calculation of the minimum liquid pressure $P_{l\min}$ requires knowledge of the vapor pressure at the evaporator end, the liquid and vapor pressure drops, and the location of the *wet point* (the point where the vapor and liquid pressures are equal). The relationships involved are illustrated in Figure 5.7.

Table 5.2. Symbols used in calculational procedure for capillary pumping limit

Symbol	Definition
α	Ratio of vapor and liquid flow channel areas
α_b	Angle between gravity and L_b
α_{de}	Ratio of vapor flow channel area to liquid flow channel area in distribution wick
β_b	Angle between external acceleration and L_b
γ	Ratio of maximum heat transport rate for laminar flow to total heat transport rate
ΔP_c	Capillary pressure
ΔP_v	Total pressure drop in vapor
ΔP_{vf}	Frictional pressure drop in vapor
ΔP_{vm}	Momentum pressure drop in vapor
ΔP_l	Total pressure drop in liquid
ΔP_{la}	Gravity/acceleration pressure drop in liquid
ΔP_{lf}	Frictional pressure drop in liquid
ΔP_{lfd}	Frictional pressure drop in liquid through distribution wick
ε_{po}	Ratio of liquid pressure drops through porous and open wicks
ξ	Axial heat flux
ξ_c	Axial heat flux at capillary pumping limit
ρ	Density
a_f through f_f, i_f, j_f	Fluid coefficients
g	Acceleration of gravity
$g(\gamma)$	Function of γ
n	External acceleration, number of g's
A	Cross-sectional area of flow channel
B	Distance around boundary of flow channel
D_c	Diameter of curvature
D_h	Hydraulic diameter
K	Friction factor for porous wick
L_a	Length of adiabatic section
L_b	Straight-line distance between lowest and highest points of heat pipe
L_c	Condenser length
L_e	Evaporator length
L_p	Heat pipe length
N	Number of arteries
P_v	Vapor pressure
Q	Heat transport rate
Q_c	Capillary pumping limit
T_v	Vapor temperature
a	Subscript denoting gravity/acceleration pressure drop
d	Subscript denoting distribution wick
f	Subscript denoting frictional pressure drop
m	Subscript denoting momentum pressure drop
v	Subscript denoting heat pipe vapor
l	Subscript denoting heat pipe liquid

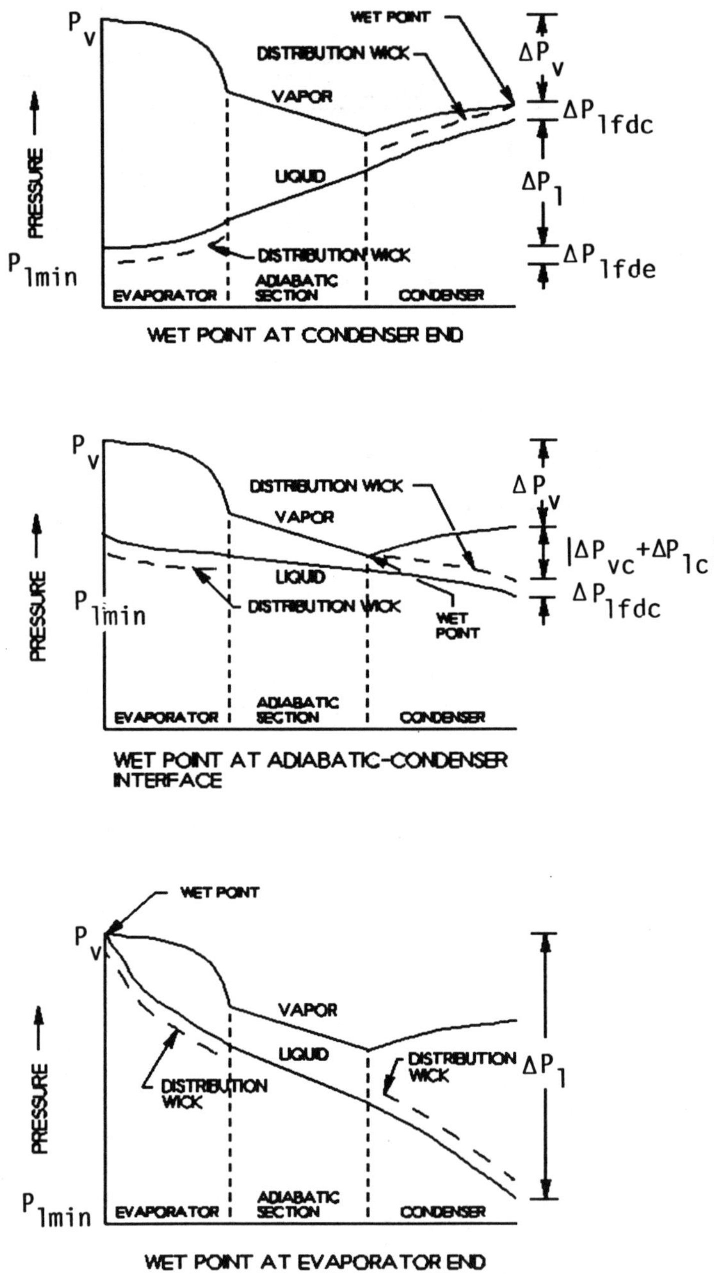

Figure 5.7. Vapor and liquid pressure distributions in a heat pipe.

The dashed lines are applicable only when an artery-type wick is present, and denote the liquid pressure in the distribution wick at locations equidistant from the arteries. The liquid pressure curves then refer to the liquid pressure in the arteries. The distribution wick pressure is greater than the artery pressure in the condenser, and less than the artery pressure in the evaporator.

As is evident from Figure 5.7, the minimum liquid pressure $P_{l\min}$ depends on whether there is a net liquid pressure drop ($\Delta P_l > 0$) or rise ($\Delta P_l < 0$), whether the absolute magnitude of a liquid pressure rise is larger or smaller than the vapor pressure drop ΔP_v, and the magnitude of the distribution wick pressure drop ΔP_{lfd} for artery-type wicks.

When there is a net liquid pressure drop, $P_{l\min}$ occurs at the evaporator end (in the distribution wick for an artery-type wick). When there is a net liquid pressure rise, $P_{l\min}$ occurs at the condenser end (in the arteries for an artery-type wick).

With an artery-type wick, it is also possible for $P_{l\min}$ to occur in the distribution wick at the juncture of the evaporator and adiabatic sections. This situation could occur with the middle graph of Figure 5.7 if the liquid pressure rise in the arteries were small compared to the distribution wick pressure drop in the evaporator.

Equations for $P_{l\min}$ are now given for three cases: (A) a net liquid pressure drop with the wet point at the condenser end, (B) a net liquid pressure rise whose absolute value is less than the vapor pressure drop, with the wet point at the condenser end and negligible vapor pressure recovery in the condenser, and (C) a net liquid pressure rise whose absolute value is greater than the vapor pressure drop, with the wet point at the evaporator end. The equations were deduced from examination of Figure 5.7, along with the concurrent use of Eqs. (5.24), (5.26), and (5.27).

A. $\Delta P_l \geq 0$:

$$P_{l\min} = P_v - (\Delta P_v + \Delta P_l + Z\Delta P_{lfd}) \tag{5.58a}$$

B. $\Delta P_l < 0$ and $|\Delta P_l| < \Delta P_v$:

$$P_{l\min} = P_v - [\Delta P_v + Z\Delta P_{lfd}/(1 + L_c/L_e)] \tag{5.58b}$$

C. $\Delta P_l < 0$ and $|\Delta P_l| > \Delta P_v$:

$$P_{l\min} = P_v + \Delta P_l \tag{5.58c}$$

In Eqs. (5.58), $Z = 1$ for an artery-type wick. Otherwise, $Z = 0$. Also, ΔP_{lfd} denotes the total pressure drop in the distribution wick (the sum of the pressure drop ΔP_{lfde} in the evaporator section and the pressure drop ΔP_{lfdc} in the condenser section of the distribution wick).

Case A is represented by the top graph of Figure 5.7, and case C by the bottom graph.

Case B would be represented by the middle graph if there were negligible recovery of vapor pressure in the condenser, and the wet point were at the condenser end instead of at the condenser-adiabatic section interface.

Since the vapor pressure drop would not ordinarily exceed 10%, it is highly unlikely that the minimum liquid pressure for the conditions depicted in the middle graph would be negative. Then the case B calculation, even though overestimating the minimum liquid pressure, will still give a reasonable indication of whether the heat pipe liquid is likely to be in tension. The alternate approach of accounting for vapor and liquid pressure recovery in the condenser will also be inaccurate, because actual vapor pressure recovery in the condenser is not known with certainty.

Temperature Drop along Heat Pipe Length. Since the heat pipe vapor is nominally saturated, its temperature is a direct function of the vapor pressure. Although knowledge of the temperature drop along the heat pipe is not required in order to determine the capillary pumping limit, it is needed to establish the extent of isothermality during heat transport. The relationship between the temperature drop along the heat pipe length and the vapor pressure drop will now be derived.

The relation between a pressure change ΔP_v in a saturated vapor and its associated temperature change ΔT_v can be approximated by the Clausius-Clapeyron equation in the form (Ref. 5.2)

$$\Delta P_v = \rho_v \kappa \Delta T_v / T_{vm} \tag{5.59}$$

where T_{vm} is the mean absolute vapor temperature, ρ_v is the vapor density at T_{vm}, and κ is the heat of vaporization. The density may be approximated using the perfect gas equation,

$$\rho_v = P_v/(R_g T_{vm}) \tag{5.60}$$

where P_v is the vapor pressure and R_g is the gas constant of the heat pipe vapor. The gas constant may be represented as

$$R_g = R_u/M \tag{5.61}$$

where R_u is the universal gas constant and M is the molecular weight of the vapor. Upon substituting Eqs. (5.60) and (5.61) into Eq. (5.59),

$$\Delta P_v = \frac{P_v\, M\, \kappa\, \Delta T_v}{R_u\, T_{vm}^2} \tag{5.62}$$

Since

$$\%\Delta P_v = 100(\Delta P_v / P_v)$$

where $\%\Delta P_v$ is the percentage drop in vapor pressure, Eq. (5.62) may be written

$$\frac{\Delta T_v}{\%\Delta P_v} = \frac{R_u\, T_{vm}^2}{100 M \kappa} \tag{5.63}$$

The universal gas constant has the value 1.985 Btu/lb-°R (8.314 J/g-°K). Then Eq. (5.63) may be written

$$\frac{\Delta T_v}{\%\ \Delta P_v} = 0.01985\, \frac{T_{vm}^2}{\kappa\, M} \tag{5.63a}$$

where T_{vm} = °R and κ is expressed in Btu/lb.

Alternately, when T_{vm} =°K, and κ = J/kg, Eq. (5.63) takes the form

$$\frac{\Delta T_v}{\%\ \Delta P_v} = 83.14\, \frac{T_{vm}^2}{\kappa\, M} \tag{5.63b}$$

Representative values of the temperature drop per percent drop in vapor pressure are given in Table 4.1. This figure varies over the range of 0.4 to 2.4 °R/%ΔP_v (0.2 to 1.3 °K/%ΔP_v) for the listed heat pipe fluids. It is evident that the vapor pressure drop should be limited to a few percent or less for relatively isothermal heat transport.

Procedure for Large Vapor Pressure Drop. If the vapor pressure drop is larger than about 10%, the density of the heat pipe vapor can no longer be considered constant over the heat pipe length, and the heat transport process may be significantly nonisothermal.

The basic assumption of incompressible, constant-density vapor flow made in developing the capillary pumping limit equations is then no longer valid. The predicted vapor pressure drop will be too low, the capillary pumping limit will be too high, and the diameter of curvature will be too large. This situation may be encountered when the heat transport rate is greater than 20–30% of the sonic limit.

The following procedure can be used to obtain more accurate capillary pumping limit results when the percentage vapor pressure drop is large:

1. Calculate the vapor temperature drop corresponding to %ΔP_v from Eq. (5.63), substituting T_v for T_{vm} in that equation.
2. Calculate the mean vapor temperature T_{vm} as

$$T_{vm} = T_v - \Delta T_v/2$$

3. Repeat the capillary pumping limit calculations, using T_{vm} as the vapor temperature.
4. Repeat steps 1 through 3 until the percentage change in ΔP_v following a recalculation is less than some arbitrarily established criterion (10% or less, for example). The last repetition may then be considered as a reasonably valid approximation to the capillary pumping limit.

In practice, one or two iterations may suffice for an adequate estimate of heat pipe characteristics at the capillary pumping limit.

Simplified Form of Capillary Pumping Limit. The capillary pumping capability of heat pipes with liquids characterized by low values of surface tension and heat of vaporization, such as ammonia and alcohol, is relatively poor. Low-pressure-drop, low-fluid-velocity designs are required for these heat pipes to achieve adequate heat transport rates.

Heat pipes of this type are frequently operated at or below ambient temperature in the weightless environment of space. In such cases, laminar flow conditions usually prevail (γ_v and γ_l are greater than unity), the vapor pressure drop may not be significant, and there is no gravity/acceleration liquid pressure drop.

For these conditions, the capillary pumping limit calculations reduce to the single equation

$$Q_c = \left[\frac{D_{hl}^2 A_l}{4\,\varepsilon_{po}\,D_p\,(L_p + L_a)}\right]\left(\frac{\sigma\,\kappa\,\rho_l}{\mu_l}\right) = G_p\,P_{ht} \tag{5.64}$$

where G_p is solely dependent on heat pipe geometry and P_{ht} is solely a function of temperature-dependent properties of the heat pipe liquid.

The parameter P_{ht} will be recognized as the *liquid heat transport parameter* that was defined in Section 3.1.2. It is evident from Eq. (5.64) that, for a given heat pipe geometry, the capillary pumping limit varies directly as the liquid heat transport parameter.

Also, for a given heat pipe liquid, operating temperature, and wick design, it can be seen that Q_c varies inversely with the sum of the heat pipe length L_p and the adiabatic length L_a. Therefore, the product $Q_c(L_p + L_a)$ (or Q_cL_p if there is no adiabatic section) is constant. For this reason, the heat transport capability of low-temperature heat pipes designed for space applications is usually expressed in W-cm or W-m.

5.2. SONIC LIMIT

The sonic limit is the highest possible heat transport rate that can be sustained in a heat pipe for a specific vapor temperature at the evaporator end of the heat pipe. The sonic limit is reached when the vapor leaving the evaporator (or the adiabatic section if present) attains sonic velocity. At that point the vapor flow is said to be choked, and the vapor flow and associated heat transport rate are at a maximum.

The sonic limit is a strongly increasing function of temperature. An increase in the heat input rate beyond what can be transported at the sonic limit for a given vapor temperature will cause the temperature to rise to a level at which the sonic limit becomes equal to the new heat input rate.

Since there is a substantial temperature drop along the evaporator length at and in the vicinity of the sonic limit, heat pipes are usually designed to operate well below the sonic limit. However, the sonic limit is encountered during the startup of liquid metal heat pipes. Heat pipe operation at the sonic limit during startup is treated in Chapter 7.

This section considers the flow of vapor through the evaporator and adiabatic section of a heat pipe when the vapor velocity is significant compared to sonic velocity. Equations are developed for the associated heat transport rates, including the sonic limit. The treatment is based on work reported in Refs. 5.3–5.9.

The key feature of heat pipe heat transport at the sonic limit is that the flowing vapor can no longer be considered as incompressible. Large decreases in pressure, temperature, and density occur as vapor is accelerated to sonic velocity at the exit of the evaporator (or adiabatic zone if present).

Friction has the same effect on compressible flow in a constant-area duct as does heat addition, causing the flow to accelerate while pressure, temperature, and density decrease (Ref. 5.10). Therefore, when friction is significant, less heat is required to accelerate the vapor in a heat pipe to sonic velocity. The sonic limit is then smaller than would be the case if friction were not present.

The presence of an adiabatic section increases the length of duct over which the heat pipe vapor is exposed to friction. Consequently, the vapor continues to accelerate after leaving the evaporator, and reaches its maximum velocity at the exit of the adiabatic section. Sonic velocity is then reached at the exit of the adiabatic section rather than the evaporator exit, and the sonic limit is below what would be the case if the adiabatic section were not present.

The heat transport rate is equal to the product of the mass evaporation rate of the heat pipe liquid and its mean heat of vaporization over the evaporator length. From mass flow rate continuity requirements, the rate at which mass enters the vapor passage by evaporation over the evaporator length is equal to the vapor mass flow rate at the evaporator exit. If an adiabatic section is present, the vapor mass flow rate in the adiabatic section is constant and equal to that leaving the evaporator.

We may then write for the heat transport rate Q,

$$Q = \rho_{ve} V_{ve} A_{ve} \kappa_m \tag{5.65}$$

where the subscript e denotes conditions at the exit of the evaporator or adiabatic section, ρ_v and V_v are the vapor density and velocity averaged over the vapor passage flow area A_{ve}, and κ_m is the heat of vaporization at the vapor passage outer boundary averaged over the evaporator length.

The Mach number Ma is defined as

$$\mathrm{Ma} = V_v / V_s \tag{5.66}$$

where V_s is the sonic velocity. Using the subscript e to denote Ma and V_s at the exit of the evaporator or adiabatic section, Eq. (5.65) can be expressed as

$$Q = \rho_{ve} \mathrm{Ma}_e V_{se} A_{ve} \kappa_m \tag{5.67}$$

When $\mathrm{Ma}_e = 1$, the exit velocity is equal to the sonic value, and the heat transport rate is equal to the sonic limit Q_s. Equation (5.67) then becomes

$$Q_s = \rho_{vs} V_{se} A_{ve} \kappa_m \tag{5.68}$$

where ρ_{vs} denotes the exit vapor density at the sonic limit.

Equations (5.67) and (5.68) are basic equations that relate the heat transport rate in a heat pipe to properties of the heat pipe vapor.

Desired forms for Q and Q_s are found by expressing the exit density and velocity of the heat pipe vapor in terms of known conditions at the beginning of the evaporator. This requires application of the principles of momentum and energy conservation to the vapor flow, along with an appropriate equation of state. The vapor is usually assumed to act as a perfect gas, which is injected into the vapor passage from the flow duct walls.

5.2.1. Vapor Pressure Relationship

The change in pressure of the heat pipe vapor as it flows through the evaporator is obtained by applying the conservation of momentum principle in the same manner as was employed for an incompressible fluid in Section 4.2. However, the vapor density is now considered to be variable, and any force on the vapor due to gravity/acceleration is assumed to be negligible. The situation is then as depicted in Figure 5.8.

For simplicity, the flow passage in Figure 5.8, as well as in Figures 5.9 and 5.10, is depicted as circular with diameter D_v. The analysis of these figures applies to other flow passage shapes as well, if D_v is interpreted as the hydraulic diameter. (See Section 4.5.)

Equating the rate of axial momentum flow leaving the differential vapor passage segment to the sum of the axial momentum flow rate entering the segment plus the net axial force acting on the segment, we obtain

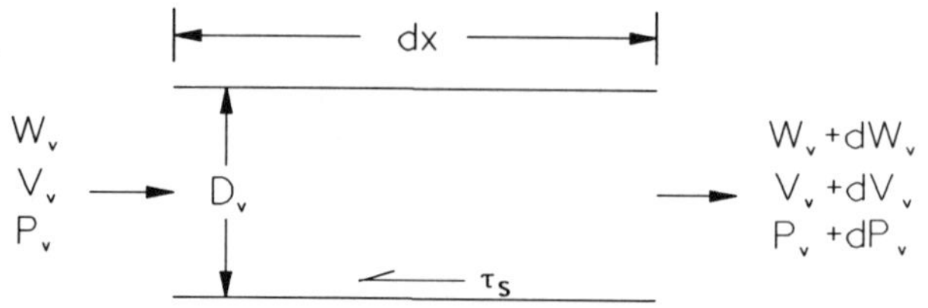

Figure 5.8. Momentum flow of a variable-density vapor.

$$(W_v + dW_v)(V_v + dV_v) = W_vV_v + (\pi/4)D_v^2P_v - (\pi/4)D_v^2(P_v - dP_v) - \tau_s\pi D_v dx \tag{5.69}$$

A momentum factor Ω is not included in the momentum terms, as was done for the incompressible flow analysis. Here, both the velocity and density vary over the flow cross section, and the integral of the momentum flow over the cross section is not readily determined. Subsequent sonic limit calculations indicate that ignoring the momentum factor does not lead to serious error.

Upon simplifying and assuming that the product of differentials is zero, Eq. (5.69) reduces to

$$d(W_vV_v) = (\pi/4)D_v^2 dP_v - \tau_s\pi D_v dx \tag{5.70}$$

Noting that

$$W_v = (\pi/4)D_v^2\rho_vV_v$$

Eq. (5.70) takes the form

$$dP_v = (4/D_v)\tau_s dx + d(\rho_vV_v^2) \tag{5.71}$$

Integrating Eq. (5.71) over the sum of the evaporator and adiabatic lengths L_e and L_a to obtain the total vapor pressure drop ΔP_v, and noting that $V_v = 0$ at $x = 0$,

$$\Delta P_v = \int_0^{L_e+L_a} \left(\frac{4}{D_v}\right)\tau_s\, dx + \rho_{ve}V_{ve}^2 = \Delta P_{vf} + \Delta P_{vm} \tag{5.72}$$

where ΔP_{vf} denotes the frictional term, ΔP_{vm} denotes the momentum term, and the subscript e denotes conditions at $x = L_e + L_a$.

As was shown by Eq. (4.13) in Section 4.2.1, ΔP_{vf} is also expressible as

$$\Delta P_{vf} = \int_0^{L_e+L_a} ff\,\rho_v \frac{V_v^2}{2D_v}\, dx \tag{5.73}$$

Letting FM be equal to the ratio of frictional and momentum pressure drops $\Delta P_{vf}/\Delta P_{vm}$, Eq. (5.72) can be written in the form

$$\Delta P_v = P_{vi} - P_{ve} = (1 + \mathrm{FM})\rho_{ve}V_{ve}^2 \tag{5.74}$$

where P_{vi} is the vapor pressure at $x = 0$ (the beginning of the evaporator).

From Eq. (5.66), the velocity V_{ve} can be expressed in terms of the sonic velocity V_{se} by the relation

$$V_{ve} = \mathrm{Ma}_eV_{se} \tag{5.75}$$

On the assumption that the vapor behaves as a perfect gas, the following equations apply:

$$\rho_{ve} = P_{ve}/(R_g T_{ve}) \tag{5.76}$$

$$V_{se} = \sqrt{k_s R_g T_{ve}} \tag{5.77}$$

where R_g is the gas constant and k_s is the ratio of specific heats at constant pressure and constant volume.

Upon substitution of Eqs. (5.75), (5.76), and (5.77) into Eq. (5.74) and simplifying, we obtain the final result for the ratio of initial vapor pressure P_{vi} to exit pressure P_{ve}.

$$P_{vi}/P_{ve} = 1 + (1 + \mathrm{FM})\mathrm{Ma}_e^2\, k_s \tag{5.78}$$

Since $\mathrm{Ma_e} = 1$ at the sonic limit,

$$P_{vi}/P_{vc} = 1 + (1 + \mathrm{FM})k_s \tag{5.79}$$

where P_{vc} is the exit vapor pressure at the sonic limit.

5.2.2. Vapor Temperature Relationship

The relationship between the initial and exit vapor temperatures is determined from an energy balance for a heat pipe vapor passage segment. (See Figure 5.9.)

The energy flow rate is the sum of the enthalpy flow rate $W_v c_p T_v$ and the kinetic energy flow rate $W_v V_v^2/2$, where c_p is the specific heat at constant pressure. The rate at which energy leaves the vapor passage segment is equal to the sum of the rates at which energy enters axially and radially.

The radially inflowing vapor is the result of vaporization of saturated liquid from the adjacent wick structure. Hence, its enthalpy is that of saturated vapor (not the heat of vaporization). The enthalpy of saturated vapor is generally much larger than its kinetic energy, or energy associated with frictional work in the vapor passage segment. Consequently, only the enthalpy of the radially inflowing vapor need be considered in the energy balance.

Taking c_p as constant, and using the mean value $T_{vs} + dT_{vs}/2$ for the temperature of saturated vapor over the length dx,

$$(W_v + dW_v)c_p(T_v + dT_v) + (W_v + dW_v)(V_v + dV_v)^2/2 = W_v c_p T_v + W_v V_v^2/2 + dW_v c_p(T_{vs} + dT_{vs}/2) \tag{5.80}$$

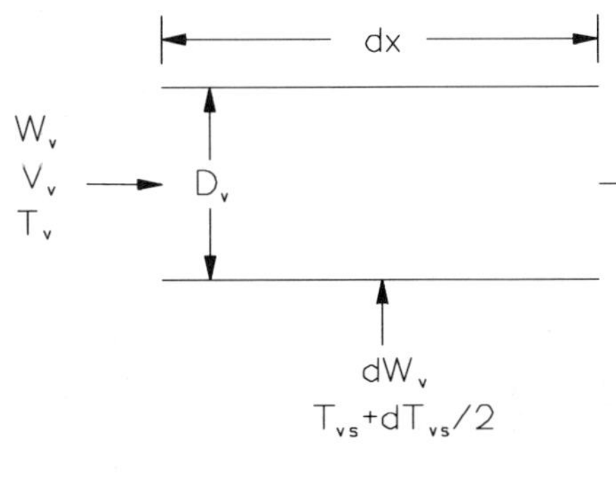

Figure 5.9. Energy flow of a variable-density vapor.

The subscript s denotes saturation conditions at the vapor passage radius.

Simplifying and letting terms involving products of differentials be equal to zero, Eq. (5.80) becomes

$$c_p d(W_v T_v) + d + (W_v V_v^2/2) = c_p T_{vs} dW_v \tag{5.81}$$

Integrating Eq. (5.81) over the length $L_e + L_a$, and noting that $V_v = 0$ at $x = 0$ and $dW_v = 0$ over the adiabatic length L_a, we obtain

$$c_p T_{vsm} W_{ve} = c_p W_{ve} T_{ve} + W_v e V_{ve}^2/2 \tag{5.82}$$

where the subscript e denotes conditions at the exit of the evaporator or adiabatic section.

In Eq. (5.82), T_{vsm} is the mean temperature over the evaporator length L_e of the saturated vapor flowing radially into the vapor passage.

Equation (5.82) may be written in the form

$$\frac{T_{vsm}}{T_{ve}} = 1 + \frac{V_{ve}^2}{2c_p T_{ve}} \tag{5.83}$$

Using Eqs. (5.75) and (5.77), and also noting that for a perfect gas

$$c_p = \left(\frac{k_s}{k_s - 1}\right) R_g \tag{5.84}$$

Eq. (5.83) takes the final form

$$\frac{T_{vsm}}{T_{ve}} = 1 + \frac{k_s - 1}{2} \mathrm{Ma}_e^2 \tag{5.85}$$

At the sonic limit, $\mathrm{Ma}_e = 1$, and Eq. (5.85) becomes

$$\frac{T_{vsm}}{T_{ve}} = \frac{k_s + 1}{2} \tag{5.86}$$

The mean saturation vapor temperature T_{vsm} over the evaporator length at the vapor passage radius can be approximated by the vapor temperature T_{vi} at the beginning of the evaporator. Therefore, Eqs. (5.85) and (5.86) are usually represented as

$$\frac{T_{vi}}{T_{ve}} = 1 + \frac{k_s - 1}{2} \mathrm{Ma}_e^2 \tag{5.85a}$$

$$\frac{T_{vi}}{T_{ve}} = \frac{k_s + 1}{2} \tag{5.86b}$$

Equation (5.85a) is the desired equation for the ratio of vapor temperatures at the beginning of the evaporator and at the distance $L_e + L_a$. Equation (5.86b) gives the same ratio at the sonic limit.

At the beginning of the evaporator, the vapor velocity is zero, and there is no radial temperature or density gradient. Therefore, at this point the vapor temperature T_{vi} is constant and equal to the saturation temperature along the vapor passage radius.

5.2.3. Vapor Density Relationship

The ratio of the mean vapor densities at the evaporator beginning and at the distance $L_e + L_a$ can be found using the perfect gas equation of state and the previously derived pressure and temperature ratios. From Eq. (5.76),

$$\frac{\rho_{vi}}{\rho_{ve}} = \left(\frac{P_{vi}}{P_{ve}}\right)\left(\frac{T_{ve}}{T_{vi}}\right) \tag{5.87}$$

Substituting Eqs. (5.78) and (5.85a) into Eq. (5.87),

$$\frac{\rho_{vi}}{\rho_{ve}} = \frac{1 + (1 + \text{FM})\,\text{Ma}_e^2\,k_s}{1 + \dfrac{k_s - 1}{2}\,\text{Ma}_e^2} \tag{5.88}$$

At the sonic limit, where $\text{Ma}_e = 1$, Eq. (5.88) becomes

$$\frac{\rho_{vi}}{\rho_{ve}} = 2\left[\frac{1 + (1 + \text{FM})\,k_s}{1 + k_s}\right] \tag{5.89}$$

Equations (5.88) and (5.89) represent the desired vapor density relationships.

5.2.4. Sonic Limit Equations

The axial heat flux will now be found as a function of the exit Mach number Ma_e and vapor conditions at the beginning of the evaporator. When $\text{Ma}_e = 1$, the axial heat flux represents the sonic limit.

First, Eq. (5.85a) is substituted into Eq. (5.77), giving

$$V_{se} = \frac{V_{si}}{\sqrt{1 + \dfrac{k_s - 1}{2}\,\text{Ma}_e^2}} \tag{5.90}$$

where V_{si} is the sonic velocity at the beginning of the evaporator.

Then, Eqs. (5.88) and (5.90) are substituted into Eq. (5.67a) to give the following expression for the axial heat flux $\xi = Q/A_v$:

$$\xi = \frac{\rho_{vi}\,V_{si}\,\kappa\,\text{Ma}_e}{1 + (1 + \text{FM})\,\text{Ma}_e^2\,k_s}\sqrt{1 + \frac{k_s - 1}{2}\,\text{Ma}_e^2} \tag{5.91}$$

When $\text{Ma}_e = 1$, the axial heat flux ξ_s at the sonic limit takes the form

$$\xi_s = \frac{\rho_{vi}\,V_{si}\,\kappa}{1 + (1 + \text{FM})\,k_s}\sqrt{\frac{1 + k_s}{2}} \tag{5.92}$$

Equation (5.92) for the sonic limit, with friction neglected (FM = 0), was first derived by Levy (Ref. 5.3). A simpler derivation, along the lines that have been presented here, was given by Deverall, Kemme, and Florschuetz (Ref. 5.5).

DeMichele and Davis (Ref. 5.7) calculated the sonic limit for water and liquid metal heat pipes on the basis of a two-dimensional analysis. Their results for frictionless flow are quite similar to the predictions of Eq. (5.92) with FM = 0.

Using Eqs. (5.76) and (5.77) for ρ_{vi} and V_{si} (replacing the subscript e with i), Eq. (5.92) becomes

$$\xi_s = \frac{P_{vi}\,\kappa}{1 + (1 + \mathrm{FM})\,k_s}\sqrt{\frac{k_s\,(1 + k_s)}{2R_g\,T_{vi}}} \tag{5.93}$$

Equation (5.93) is a convenient form for calculating the axial heat flux at the sonic limit, since it is in terms of the maximum heat pipe vapor pressure and temperature at the beginning of the evaporator.

If Eq. (5.91) is divided by Eq. (5.92),

$$\frac{\xi}{\xi_s} = \frac{[1 + (1 + \mathrm{FM})\,k_s]\ \mathrm{Ma}_e}{1 + (\,1 + \mathrm{FM})\,\mathrm{Ma}_e^{\,2}\,k_s}\sqrt{\left(\frac{2}{1 + k_s}\right)\left(1 + \frac{k_s - 1}{2}\,\mathrm{Ma}_e^{\,2}\right)} \tag{5.94}$$

Equation (5.94) gives the ratio of the exit axial heat flux to the sonic limit as a function of the exit Mach number. It may also be interpreted as the axial heat flux ratio at an axial location where the Mach number is Ma_e, when the heat pipe is operating at the sonic limit.

Since, for a given surface heat flux distribution, the axial heat flux is a known function of the axial distance x, Eq. (5.94) may also be used to determine the Mach number as a function of x. As shown by Eq. (5.98), the axial heat flux varies linearly with x when the surface heat flux is constant.

5.2.5. Relation between Axial and Surface Heat Fluxes

The relation between the axial and surface heat flux distributions (see also Section 5.4.6) is found by applying the conservation of mass principle to a vapor passage segment of length dx and diameter D_v. (See Figure 5.10.)

As was done for the incompressible flow case in Section 4.2.1, the mass flow rate leaving the segment is set equal to the rate at which mass enters the segment axially and radially. In this case, however, the variable density of the vapor is taken into account. Thus, we may write

$$G_v(\pi/4)D_v^2 + G_{vsm}\pi D_v dx = (G_v + dG_v)(\pi/4)D_v^2$$

where $G_v = \rho_v V_v$ is the axial mass flow rate per unit cross section of the vapor passage,

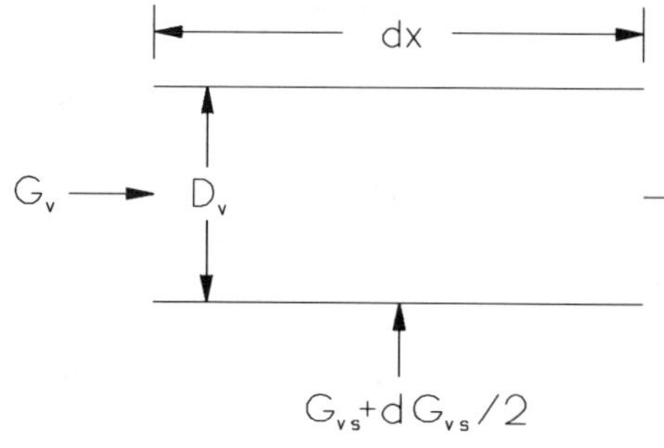

Figure 5.10. Mass flow of a variable-density vapor.

and G_{vsm} is the mean mass flow rate through the vapor passage lateral boundary per unit surface area. Noting that $G_{vsm} = G_{vs} + dG_{vs}/2$, and letting products of differentials be zero, it follows that

$$dG_v = (4/D_v)G_{vs}dx \tag{5.95}$$

where G_{vs} is the mass flow rate through the vapor passage periphery per unit surface area at x.

Now, at a given axial distance x, the heat and mass fluxes are related as follows:

$$\xi = \kappa_m G_v \quad \text{and} \quad q = \kappa G_{vs}$$

where ξ is the axial heat flux, q is the surface heat flux at the vapor passage periphery, κ is the heat of vaporization at x, and κ_m is the mean value of the heat of vaporization between 0 and x.

Then, Eq. (5.95) can be written

$$d\xi = \frac{4\ \kappa_m}{D_v\ \kappa}\ q\ dx \tag{5.96}$$

Although the heat of vaporization is a function of temperature at the evaporating surface, and hence varies with x, the ratio κ_m/k generally remains close to 1, and is assumed to be unity here. Therefore, on integrating Eq. (5.96) over x,

$$\xi = \frac{4}{D_v}\int_0^x q\ dx \tag{5.97}$$

If the surface heat flux q is constant, it follows that

$$\xi = \frac{4}{D_v}\ q\ x \tag{5.98}$$

and ξ is then a linear function of x.

5.2.6. Mach Number, Pressure, and Temperature Distributions in an Evaporator with No Friction

If there is no friction, FM = 0, and sonic velocity is reached at the evaporator exit, regardless of whether an adiabatic section is present. Then, from Eq. (5.98),

$$\xi_s = \frac{4\ q}{D_v}\ L_e$$

and

$$\frac{\xi}{\xi_s} = \frac{x}{L_e} \tag{5.99}$$

Therefore, when the surface heat flux is constant and the heat pipe is operating at the sonic limit, Eq. (5.94) also describes the relationship between the axial distance x along the evaporator and the associated Mach number. The pressure and temperature ratios may also be expressed as functions of ξ/ξ_s and x/L_e by substituting corresponding values of Ma_e in Eqs. (5.78) and (5.85a).

The use of these equations requires that the ratio of specific heats at constant pressure and volume k_s be specified. Theory predicts the following values for k_s:

Molecule	k_s
Monatomic	5/3
Diatomic	7/5

These values of k_s are very close to measured values for many monatomic and diatomic gases (Ref. 5.2).

The distributions of Mach number, Mach number squared, pressure ratio, and temperature ratio over the evaporator length are shown in Figures 5.11–5.14. These curves apply for a heat pipe operating at the sonic limit with a constant surface heat flux, negligible friction, and a monatomic heat pipe vapor. It is evident that the effects of compressibility, which result in a rapid increase in the slope of the curves, become pronounced only in the latter 20–30% of the evaporator length.

5.2.7. Surface Temperature Distribution in an Evaporator

Whereas the distribution of the mean vapor temperature along the evaporator length is needed to calculate the sonic limit, it is the temperature distribution along the evaporator wall that influences convective and radiative heat transfer into the heat pipe. The wall temperature is equal to the temperature at the wick inner surface (the liquid-vapor interface) plus the temperature drop through the wall and wick. If a heat pipe section is adiabatic (thermally insulated with no heat transfer through the wall/wick), the wall and wick inner surface temperatures are the same.

Kemme (Ref. 5.6) measured wall temperatures of short adiabatic regions at the evaporator beginning and exit of sodium and cesium heat pipes. The measurements were made for a wide range of maximum evaporator temperatures, and for axial heat transport rates ranging from 50% to 100% of the sonic limit. Since the wall temperature measurement points were thermally insulated, the temperatures on the exterior wall surface were essentially the same as the vapor temperatures at the liquid-vapor interface.

Kemme assumed that the vapor at the interface was saturated, and hence that its temperature was a unique function of the vapor pressure. He then obtained the vapor pressures corresponding to the measured wall temperatures, and plotted the ratio of initial and exit evaporator pressures as a function of the ratio of the axial heat transport rate to the sonic limit. The plotted points were found to lie on or very close to the curve of P_{vi}/P_{ve} versus evaporator length shown in Figure 5.13, which was calculated for a monatomic vapor.

These results indicate that sodium and cesium behave essentially as monatomic vapors in a heat pipe. Additionally, the vapor pressure at a given point along the evaporator appears to be constant over the vapor passage radius, and equal to the saturation pressure at the liquid-vapor interface. It follows that the vapor at the interface is saturated.

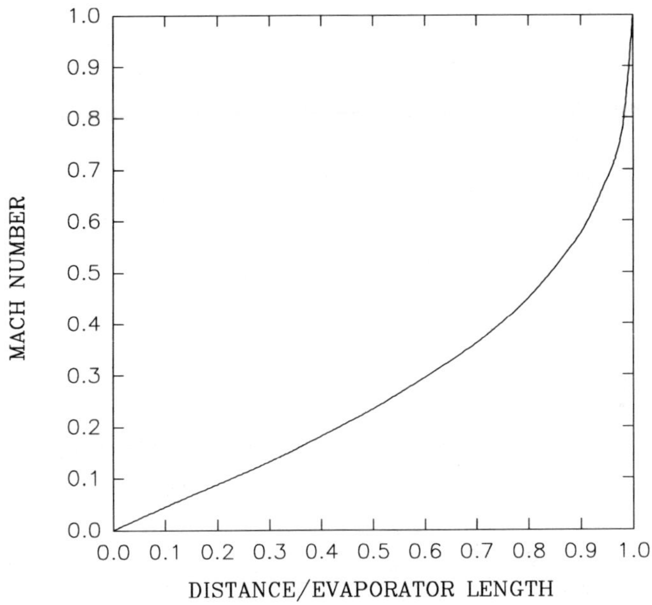

Figure 5.11. Vapor Mach number versus distance along an evaporator at the sonic limit.

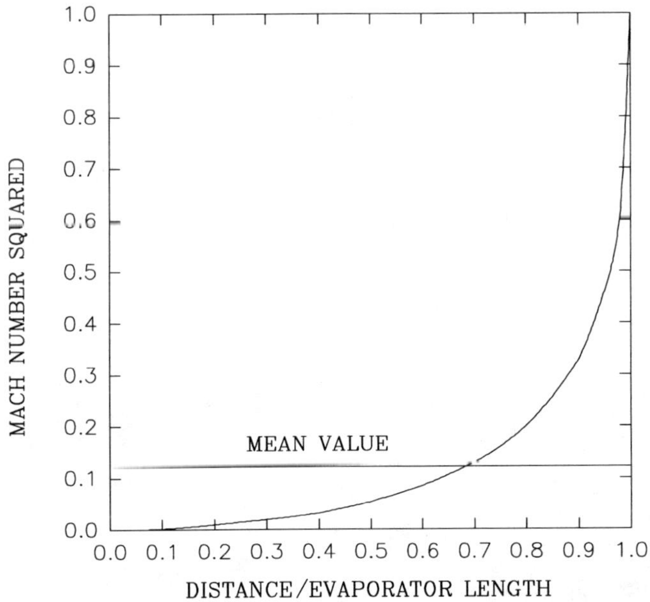

Figure 5.12. Square of vapor Mach number versus distance along an evaporator at the sonic limit.

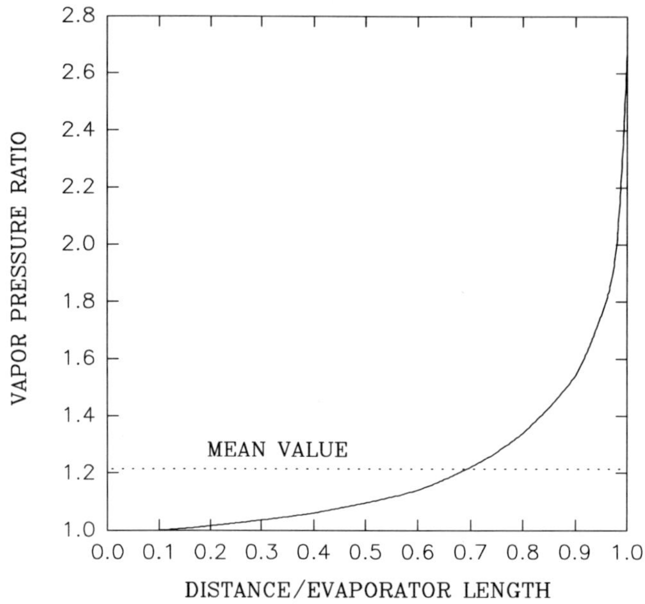

Figure 5.13. Variation of pressure ratio along an evaporator length at the sonic limit.

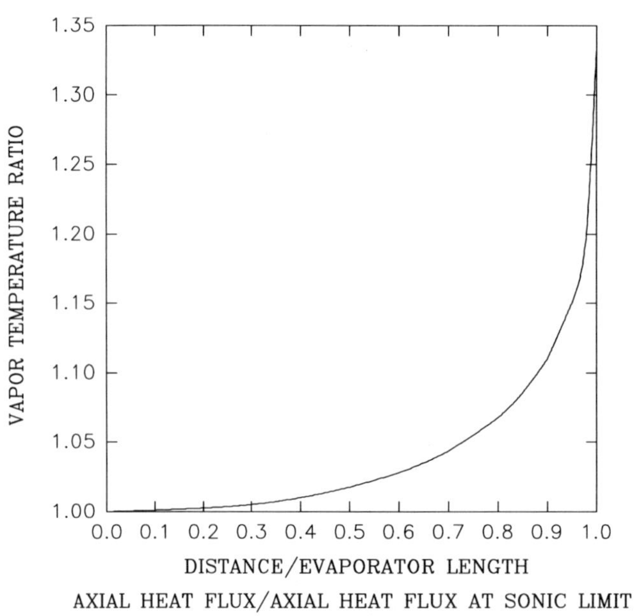

Figure 5.14. Variation of temperature ratio with evaporator position or axial heat flux.

At the beginning of the evaporator, the vapor velocity is zero, and there is no temperature gradient along the vapor passage radius. Consequently, the vapor at that point can be presumed to be saturated along the radius, so that $T_{vi} = T_{vsi}$, where T_{vsi} is the saturation temperature at the beginning of the evaporator.

The interface vapor temperature (i.e., the saturation temperature) can then be found as a function of the axial heat transport ratio ξ/ξ_s and the initial vapor temperature T_{vi} by the following procedure:

1. Specify $T_{vi} = T_{vsi}$ and ξ/ξ_s.
2. Find the saturation pressure P_{vsi} corresponding to T_{vsi} from tables or an equation relating the vapor pressure to the saturation vapor temperature.
3. Find the Mach number corresponding to ξ/ξ_s from Eq. (5.94).
4. Calculate the pressure $P_{vs} = P_{ve}$ from Eq. (5.78).
5. Find the saturation temperature T_{vs} corresponding to P_{vs} from tables or an equation relating the vapor pressure to the saturation vapor temperature. This is the T_{vs} that corresponds to the specified T_{vi} and ξ/ξ_s.

Drop in Saturation Temperature versus Mach Number. A direct expression for the drop in saturation vapor temperature T_{vs} as a function of Mach number along the evaporator will now be derived.

The pressure-temperature relationship for a saturated vapor can be represented as

$$\log_{10} P_v = A - \frac{B}{T_v} \tag{5.100}$$

where A and B are constants for a specific vapor (Refs. 5.11–5.14), T_v is in °R, and P_v is in psia. When T_v is in °K and P_v is in N/m^2, Eq. (5.100) takes the form

$$\log_{10} P_v = A + 3.8385 - 0.5556 \frac{B}{T_v} \tag{5.100a}$$

Values of A and B for a few liquid metals are given in Table 5.3.

Solving Eq. (5.100) for T_v,

$$T_v = \frac{B}{A - \log_{10} P_v} \tag{5.101}$$

Evaluating Eq. (5.101) at $L_e + L_a$, where $T_v = T_{vs}$ and $P_v = P_{vs}$, and using Eq. (5.78),

Table 5.3. Constants in the pressure-temperature equation for saturated vapor [Eq. (5.100)]

Vapor	A	B	Reference
Lithium	6.1765	14,516	5.11
Sodium	5.6883	9,396	5.12
Potassium	5.3157	7,707	5.13
Cesium	5.0948	6,680	5.14

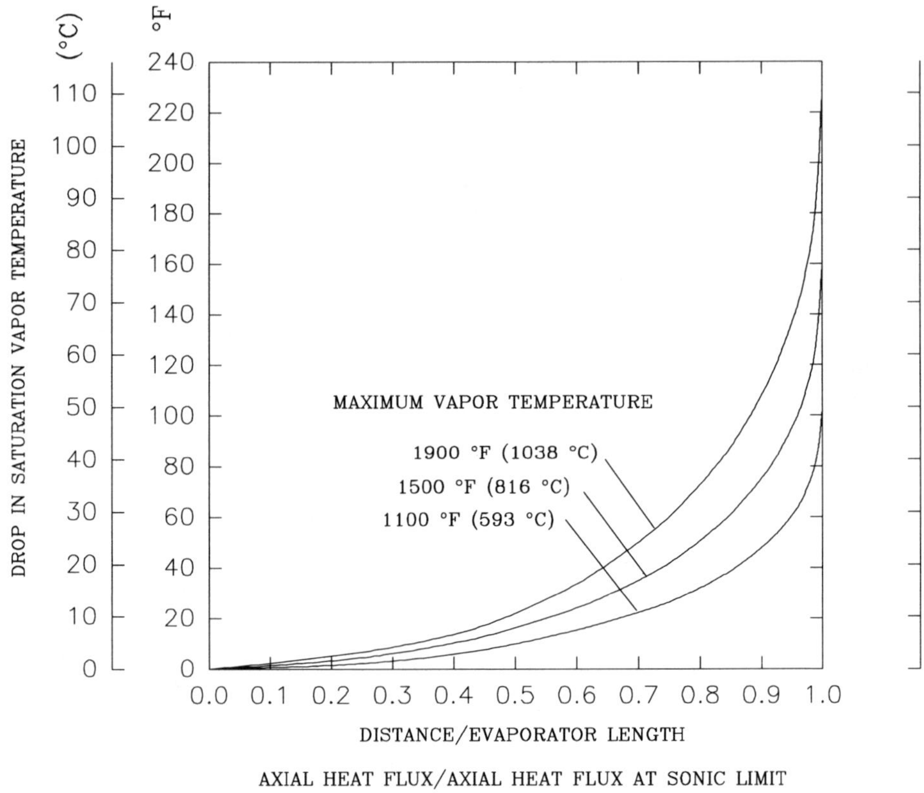

Figure 5.15. Distribution of saturation vapor temperature drop along the evaporator of a sodium heat pipe.

$$T_{vs} = \frac{B}{A - \log_{10} P_{vi} + \log_{10}[1 + (1 + \text{FM})\,\text{Ma}_e^2\, k_s]}$$

Substituting Eq. (5.100) for $\log_{10} P_{vi}$ in the above equation, with $P_v = P_{vi}$ and $T_v = T_{vi}$, yields the result

$$\frac{T_{vs}}{T_{vi}} = \frac{B}{B + T_{vi}\,\log_{10}\,[1 + (1 + \text{FM})\,\text{Ma}_e^2\, k_s]} \tag{5.102}$$

Letting $\Delta T_{vs} = T_{vi} - T_{vs}$, Eq. (5.102) can be rewritten in the form

$$\frac{\Delta T_{vs}}{T_{vi}} = 1 - \frac{B}{B + T_{vi}\,\log_{10}\,[1 + (1 + \text{FM})\,\text{Ma}_e^2\, k_s]} \tag{5.103}$$

When ΔT_{vs} and T_{vi} are in °K, T_{vi} in the second term on the right side of Eq. (5.103) should be multiplied by 1.8.

Equation (5.103) gives the drop in saturation vapor temperature along the evaporator as a function of the initial (maximum) vapor temperature and the exit Mach number. For a constant-surface heat flux, this is also the temperature drop along the external wall surface.

Since, from Eq. (5.94), Ma_e is a function of the ratio of the axial heat flux to the axial sonic limit ξ/ξ_s, ΔT_{vs} is also a function of T_{vi} and ξ/ξ_s. When $\mathrm{Ma}_e = 1$, the saturation temperature drop is that corresponding to the sonic limit.

Equations (5.94) and (5.103) were used to generate plots of the drop in saturation vapor temperature versus the fraction of the sonic limit for a sodium heat pipe with negligible friction (FM = 0). The resulting curves are shown in Figure 5.15. As was shown to be the case for the Mach number and pressure ratio distributions, the plot of ΔT_{vs} versus ξ/ξ_s is also a plot of ΔT_{vs} versus x/L_e if the surface heat flux is constant and the heat pipe is operating at the sonic limit.

The drop in saturation temperature is seen to increase with both distance along the evaporator and the maximum vapor temperature. At the sonic limit, ΔT_{vs} varies from 103 to 228°F (57 to 127°C) as T_{vi} increases from 1100 to 1900°F (593 to 1038°C). Two-thirds of this temperature drop occurs within 20% of the evaporator exit. It is evident that the axial heat flux should not exceed 20–30% of the sonic limit for reasonably isothermal heat transport [$\Delta T_{vs} < 5$–10°F (3–6°C)] in the evaporator.

Drop in Mean Vapor Temperature at the Sonic Limit. The drop in the mean vapor temperature at the sonic limit $\Delta T_{vc} = T_{vi} - T_{ve}$, as obtained from Eq. (5.86b), is given by

$$\frac{\Delta T_{vc}}{T_{vi}} = \frac{k_s - 1}{k_s + 1}$$

In Figure 5.16, ΔT_{vc} and ΔT_{vs} at the sonic limit are shown as a function of the

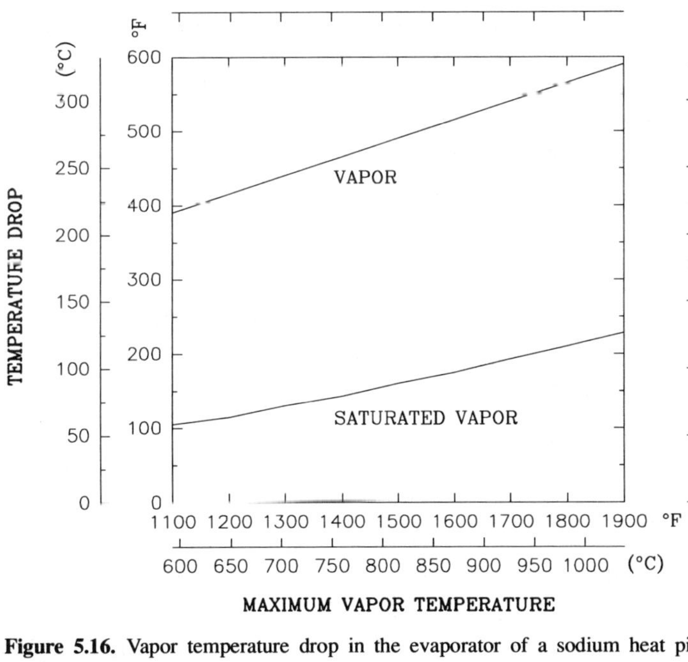

Figure 5.16. Vapor temperature drop in the evaporator of a sodium heat pipe at the sonic limit versus maximum vapor temperature.

maximum vapor temperature T_{vi} for a sodium heat pipe. The drop in the mean vapor temperature is several hundred degrees greater than the drop in the saturation temperature. These figures indicate that the bulk of the heat pipe vapor becomes highly supersaturated as it expands through the evaporator. Additionally, a radial temperature drop on the order of several hundred degree develops between the saturated vapor at the liquid-vapor interface and the vapor in the interior of the vapor passage.

The high degree of supersaturation is a direct consequence of the assumption that the heat pipe vapor acts as a perfect gas. If the entering saturated vapor were to remain in thermodynamic equilibrium throughout its expansion in the evaporator, its temperature drop would correspond to ΔT_{vs}, and some of the vapor would have condensed into liquid droplets.

Sonic limit calculations and experimental data are compared in Section 5.2.10. The good correlation between theory and experiment based on the perfect gas model indicates that the evaporation and expansion of vapor in a heat pipe evaporator is a highly nonequilibrium process. Although some condensation of liquid droplets may occur, the process does appear to be characterized by a high degree of supersaturation and large radial temperature gradients.

5.2.8. Sonic Limit with Friction

Levy and Chow (Ref. 5.4) analyzed the sonic limit of sodium heat pipes, taking into account the effects of friction, condensation of vapor into liquid droplets during the vapor expansion process, and the presence of varying proportions of diatomic molecules (dimers) in the nominally monatomic sodium vapor. Depending on the temperature, roughly 5–10% of the sodium molecules are diatomic under equilibrium conditions (Ref. 5.7). Levy and Chow found that the effect on the sonic limit of condensation and dimer concentration is small compared to the effect of friction. Frictional effects on the sonic limit were also considered by DeMichele and Davis (Ref. 5.7).

We now proceed to develop a simplified method for predicting the sonic limit when friction is present. As can be seen from Eq. (5.94), the effect of friction on the sonic limit is established by FM, the ratio of frictional and momentum pressure drops. Expressions for FM at the sonic limit will now be obtained for the cases of laminar and turbulent flow.

Using Eqs. (5.72) and (5.73), FM may be expressed as

$$\mathrm{FM} = \frac{\displaystyle\int_0^{L_e + L_a} \frac{f \rho_v V_v^2}{2 D_v} dx}{\rho_{ve} V_{ve}^2} \tag{5.104}$$

Laminar Flow. For laminar flow, f from Eq. (4.20) is substituted into Eq. (5.104) to obtain

$$\mathrm{FM}_l = \frac{32\,\mu_{vm}\,(L_e + L_a)\,V_{vm}}{D_v^2\,\rho_{ve}\,V_{ve}^2} \tag{5.105}$$

where the subscript l denotes laminar flow, and μ_{vm} and V_{vm} are values of μ_v and V_v averaged over the length $L_e + L_a$.

Using Eqs. (5.66) and (5.76)–(5.78), Eq. (5.105) can be expressed as

$$\mathrm{FM}_l = y_1/(1 - y_1) \tag{5.106}$$

where

$$y_1 = \left[\frac{32\,\mu_{vm}\,(L_e + L_a)\,\mathrm{Ma}_m\,(1 + k_s)}{D_v^2\,P_{vi}}\right]\sqrt{\frac{R_g\,T_{vm}}{k_s}} \tag{5.107}$$

Equation (5.106), with y_1 defined by Eq. (107), is the desired equation for FM_l. Its use requires that the mean values μ_{vm}, Ma_m, and T_{vm} be known. The averages μ_{vm} and T_{vm} can be approximated closely by μ_{vi} and T_{vi}. The average Mach number for the evaporator can be estimated by finding the average of the curve of Mach number versus x/L_e for no friction (FM = 0), as shown in Figure 5.11. Numerical integration of this curve using Simpson's rule yields $\mathrm{Ma}_m = 0.280$.

When the heat pipe includes an adiabatic section, sonic velocity in the vapor is reached at the end of the adiabatic section. The additional pressure drop in the adiabatic section results in a further decrease in the sonic limit. Ma_m then represents the mean Mach number averaged over the length $L_e + L_a$. Its value can be conservatively estimated by assuming that $\mathrm{Ma}_m = 0.280$ in the evaporator and 1 in the adiabatic section. Then

$$\mathrm{Ma}_m = \frac{0.28\,L_e + L_a}{L_e + L_a} \tag{5.108}$$

Using Eq. (5.108) in (5.107), y_1 for laminar flow takes the final form

$$y_1 = \left[\frac{32\,\mu_{vi}\,(0.28\,L_e + L_a)\,(1 + k_s)}{D_v^2\,P_{vi}}\right]\sqrt{\frac{R_g\,T_{vi}}{k_s}} \tag{5.109}$$

Turbulent Flow. If the axial Reynolds number at the evaporator exit exceeds the maximum value for laminar flow, turbulent flow can be expected in the latter part of the evaporator and in the adiabatic section. Although both laminar and turbulent flow will then exist in the evaporator, for purposes of estimating the sonic limit with friction the evaporator flow will be assumed to be all laminar or all turbulent.

In Section 5.1.1, it was established that in a heat pipe evaporator the flow would be considered turbulent beyond the point where $\mathrm{Re}_a > 1600$, and laminar prior to that point. In the turbulent flow regime, the friction factor f was assumed to have the constant value of 0.04.

Here, it will be assumed that, when $\mathrm{Re}_a > 1600$ at the evaporator exit, turbulent flow exists in the adiabatic section. Otherwise, the adiabatic section flow will be considered laminar. In the evaporator, flow will be considered all turbulent if $\mathrm{Re}_a > 5000$, and all laminar otherwise.

The ratio of frictional and momentum pressure drops at the sonic limit for turbulent flow FM_t can be obtained from Eq. (5.104) with $f = 0.04$ and the use of Eqs. (5.75)–(5.77). FM_t then takes the form

$$FM_t = 0.02\left(\frac{L}{D_v}\right)\left(\frac{P_{vi}}{P_{ve}}\right)_m (Ma^2)_m \tag{5.110}$$

where the subscript m denotes values averaged over the appropriate length L. If the evaporator flow is turbulent, the appropriate length $L = L_e + L_a$. If the evaporator flow is laminar, then $L = L_a$.

The average values of the pressure ratio and square of the Mach number over the evaporator length can be approximated by the averages of the curves of Figures 5.13 and 5.12 for frictionless flow. These values, based on numerical integration with Simpson's rule, are

$$(P_{vi}/P_{ve})_m = 1.210$$

$$(Ma^2)_m = 0.126$$

Over the adiabatic length, the averages of these parameters may conservatively be taken as unity.

The value of the parameters over the total turbulent flow length may then be found from the following equations:

$$\left(\frac{P_{vi}}{P_{ve}}\right)_m = \frac{1.21\,L_e + L_a}{L_e + L_a} \tag{5.111}$$

$$(Ma^2)_m = \frac{0.126\,L_e + L_a}{L_e + L_a} \tag{5.112}$$

When turbulent flow is restricted to the adiabatic section, L_e is taken to be zero, and Eqs. (5.111) and (5.112) reduce to 1.

Substitution of Eqs. (5.111) and (5.112) into Eq. (5.110) gives the final desired form of FM_t:

$$FM_t = \left[\frac{0.02}{D_v\,(L_e + L_a)}\right](1.21\,L_e + L_a)\,(0.126\,L_e + L_a) \tag{5.113}$$

Summary of FM Equations. The use of the equations for FM_l and FM_t is summarized as follows:

$FM = FM_l$ $\quad Re_a < 1600$

$FM = FM_l\,(L_a = 0) + FM_t\,(L_e = 0)$ $\quad 1600 \le Re_a \le 5000$

$FM = FM_t$ $\quad Re_a > 5000$

5.2.9. Axial Reynolds Number at Sonic Limit

At the sonic limit, the axial Reynolds number Re_{as} takes the form

$$\text{Re}_{as} = \frac{\rho_{ve}\, V_{se}\, D_v}{\mu_{ve}} \tag{5.114}$$

where V_{se} is the sonic velocity at the exit of the evaporator.

The ratio V_{se}/μ_{ve} is relatively insensitive to temperature, and hence may be evaluated at the maximum vapor temperature T_{vi}. Letting $V_{si} = (k_s R_g T_{vi})^{1/2}$ and using Eq. (5.89) for ρ_{ve}, Eq. (5.114) may be written in terms of properties at the beginning of the evaporator as

$$\text{Re}_{as} = \frac{\rho_{vi}\, D_v\, (1 + k_s)\, \sqrt{k_s\, R_g\, T_{vi}}}{2\, \mu_{vi}\, [1 + (1 + \text{FM})\, k_s]} \tag{5.115}$$

5.2.10. Comparison with Experimental Sonic Limit Data

Kemme (Ref. 5.6) measured the sonic limit of sodium, potassium, and cesium heat pipes at several different maximum vapor temperatures. His results for the sodium heat pipe are reported here and compared with calculated values of the sonic limit. The length of the heat pipe evaporator was 14 cm, and the vapor passage diameter was 1.14 cm. There was an adiabatic section of undefined, but apparently negligible, length. The L_e/D_v ratio was 12.28.

The sonic limit ξ_s for this heat pipe was calculated for frictional and frictionless flow cases. In Table 5.4, the measured and calculated values of ξ_s are shown as a function of the maximum vapor temperature T_{vi}. The percent deviations of the calculated values from the measured values are given in brackets. The axial Reynolds number, calculated from Eq. (5.115) with FM = 0, is also given in the table. Since $\text{Re}_{as} < 5000$ and $L_a = 0$, the laminar friction parameter applies at all temperatures.

The calculated sonic limits are larger than the experimental values. For the frictionless case (FM = 0), the difference is almost 50% at the lowest temperature, and decreases to 5–10% at the highest temperatures. When friction is taken into account, the

Table 5.4. Comparison of calculated and experimental sonic limits for sodium heat pipes

T_{vi}, °F (°C)	P_{vi}, psia (N/m²)	Re_{as}	ξ_{sx}, Btu/in²-sec (W/m²)	ξ_{so}, Btu/in²-sec (W/m²)	ξ_{sf}, Btu/in²-sec (W/m²)
932	0.087	422	1.13	1.65 [46.6]	1.28 [13.4]
(500)	(600)		(1.84)	(2.70)	(2.09)
1022	0.223	1071	3.54	4.05 [14.5]	3.70 [4.5]
(550)	(1538)		(5.79)	(6.63)	(6.05)
1112	0.514	2212	8.26	8.97 [8.7]	8.62 [4.4]
(600)	(3544)		(1.35)	(1.47)	(14.1)
1202	1.084	4919	17.4	18.2 [4.2]	17.8 [2.2]
(650)	(7474)		(28.5)	(29.7)	(29.1)

Note: ξ_{sx} = experimental sonic limit; ξ_{so} = calculated sonic limit (FM = 0); ξ_{sf} = calculated sonic limit with friction.

calculated sonic limit agrees with experimental values to within 5%, except at the lowest temperature, at which the difference is 13%.

The difference between ξ_{sf} and ξ_{sx} may be attributed to not including the effect of the small adiabatic section and the evaporator momentum factor, approximations made in evaluating temperature-dependent properties and the friction parameter FM, and experimental error.

The calculated sonic limit with friction for an evaporator length of 140 cm is 55% of the measured value ξ_{sx} at T_{vi} = 1112°F (600°C). Thus, friction can substantially reduce the sonic limit in long evaporator (and adiabatic) sections.

For most design situations, a sodium heat pipe would probably operate at a vapor pressure of 1 psia (6895 N/m^2) or higher, in order to have a reasonably large axial heat flux and hence a reasonably small vapor passage area. The frictionless sonic limit then gives results of sufficient accuracy for design purposes, although they are somewhat on the high side.

If L_e/D_v is large, if an adiabatic section is present, or if the vapor pressure is substantially below 1 psia (6895 N/m^2), frictional effects can be significant. The inclusion of FM in the sonic limit calculations will then yield improved accuracy.

5.2.11. Simplified Equation for Sonic Limit

In Ref. 5.9, it is shown that the frictionless sonic limit can be represented as the following simple power law function of vapor pressure:

$$\xi_s = A_s P_{vi}^{\alpha_s} \tag{5.116}$$

where P_{vi} is the maximum vapor pressure in psia and ξ_s is the sonic limit in Btu/in^2-sec.

When P_{vi} is in N/m^2 and ξ_s is in W/m^2, Eq. (5.116) takes the form

$$\xi_s = 1.636 \times 10^6 \ A_s \left(\frac{P_{vi}}{6895} \right)^{\alpha_s} \tag{5.116a}$$

Values of the power law constants A_s and α_s are given in Table 5.5 for a number of heat pipe fluids.

Since most heat pipes tend to operate over a limited vapor pressure range (roughly 1 to 100 psia [6895 to 689,500 N/m^2]), the sonic limit capabilities of various heat pipe fluids are readily compared using Eq. (5.116).

Table 5.5. Power law constants for sonic limit Eqs. (5.116) and (5.116a) (Ref. 5.9)

Fluid	A_s	α_s
Lithium	39.02	0.934
Sodium	16.88	0.929
Water	15.24	0.931
Ammonia	12.74	0.889
Freon 11	4.04	0.910

For example, at a vapor pressure of 1 psia (6895 N/m^2), the sonic limit ξ_s is equal to the constant A_s. The sonic limit of lithium is then about 2½ times that of sodium, and the sonic limits of sodium and water are roughly equal.

5.3. ENTRAINMENT LIMIT

The vapor flowing through a heat pipe is in contact with its counterflowing liquid. At a sufficiently high vapor velocity, the shear stress exerted by the vapor on the adjacent liquid may be sufficient to disrupt the liquid-vapor interface. Liquid droplets may then be torn loose from the liquid surface, entrained in the vapor, and transported to the end of the condenser. Entrainment of the heat pipe liquid in this manner will be initiated at the evaporator exit, where the vapor velocity and shear stress are at a maximum.

When entrainment occurs, liquid that otherwise would flow into the evaporator to replenish vaporizing liquid instead recirculates through the condenser region. The reduced flow of liquid into the evaporator leads to dryout of the wick and subsequent overheating. The heat transport rate at which entrainment occurs is called the *entrainment limit*. Like the sonic and capillary pumping limits, it may be expressed either as a heat transport rate or as an axial heat flux.

The entrainment limit has been and remains the subject of some controversy. Questions have been raised about whether experimentally measured entrainment limits are in reality due to some other heat-transport-limiting phenomenon. In this section, entrainment limit equations are derived, and uncertainties associated with their use are discussed. As was the case for the capillary pumping limit derivation, the vapor density is assumed to be constant.

5.3.1. Weber Number Criterion for Entrainment

We now consider a circular flow duct, with vapor flowing through the center of the duct and liquid flowing along the duct periphery. The *Weber number* We is defined as the ratio of the vapor inertia force acting over the vapor space cross section (i.e., the change in the momentum flow rate as the vapor velocity increases from 0 to V_v) to the liquid surface tension force acting over the circumference of the vapor space. Thus,

$$\text{We} = \frac{\rho_v V_v^2 (\pi/4) D_v^2}{\sigma \pi D_v} = \frac{\rho_v V_v^2 D_v}{4\sigma}$$

More generally, the linear dimension in the Weber number definition, which in this case is the vapor space diameter D_v, is denoted by λ_w.

Cotter (Ref. 5.15) pointed out that when We > 1 surface waves in the liquid can be expected to grow and become unstable, thereby disrupting the liquid surface. He proposed that the heat transport rate in a heat pipe would reach a maximum value (the entrainment limit) at a Weber number equal to unity, provided that an appropriate value of the characteristic dimension λ_w could be identified for the capillary structure of the heat pipe wick.

In Cotter's definition of We, the constant 2π appears in the denominator instead of 4. Since the characteristic dimension is generally an empirically determined factor, the choice of constant is of no great importance, as long as it is used consistently in arriving at a value for λ_w and in subsequent calculations of We. For consistency with the literature, the constant 2π will be used in the Weber number definition.

In addition, for a heat pipe the momentum pressure drop term $\rho_v V_v^2$ should be multiplied by the evaporator momentum factor Ω_e to reflect more accurately the momentum pressure drop. Also, as indicated by the subscript *e*, the Weber number should be evaluated at the evaporator exit, where the momentum flow rate is greatest. Then the Weber number criterion for the onset of entrainment is expressible as

$$\mathrm{We} = \frac{\Omega_e\, \rho_{ve}\, V_{ve}^2\, \lambda_w}{2\,\pi\,\sigma} = 1 \tag{5.117}$$

Ernst (Ref. 5.16) and Kemme (Ref. 5.17) determined that the effective diameter D_p corresponding to the maximum capillary pressure in square woven screen wicks is well represented by

$$D_p = D_s + D_w \tag{5.118}$$

where D_s is the spacing between wires and D_w is the wire diameter. Further investigation by Kemme (Ref. 5.17) established that the onset of entrainment correlates well with experimental data when

$$\lambda_w = D_p = D_s + D_w \tag{5.119}$$

The entrainment limit is found by first solving for the vapor velocity V_{ve} at the evaporator exit from Eq. (5.117), yielding

$$V_{ve} = \sqrt{\frac{2\,\pi\,\sigma}{\Omega_e\, \rho_{ve}\, \lambda_w}} \tag{5.120}$$

It will be recalled that the heat transport rate Q is given by

$$Q = \kappa \rho_{ve} V_{ve} A_v$$

where V_{ve} is the vapor velocity at the evaporator exit.

Upon substituting V_{ve} from Eq. (5.120) for V_{ve}, and letting $Q = Q_e$, we obtain for the *entrainment limit*

$$Q_e = \kappa\, A_v \sqrt{\frac{\rho_{ve}}{\Omega_e}\left(\frac{2\,\pi\,\sigma}{\lambda_w}\right)} \tag{5.121}$$

In terms of the axial heat flux $\xi_e = Q_e/A_v$, Eq (5.121) for the entrainment limit takes the form

$$\xi_e = \kappa \sqrt{\frac{\rho_{ve}}{\Omega_e}\left(\frac{2\,\pi\,\sigma}{\lambda_w}\right)} \tag{5.122}$$

Kemme (Ref. 5.17) found that in vertical, gravity-assisted heat pipes (evaporator at bottom and condenser at top), the correlation of Eq. (5.122) with experimental data was improved by the addition of a hydrostatic pressure term $\rho_l g D_v$ to the capillary pressure term $2\pi\sigma/\lambda_w$. In the hydrostatic pressure term, ρ_l is the liquid density and D_v is the vapor passage diameter. Tien and Chung (Ref. 5.18) showed that the addition of a hydrostatic pressure term based on the vapor passage diameter was justified on the basis of flooding correlations for countercurrent vapor-liquid flow.

When a heat pipe is inclined from the vertical (at angle θ), and an external acceleration of n g's acts along the heat pipe axis in a direction that increases the liquid pressure in the evaporator, it is proposed that the hydrostatic pressure term be expressed in the form

$$(n + \cos\theta)\rho_l g D_v$$

Equation (5.122) then becomes

$$\xi_e = \kappa A_v \sqrt{\frac{\rho_{ve}}{\Omega_e}\left[\frac{2\pi\sigma}{\lambda_w} + (n + \cos\theta)\,\rho_l\,g\,D_v\right]}$$

The entrainment limit can then be presented in final form as

$$\xi_e = \kappa\sqrt{\frac{2\pi\rho_{ve}\sigma}{\Omega_e\lambda_w}\,(1 + z_e)} \tag{5.123}$$

where

$$z_e = \frac{\lambda_w\,(n + \cos\theta)\,\rho_l\,g\,D_v}{2\pi\sigma} \tag{5.124}$$

The value of the momentum factor Ω_e depends on whether the vapor flow exiting the evaporator is laminar or turbulent. On the basis of experimental observations, Kemme (Ref. 5.17) recommended the following values for Ω_e:

Flow	Ω_e
Laminar	$\pi^2/8$
Turbulent	2.2

The parameter z_e can be regarded as a correction factor to ξ_e. As an example, consider a sodium heat pipe with a 100 mesh screen wick (λ_w = 254 microns), which operates at 1500°F (816°C). If $n = 0$, $\theta = 0$ deg, and $D_v = 0.6$ in., $z_e = 0.0377$. Had 16 mesh screen (λ_w = 1590 microns) been used for the wick, the value of z_e would have been 0.236. In general, z_e may be considered as negligible unless the screen wick is very coarse, the heat pipe diameter is large, and/or the heat pipe is subjected to significant external acceleration.

If the vapor pressure drop in the evaporator is significant, ρ_{ve} should be obtained from Eq. (5.88), with the exit Mach number Ma_e selected such that the axial heat fluxes calculated from Eqs. (5.91) and (5.123) are equal. In most cases, the vapor pressure drop in the evaporator will be small enough that ρ_v corresponding to T_{vi} may be used in Eq. (5.123) to calculate ξ_e.

The entrainment limit as derived above is based on Cotter's assumption that when We > 1 any wavelike disturbances on the liquid surface will become unstable, increasing in amplitude until liquid droplets form and are entrained in the heat pipe vapor (Ref. 5.15). Questions have been raised as to whether this type of behavior actually develops at the liquid surface within the pores of a capillary wick structure, particularly in horizontal heat pipes (Refs. 5.17, 5.18).

If the experimental heat pipe dryout limit predicted by Eq. (5.123) is not the result of entrainment, then this limit must be due to some other heat transport–disrupting mechanism. In assessing this possibility, an alternate form of the entrainment limit will now be considered that is based on the shear stress acting on the liquid surface of the wick pores.

5.3.2. Vapor Shear Stress Criterion for Entrainment

Figure 5.a shows an idealized picture of the forces acting on a segment of liquid surface. The liquid surface is assumed to lie within a square pore with sides of length equal to the effective pore size D_p.

The vapor shear stress acting on the surface of area D_p^2 is resisted by surface tension acting over the length D_p. Thus,

$$\tau_s D_p^2 = \sigma D_p$$

or

$$\tau_s = \sigma / D_p \tag{5.125}$$

From Eq. (4.14),

$$\tau_s = (f/8)\rho_v V_v^2$$

where f is the friction factor. On substituting this value of τ_s in Eq. (5.125) and solving for V_v,

$$V_v = \sqrt{\frac{8\,\sigma}{f\,\rho_v\,D_p}}$$

Again, note that

$$Q = \kappa \rho_{ve} V_{ve} A_v$$

Substituting V_v from the previous equation for V_{ve} and letting $\rho_v = \rho_{ve}$, $Q = Q_e$, and $\xi_e{}^* = Q_e/A_v$, we obtain

$$\xi_e^* = \kappa\,\frac{\sqrt{8\,\rho_{ve}\,\sigma}}{f\,D_p} \tag{5.126}$$

On the assumption that the vapor flow at the evaporator exit is turbulent, a value of 0.04 can be used for f. (See Section 4.2.3.) Then, dividing Eq. (5.126) by Eq. (5.123),

Figure 5.a. Force balance on the liquid-vapor interface in wick pores.

using the value of 2.2 for Ω_e with turbulent flow (see Section 5.3.1), and letting $z_e = 0$, and $D_p = \lambda_w$,

$$\frac{\xi_e^*}{\xi_e} = \sqrt{\frac{8/f}{2\pi/2.2}} = 8.37$$

Thus, the entrainment limit based on a balance of shear and surface tension forces is seen to be more than eight times that based on the Weber number criterion.

5.3.3. Capillary Pumping Limit for Interior Wet Point

If ξ_e* indeed represents a true entrainment limit, then it must be presumed that a different phenomenon is responsible for the heat transport limit ξ_e that has been correlated with experimental dryout data. The nature of this phenomenon can be clarified by reference to the heat pipe pressure diagram of Figure 5.17.

This pressure diagram is representative of a horizontal heat pipe with no adiabatic section and negligible frictional pressure drops. Notice that the wet point (where the liquid and vapor pressures are equal) is located within the heat pipe at the evaporator exit. The maximum difference between the vapor and liquid pressures $P_v - P_l$ occurs at the evaporator beginning, and is equal to the vapor momentum pressure drop ΔP_{vm}.

The capillary pumping limit is reached when the maximum vapor-liquid pressure difference is equal to the maximum available capillary pressure in the wick pores. In this case,

$$\Omega_e \rho_{ve} V_{ve}^2 = 4\sigma / D_p$$

If V_{ve} from this equation is substituted into the heat transport rate expression

$$Q = \kappa \rho_{ve} V_{ve} A_v$$

we obtain for the capillary pumping limit

$$\xi_c = \frac{Q}{A_v} = \kappa \sqrt{\frac{4\,\rho_{ve}\,\sigma}{\Omega_e\, D_p}} \tag{5.127}$$

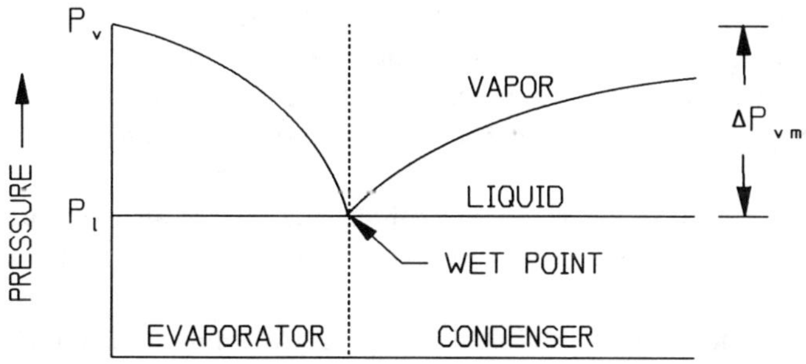

Figure 5.17. Pressure diagram for horizontal heat pipe (no adiabatic section, negligible friction).

Assuming that $D_p = \lambda_w$, the axial heat flux limit given by Eq. (5.127) is 80% of ξ_e from Eq. (5.123) for a horizontal heat pipe ($z_e = 0$).

Thus, it appears that ξ_e may really be a capillary pumping rather than an entrainment limit. This capillary pumping limit develops when the wet point occurs at some location along the heat pipe length, instead of at the evaporator or condenser end. The fact that the characteristic dimension used in the entrainment limit given by Eq. (5.123) is equal to the effective pore size for maximum capillary pressure lends credibility to this assertion.

The sound of liquid droplets impinging on the end of the condenser is characteristic of the entrainment phenomenon. In experiments in which an entrainment limit was first identified (Ref. 5.19), liquid impingement sounds were frequently heard *after* dryout and overheating of the evaporator had taken place. This observation indicates that entrainment of liquid droplets by the heat pipe vapor may have been a consequence of dryout, rather than its cause.

If the wet point occurs at some point along the heat pipe length when there is a significant hydrostatic pressure rise, as in gravity-assisted heat pipes, calculation of the capillary pumping limit can become quite complicated. Busse and Kemme (Ref. 5.20) give a perceptive, but quite involved, analysis of this situation.

5.3.4. Additional Considerations

From the foregoing discussion, it is evident that the concept of the entrainment limit is controversial and subject to considerable interpretation. An entrainment-produced dryout may occur at heat transport rates several times greater than those predicted by the entrainment limit equation based on a Weber number of unity. The latter equation may actually be predicting a dryout limit due to inadequate capillary pumping capability rather than entrainment of liquid droplets in the vapor stream.

For design purposes, the use of Eq. (5.123) is recommended for calculating the "entrainment" limit, as a means of identifying a possible capillary pumping limitation on heat pipe heat transport. Such a limitation might not be apparent from the usual capillary pumping limit calculation, which assumes that the wet point is located at the evaporator or condenser end of the heat pipe.

In situations in which the calculated entrainment limit imposes serious restrictions on the allowable heat transport rate, closer examination of the heat pipe pressure diagram is warranted, to establish whether a wet point does in fact exist between the heat pipe extremities. If such a wet point does not exist, the entrainment limit calculated from Eq. (5.123) may be spurious, and may not represent a heat pipe limit at all. On the other hand, if an interior wet point is identified, an alternate capillary pumping limit calculation based on a detailed analysis of the pressure diagram may be required.

The following additional important point related to capillary pumping limits associated with interior wet points is brought out in Ref. 5.20. When an increase in the heat transport rate causes the difference between vapor and liquid pressures at a location to the right of the wet point to exceed the maximum available capillary pressure, capillary pumping does not necessarily stop.

Instead, as liquid recedes into the heat pipe wick, the liquid flow area decreases, producing an increase in the liquid pressure drop. The liquid pressure to the right of the wet point will then rise, causing the vapor-liquid pressure difference to decrease until a balance with the maximum available capillary pressure is restored.

This process may result in a partial dryout over the portion of the wick circumference to the right of the wet point, but the heat pipe region to the left of the wet point, including the evaporator, will not be affected. Therefore, capillary pumping can be expected to continue at the higher heat transport rate.

However, if the wet point occurs at the evaporator end of the heat pipe, the partial dryout region could extend over the entire heat pipe length. Evaporator dryout and the cessation of capillary pumping would then be expected, if the calculated capillary pumping limit were exceeded.

The region to the right of an interior wet point includes the condenser. If the heat transport rate should exceed the calculated capillary pumping limit, any resulting partial dryout would reduce the effective condenser area for the transfer of heat to an adjacent heat sink. The heat pipe temperature would then rise above design levels to offset the reduced heat transfer area. Therefore, although capillary pumping may continue at the higher heat transport rate, the calculated capillary pumping limit should still be accepted as a valid limiting condition for this situation.

The occurrence of the maximum vapor-liquid pressure difference to the right of the wet point is most likely in a gravity-assisted heat pipe, or in a heat pipe subjected to external acceleration that acts to facilitate the return of condensate to the evaporator.

Otherwise, the maximum vapor-liquid pressure difference will most likely occur to the left of the wet point, in the evaporator. An increase in liquid pressure drop caused by recession of the liquid into the wick pores will then cause a further increase in the vapor-liquid pressure difference and the cessation of capillary pumping. In this instance, a capillary pumping limit will definitely be reached when the pressure difference equals the maximum available capillary pressure.

5.3.5. Simplified Equation for Entrainment Limit Parameter

Eq. (5.123) for the entrainment limit can be written in the form

$$\xi_e = \left[\frac{2\,\pi}{\Omega_e\,\lambda_w} (1 + z_e) \right]^{1/2} \phi_e \tag{5.128}$$

where

$$\phi_e = \kappa\,(\rho_{ve}\,\sigma)^{1/2} \tag{5.129}$$

The entrainment limit parameter ϕ_e is a function of thermophysical properties of the heat pipe fluid, all of which are temperature-dependent. Although the parameter z_e is

Table 5.6. Power law constants for entrainment limit parameter ϕ_e

Fluid	A_e	α_e
Lithium	506.4	0.383
Sodium	140.2	0.384
Water	87.28	0.400
Ammonia	64.18	0.306

also a function of fluid thermophysical properties, its temperature dependence is minimal. Also, in many cases z_e will be small enough to be neglected.

In Ref. 5.9, it is shown that, when $P_{ve} \approx P_v$ (the maximum vapor pressure, corresponding to T_{vi}), ϕ_e is expressible as a simple power law function of the vapor pressure P_v in the form

$$\phi_e = A_e P_v^{\alpha_e} \tag{5.130}$$

where A_e and α_e are constants for a particular heat pipe fluid. P_v is in psia and ϕ_e is in micron$^{1/2}$-Btu/in^2-sec. With these units, λ_w in Eq. (5.128) is expressed in microns and ξ_e is in Btu/in^2-sec.

If P_v is in N/m^2 and ϕ_e is in micron$^{1/2}$-W/m^2, then ξ_e is in W/m^2. Equation (5.130) then takes the form

$$\phi_e = 1.636 \times 10^6 A_e \left(\frac{P_v}{6895}\right)^{\alpha_e} \tag{5.130a}$$

Values of the power law constants are given in Table 5.6 for some heat pipe fluids. The vapor pressure corresponding to a particular heat pipe temperature can be found from saturation vapor pressure-temperature tables or equations. (See Section 5.2.7.) Once ϕ_e has been determined from Eq. (5.130) for a particular heat pipe fluid and vapor pressure, the entrainment limit is readily obtained from Eq. (5.128).

Equation (5.130) provides a convenient basis for comparing the entrainment limit of various heat pipe fluids at a given vapor pressure in a given heat pipe. At a vapor pressure of 1 psia, the entrainment limit is directly proportional to A_e. Then, the entrainment limit of a heat pipe with lithium is 3.6 times that for sodium, almost 6 times that for water, and almost 8 times that for ammonia.

5.4. BOILING LIMIT (Refs. 5.21, 5.22)

The transfer of heat into a heat pipe produces a temperature gradient across the wick. At the interface between the wick and the vapor space, the liquid temperature in the wick is equal to or greater than that of the adjacent saturated vapor. The liquid temperature rises steadily with distance from the interface, reaching a maximum value at the outer wick surface. The temperature of the liquid is then always equal to or greater than the saturation temperature. Hence, the liquid is superheated over the entire wick thickness, with the highest superheat occurring at the heat pipe wall.

When the superheat exceeds some critical value, vapor bubbles at nucleation sites in the heat pipe wall will break loose, initiating the boiling process. The heat transfer rate corresponding to the incipient boiling condition is called the boiling limit, and is usually expressed as a surface heat flux (Btu/ft^2-sec or W/m^2).

For reasons that have been enumerated in Section 2.3.2, the onset of boiling in the heat pipe liquid is generally to be avoided, although under some circumstances heat transport can continue in the boiling mode.

In this section, expressions for the boiling limit are derived, and additional information relative to use and interpretation of the boiling limit is presented. The derivation proceeds first by finding an expression for the wick temperature drop at incipient boiling, and then by substituting this expression into the heat conduction equation to obtain the associated heat transfer rate.

5.4.1. Wick Temperature Drop at Incipient Boiling

The temperature and pressure distribution across the wick thickness and in the adjacent vapor space are indicated in Figure 5.b.

The liquid temperature decreases from T_w at the wall to T_i at the liquid-vapor interface, and drops further across the interface to the temperature T_s of the saturated vapor. The vapor pressure corresponding to the fluid temperature drops from P_{vw} at the wall to P_{vi} at the interface to P_{vs} in the saturated vapor. The fluid pressure is constant and equal to the liquid pressure P_l across the wick thickness, and rises to the saturation pressure P_{vs} in the vapor. The saturation temperature T_{ls} associated with the liquid pressure is also constant over the wick thickness, and rises to the saturation temperature T_s in the vapor.

At the wall, vapor is always present in microcavities in the wall surface. These microcavities are called *nucleation sites*. Noncondensable gas may also be present. The vapor/gas pressure in the nucleation sites is higher than that of the adjacent liquid, the pressure difference being balanced by capillary pressure. The incipient boiling condition exists when the vapor/gas-liquid pressure difference is equal to the maximum capillary pressure that can be supported in the nucleation sites, or when

$$P_{vw} - P_l = 2\sigma(T_w)/r_n \tag{5.131}$$

In Eq. (5.131), $\sigma(T_w)$ is the liquid surface tension at T_w, and r_n is the smallest equivalent nucleation site radius. The equivalent nucleation site radius is equal to the actual nucleation site radius divided by the cosine of the wetting angle. [See Eq. (2.1).] In this book, the equivalent nucleation site radius is referred to as simply the nucleation site radius, unless the effect of the wetting angle is specifically being addressed.

If a significant quantity of noncondensable gas is present in the microcavities, the partial pressure of the gas will contribute to the total cavity pressure. The vapor pressure, and hence the superheat at incipient boiling, will then be smaller than if noncondensable gas were not present.

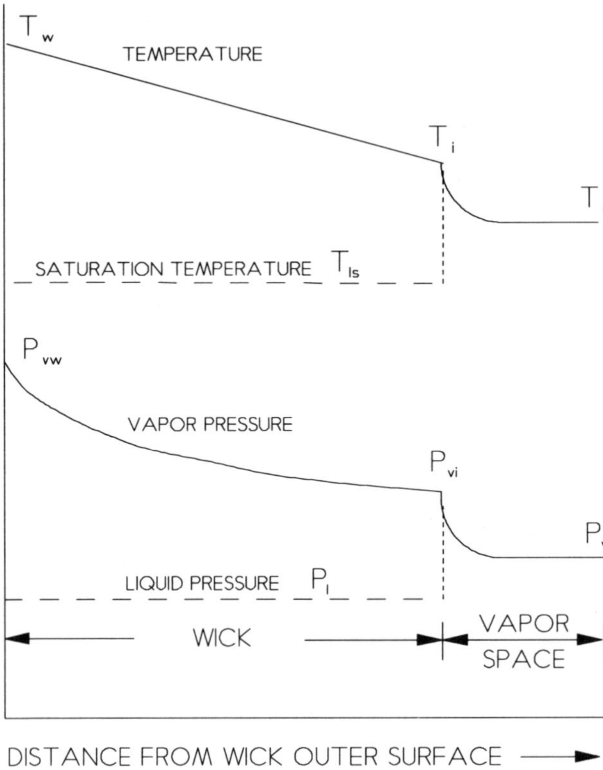

Figure 5.b. Pressure and temperature distributions across the thickness of a heat pipe wick.

At the liquid-vapor interface of the wick, the difference between vapor and liquid pressures is again balanced by capillary pressure, according to the equation

$$P_{vi} - P_l = 2\sigma(T_i)/r_c \tag{5.132}$$

where $\sigma(T_i)$ is the liquid surface tension at T_i and r_c is the minimum radius of curvature of the liquid-vapor interface in the wick pores at incipient boiling. The minimum radius of curvature r_c is established from capillary pumping requirements, and hence depends on both the heat transport rate at incipient boiling Q_b and the heat pipe internal geometry.

Subtraction of Eq. (5.132) from Eq. (5.131) gives the vapor pressure drop across the wick thickness at incipient boiling, or

$$P_{vw} - P_{vi} = \frac{2\,\sigma(T_w)}{r_n}\left[1 - \frac{\sigma(T_i)\,r_n}{\sigma(T_w)\,r_c}\right] \tag{5.133}$$

The associated wick temperature drop at incipient boiling $T_w - T_i$ can be found from Eq. (5.133) for a specified T_i, r_n, and r_c by an iterative process. First, P_{vi} corresponding to T_i is obtained from vapor pressure versus temperature tables or equations. Then, a value is assumed for the wall temperature T_w, P_{vw} is calculated from Eq. (5.133), and a new value for T_w corresponding to P_{vw} is obtained, again from vapor pressure tables or equations. If the new and initial values of T_w do not agree within some specified percentage, they are averaged, and the calculation is repeated with the averaged value of T_w. The process is repeated until the latest and previously calculated values of T_w agree within an acceptably small percentage.

If the temperature dependence of the surface tension is neglected, P_{vw} can be solved directly from Eq. (5.133), and T_w obtained immediately from vapor pressure tables or equations.

Alternate Method. A simpler, somewhat less accurate method for calculating $T_w - T_i$ utilizes Eq. (5.59), which is the Clausius-Clapeyron equation relating pressure and temperature changes of a saturated fluid. Repeating Eq. (5.59),

$$dP_v = \frac{\rho_v \kappa}{T} dT \tag{5.59}$$

where T denotes the absolute temperature corresponding to the vapor pressure P_v.

Upon integrating this equation over the vapor pressure and temperature ranges between the wick-vapor space interface (subscript i) and the wall-wick interface (subscript w),

$$P_{vw} - P_{vi} = \left(\frac{\rho_v \kappa}{T}\right)_m (T_w - T_i) \tag{5.134}$$

where the subscript m denotes the average value of $\rho_v \kappa / T$ over the indicated temperature range.

If $P_{vw} - P_{vi}$ is taken to be the vapor pressure drop across a heat pipe wick at incipient boiling, then $T_w - T_i$ is the associated wick temperature drop.

Upon substituting Eq. (5.133) in Eq. (5.134) and solving for $T_w - T_i$,

$$T_w - T_i = \frac{2\,\sigma(T_w)}{(\rho_v\,\kappa/T)_m\, r_n}\left[1 - \frac{\sigma(T_i)}{\sigma(T_w)}\frac{r_n}{r_c}\right] \tag{5.135}$$

If the following assumptions are made,

$$\sigma(T_w) = \sigma(T_i) = \sigma(T_m)$$

$$\left(\frac{\rho_v \kappa}{T}\right)_m = \left(\frac{\rho_v \kappa}{T}\right)_{T=T_m}$$

where $T_m = (T_w + T_i)/2$, Eq. (5.135) can be written in the simplified form

$$\Delta T_{wb} = T_w - T_i = 2\beta(T_m)/r_{ne} \tag{5.136}$$

where

$$\beta\,(T_m) = \left(\frac{\sigma\, T}{\rho_v\, \kappa}\right)_{T=T_m} \tag{5.137}$$

$$r_{ne} = \frac{r_n}{1 - r_n/r_c} \tag{5.138}$$

and ΔT_{wb} is the wick temperature drop at incipient boiling.

The variable β, which is a function of the mean wick temperature and the heat pipe fluid, is called the *boiling limit parameter.* From Eq. (5.136), if r_{ne} is expressed in microns and ΔT_{wb} in °F or °R, the unit of β is microns-°R. If ΔT_{wb} is in °C or °K, β has the unit microns-°K.

The variable r_{ne} may be considered as an *effective nucleation site radius*. It should be noted from Eq. (5.136) that, when r_{ne} = 2 microns, the wick temperature drop at incipient boiling is equal to β.

Eq. (5.136) predicts a higher wick temperature drop at incipient boiling than the more accurate Eq. (5.133). For wick temperature drops up to 200–300 °F (111–167°C), the error is generally no more than 10–12%. Eq. (5.136) is considered adequate for design purposes, and will be used in calculating the boiling limit in this book. Where greater accuracy is desired, recourse to Eq. (5.133) is recommended.

Simplified Equation for Boiling Limit Parameter. The boiling limit parameter β is expressible as the following simple power law function of the vapor pressure (Ref. 5.9):

$$\beta = A_b P_v^{\alpha_b} \tag{5.139}$$

where P_v is evaluated at the mean wick temperature, and A_b and α_b are constants for a particular heat pipe fluid. The vapor pressure P_v is in psia, and the boiling parameter β is in microns-°R. The constants for a number of heat pipe fluids are given in Table 5.7.

If P_v is in N/m^2 and β is in microns-°K, then Eq. (5.139) takes the form

$$\beta = \left(\frac{A_b}{1.8}\right)\left(\frac{P_v}{6895}\right)^{\alpha_b} \tag{5.139a}$$

The variation of β with the mean wick temperature and associated vapor pressure in a sodium heat pipe is shown below.

T_m, °F (°C)	P_v, psia (N/m^2)	β, microns-°R (microns-°K)
1300 (704)	2.24 (15,440)	1342 (746)
1400 (760)	4.33 (29,860)	735 (408)
1500 (816)	7.84 (54,060)	429 (238)
1600 (871)	18.40 (126,900)	263 (146)

Table 5.7. Power law constants for boiling limit parameter β [Eqs. (5.139) and (5.139a)]

Fluid	A_b	α_b
Lithium	6336	–0.893
Sodium	2792	–0.910
Water	337.1	–0.907
Ammonia	172.8	–0.992

It is evident that β, and hence the wick temperature drop at incipient boiling, falls off sharply as the mean wick temperature increases. Since the boiling limit is proportional to β (see Section 5.4.2), it is also a sharply decreasing function of temperature.

However, even at 1600°F (871°C), the boiling limit parameter for sodium is large enough to require a wick temperature drop of 263°F (146°C) for the initiation of boiling when $r_{ne} = 2$ microns. High wick superheats at boiling inception and high boiling limits are characteristic of liquid metal heat pipes.

The wick temperature drops for incipient boiling at a given vapor pressure for various heat pipe fluids are readily compared by using Eq. (5.139). At a vapor pressure of 1 psia (6895 N/m^2), ΔT_{wb} is directly proportional to A_b. Then, ΔT_{wb} of a heat pipe with lithium is 2.3 times that for sodium, almost 19 times that for water, and almost 37 times that for ammonia.

5.4.2. Surface Heat Flux at Incipient Boiling

Using the one-dimensional steady-state heat conduction equation and Eq. (5.136), the heat flux at the outer surface of a uniformly heated heat pipe wick at incipient boiling q_b is expressible as

$$q_b = C_{g1} \frac{k_w}{t_{\text{wick}}} \Delta T_{wb} = C_{g1} \frac{k_w}{t_{\text{wick}}} \left[\frac{2\,\beta(T_m)}{r_{ne}} \right] \qquad (5.140)$$

where k_w is the mean thermal conductivity of the wick, t_{wick} is the wick thickness, ΔT_{wb} is the wick temperature drop at incipient boiling, and C_{g1} is a correction factor that accounts for the coordinate system being used.

C_{g1} was derived from standard formulations for one-dimensional conductive heat transfer in rectangular, cylindrical, and spherical coordinates (Ref. 5.23). Methods for calculating the wick thermal conductivity are presented in Chapter 6. Eq. (5.140) represents the boiling limit evaluated at the outer wick surface. It can be referenced to the outer heat pipe surface by appropriate geometrical corrections.

At the liquid-vapor interface, the surface heat flux for incipient boiling q_{bi} can be expressed as

$$q_{bi} = h_i(T_i - T_s) = C_{g2} q_b \qquad (5.141)$$

where h_i is the interface heat transfer coefficient and C_{g2} is a correction factor that accounts for the radial variation of the surface heat flux in cylindrical and spherical geometries.

The correction factors C_{g1} and C_{g2} are given in Table 5.8, and C_{g1} is plotted against t_{wick}/r_o in Figure 5.18. The outer wick radius is denoted by r_o, and the inner wick radius by r_i.

It is evident that the boiling limit for a given ΔT_{wb}, expressed as a surface heat flux, is smaller for geometries with curved surfaces, the more so for spherical geometry.

Figure 5.18 is also useful for establishing the wick thickness/radius ratio at which rectangular geometry is a good approximation to geometries with curved coordinates. With cylindrical geometry, if $t_{\text{wick}}/r_o < 0.1$, rectangular geometry can be used to

Table 5.8. Geometrical correction factors for surface heat flux

Coordinate system	C_{g1}	C_{g2}
Rectangular	1	1
Cylindrical	$-(t_{wick}/r_o)/\ln[1-(t_{wick}/r_o)]$	r_o/r_i
Spherical	$1-(t_{wick}/r_o)$	$(r_o/r_i)^2$

calculate the surface heat flux with an error of less than 5%. With spherical geometry, the error margin will be the same if $t_{wick}/r_o < 0.05$.

When the surface heat flux is specified at the outer surface of the heat pipe envelope, as is usually the case, the boiling limit should also be referenced to the outer heat pipe surface. The correction is of the same form as C_{g2}, except that the outer heat pipe radius r_w is used in place of r_i for curvilinear geometries. Thus, letting the boiling limit at the outer heat pipe surface be q_{bw}, we have

$$q_{bw} = q_b \qquad \text{(rectangular geometry)}$$

$$q_{bw} = (r_o/r_w)q_b \qquad \text{(cylindrical geometry)}$$

$$q_{bw} = (r_o/r_w)^2 q_b \qquad \text{(spherical geometry)}$$

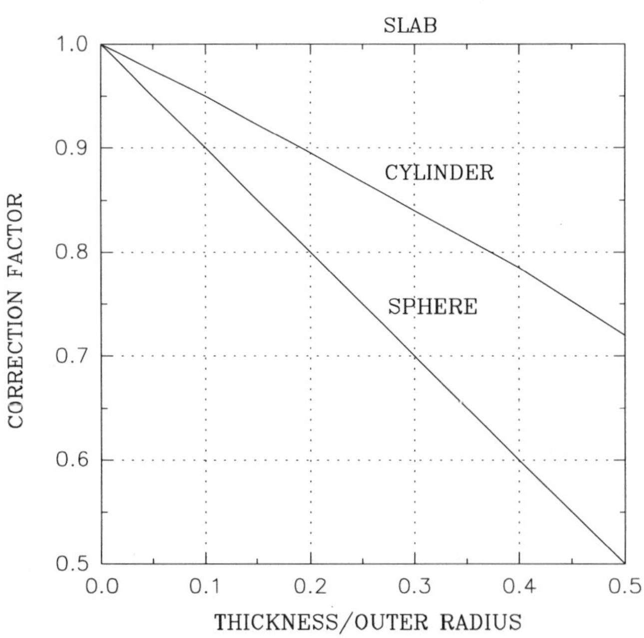

Figure 5.18. Surface heat flux correction factors C_{g1} for various geometries.

Nonuniform Surface Heat Flux. Equation (5.140) is applicable when the incident heat flux over the evaporator surface is uniform. If the incident heat flux is a function of position on the evaporator surface, the wick temperature drop will be greatest at the point where the heat flux is a maximum. Boiling will be initiated when the wick temperature drop at that point is equal to that for incipient boiling, ΔT_{wb}.

Where the incident heat flux is a maximum, some of the heat will be diffused laterally through the heat pipe wall and wick. A higher surface heat flux will then be necessary to produce a given temperature drop across the wick thickness. The boiling limit is then greater than that predicted by Eq. (5.140). If the incident heat flux is concentrated over an area whose linear dimensions are comparable to or smaller than the wall-plus-wick thickness, lateral diffusion of the incident heat flux will be extensive. The boiling limit at the maximum heat flux location may then be far greater than that predicted by Eq. (5.140).

In general, a two- or three-dimensional heat transfer analysis will be required to establish the wick temperature distribution and boiling limit when the surface heat flux over the evaporator surface is nonuniform. However, if the peak-to-average heat flux ratio is moderate, and the dimensions of the surface area over which the heat flux is elevated are a large multiple of the wall-plus-wick thickness, the effect of lateral thermal diffusion on the boiling limit will be minimal. Equation (5.140) can then be used to calculate the boiling limit at the maximum heat flux location with reasonable accuracy.

Surface Heat Flux at Incipient Boiling in Terms of Total Temperature Drop. In the design of heat pipes, the vapor temperature T_s is usually specified rather than the liquid-vapor interface temperature T_i. Therefore, it is preferable to express the surface heat flux at incipient boiling q_b in terms of the total temperature drop $T_w - T_s$ across the wick and the liquid-vapor interface, instead of just the wick temperature drop $T_w - T_i$. The temperature difference $T_w - T_s$ will be referred to as the *wick superheat for incipient boiling*. (The actual superheat at the heat pipe wall is $T_w - T_{s'}$, where $T_{s'}$ is the saturation temperature corresponding to the liquid pressure in the heat pipe wick.)

From Eqs. (5.140) and (5.141),

$$T_i - T_s = \frac{C_{g2}}{h_i}\, q_b = C_{g1}\, C_{g2} \left(\frac{k_w}{h_i\, t_{\text{wick}}} \right) \left(\frac{2\beta}{r_{ne}} \right) \tag{5.142}$$

Then, adding this equation to Eq. (5.136) and simplifying, we obtain an expression for the wick superheat at incipient boiling $Tw - T_s$:

$$T_w - T_s = \left[1 + C_{g1}\, C_{g2} \left(\frac{k_w}{h_i\, t_{\text{wick}}} \right) \right] \left(\frac{2\beta}{r_{ne}} \right) \tag{5.143}$$

From Eq. (5.140), the corresponding surface heat flux at incipient boiling q_b is then

$$q_b = \frac{C_{g1}\, (k_w/t_{\text{wick}})\, (T_w - T_s)}{1 + C_{g1}\, C_{g2}\, [k_w/(h_i\, t_{\text{wick}})]} \tag{5.144}$$

If the interface thermal resistance $1/h_i$ is zero and the radius of curvature r_c is infinite, Eq. (5.143) reduces to

$$T_w - T_s = T_w - T_i = 2\beta/r_n \tag{5.145}$$

and Eq. (5.144) reverts back to Eq. (5.140).

Equation (5.145) is also a good approximation to the wick superheat for incipient boiling in a heat pipe for which $C_{g1}C_{g2}(k_w/h_i t_{\text{wick}}) << 1$ and $r_n/r_c << 1$. It also represents the superheat that would be required to initiate boiling at a surface in a completely enclosed, saturated liquid with negligible interface thermal resistance.

Since information is required about the nucleation site radius and interface heat transfer coefficient before incipient boiling conditions can be calculated, these parameters will now be discussed.

5.4.3. Nucleation Site Radius

The *equivalent nucleation site radius* r_n is equal to the actual radius of a nucleation site divided by the cosine of the contact angle, and is strongly influenced by the wettability of the heat pipe wick and wall surfaces. Good wettability favors a small contact angle, a smaller equivalent nucleation site radius, and greater resistance to boiling.

Good wettability can be attributed to the capability of a liquid to dissolve or chemically remove oxides and other impurities from a solid surface. A heat pipe liquid with this capability can penetrate microscopic cavities to smaller radii more readily, further reducing the equivalent nucleation site radius.

Since it is the equivalent nucleation site radius that determines the maximum capillary pressure available in a surface pore, and the wetting angle is usually not known with precision, the equivalent nucleation site radius is referred to here as simply the nucleation site radius. The "nucleation site radius" of a given surface may vary considerably if the wetting angle is significantly influenced by the heat pipe liquid and wall/wick materials selected and/or the presence of surface impurities.

Effective Nucleation Site Radius in Heat Pipes. The effective nucleation site radius r_{ne} in heat pipes accounts for the influence of the radius of curvature in the wick pores on incipient boiling. It can be calculated from Eqs. (5.143) and (5.144), using experimental data for the vapor temperature and the heat transfer rate at which dryout due to boiling is observed. In almost all cases, the term involving the interface thermal resistance $1/h_i$ may be neglected.

If $r_n/r_c << 1$, as is usually the case, Eq. (5.138) shows that the nucleation site radius r_n may be assumed to be equal to r_{ne}. Otherwise, r_n must be calculated for values of r_{ne} and r_c at dryout from Eq. (5.138). The radius of curvature r_c would then have to be found from the capillary pumping equation.

If the dryout condition is definitely responsible for the onset of boiling, the calculated r_n will be a reasonably accurate estimate of the actual nucleation site radius. If the dryout is the result of some other limiting phenomenon, such as liquid entrainment or interruption of capillary pumping, then the boiling limit will be greater than the experimentally observed dryout limit. In this event, the calculated r_n will be larger than the actual r_n, and may then be regarded as an upper limit on the nucleation site radius.

The nucleation site radius was calculated by the author in the manner just described for a sodium-molybdenum heat pipe (Ref. 5.24) and for an ammonia-aluminum heat pipe (Ref. 5.25).

The sodium heat pipe was of the two-layer type, in which a cylinder of fine-pored screen in contact with the cylindrical heat pipe wall along a single axial line defined a crescent-shaped open liquid flow channel. A boiling limit of 84.5 Btu/ft^2-sec (9.60×10^5 W/m^2) was observed at a vapor temperature of 2035°F (1113°C), and one of 68.4 Btu/ft^2-sec (7.77×10^5 W/m^2) at 2124°F (1162°C). Using the maximum thickness of the flow channel plus the thickness of the screen as the wick thickness, a nucleation site radius of 1.6 to 1.7 microns was calculated.

The ammonia heat pipe was of the monogroove type, with a single liquid flow channel located outside the vapor passage and a distribution wick on the vapor passage wall consisting of fine circumferential grooves. The evaporator section was confined to the quadrant of the vapor passage opposite the liquid flow channel. The surface heat flux at the inception of local dryout and a vapor temperature of 104°F (40°C) was 9.97 Btu/ft^2-sec (1.13×10^5 W/m^2). This was taken to be the boiling limit, although the heat transfer rate for this heat pipe could be increased by another 40% before widespread dryout requiring the termination of heat input was observed.

The ratio between the heat flux at the outer wall and the temperature drop from the outer wall to the vapor is referred to as a film coefficient in Ref. 5.25, although this term is more properly called the unit thermal conductance U. Using the specified value of 9900 Btu/hr-ft^2-°F (5.62×10^8 W/m^2-°K) for U to obtain the wall-to-vapor temperature drop, and subtracting the temperature drop between the wall outer surface and the root of the grooves, the mean temperature drop across the thickness of the circumferential grooves (i.e, wick) was obtained. From this data, a nucleation site radius of 0.8 microns was calculated.

Other calculations, reported in Section 2.3.2, yielded upper limits of 3 to 7 microns for a sodium–stainless steel heat pipe and 24 to 67 microns for a potassium-nickel heat pipe.

Effect of Incomplete Wetting and/or Noncondensable Gas. In grooved and porous wicks in which wettability of the heat pipe liquid is poor or excessive noncondensable gas is present, the possibility exists that crevices and corners within the wick structure could act as nucleation sites. In that event, r_n could be considerably larger, and the boiling limit considerably lower, than would be the case if the nucleation sites were confined to microcavities in the heat pipe wall. These factors could have contributed to the large apparent nucleation site radius for the potassium-nickel heat pipe.

Although the breadth and depth of surface cavities are reduced considerably on highly polished surfaces, there is not necessarily a direct correlation between surface finish and the nucleation site radius.

In a pool boiling experiment with sodium in contact with a stainless steel heating surface, the superheat for incipient boiling was found to be the same for a mirror smooth surface finish of 8 microinches (0.203 micrometers), as for a surface that had been sandblasted to a finish of 240 microinches (6.10 micrometers) (Ref. 5.26). Since the superheat for boiling was the same for each surface, the nucleation site radius can also be presumed to be the same.

Effect of Heat Pipe Conditioning. Heat pipes are usually conditioned prior to actual use by operation at low power and elevated temperature for a period of hours or days. The purpose of this conditioning process is to assure thorough wetting of the wick and wall surfaces by the heat pipe liquid, and maximum removal of any residual noncondensable gas from the surface cavities. The nucleation site radius is then minimized, and heat transport capability (to the extent that it is restricted by the boiling limit) is maximized.

In situations in which a heat pipe is required to operate at very high surface heat fluxes and relatively high vapor pressures, continuance of the conditioning process over a period of several weeks or more may be advisable to minimize the nucleation site radius and maximize the boiling limit.

Pool and forced convection boiling experiments were conducted with sodium, both over a period of several weeks and for several tens of test runs over shorter intervals. Results indicated an increase in the superheat for incipient boiling by a factor of three or more between early and later test runs (Ref. 5.26).

Nucleation site radii may also be reduced, and incipient boiling superheats increased, by forcing pressurized liquid into the microscopic surface cavities prior to operation (Ref. 5.27). If the liquid is capable of dissolving impurities on the surface of cavities that have been penetrated, it will wet these surfaces and remain in place after the liquid pressure has been reduced. This technique could potentially be applied to increase the boiling limit in heat pipes by introducing a pressurized inert gas, at a temperature above the melting point of the heat pipe liquid, prior to the start of operation.

Recommended Values for Nucleation Site Radius. From the foregoing discussion, it is reasonable to assume that the nucleation site radius in properly fabricated and conditioned heat pipes lies within the range of 1 to 7 microns. A value of 3 microns is suggested for preliminary design purposes.

5.4.4. Interface Heat Transfer Coefficient

The impact of the interface thermal resistance on boiling inception conditions depends on the magnitude of the term $k_w/h_i t_{\text{wick}}$ that appears in Eqs. (5.143) and (5.144). This term is just the ratio of the resistance to heat flow across the liquid-vapor interface of a heat pipe $(1/h_i)$ to that across the wick thickness (t_{wick}/k_w) in rectangular geometry. If this ratio is much smaller than 1, the effect of the interface thermal resistance will be negligible. Otherwise, the wick superheat and surface heat flux at incipient boiling may be significantly lower than would be predicted if the interface thermal resistance were ignored (i.e., if h_i were assumed to be infinite).

In general, the larger k_w and the smaller t_{wick}, the more significant will be the interface thermal resistance. Since high thermal conductivity and small thickness are characteristic of many liquid metal heat pipe wicks, interface thermal resistance is most likely to be of concern in liquid metal heat pipes.

The magnitude of h_i is dependent on the degree of interaction between liquid molecules that evaporate into the surrounding vapor and vapor molecules that condense on the liquid surface. At low vapor pressures (below about 0.1 psia), such interactions are negligible, and h_i is predictable from gas kinetics considerations.

The frequency of molecular interactions increases at higher vapor pressures. At a sufficiently high pressure, interface heat transfer may be regarded as a diffusion process through an adjacent vapor layer with finite temperature and pressure gradients. The resistance to heat flow will then be appreciably greater, and h_i appreciably smaller, than when molecular interactions are negligible. Therefore, the interface heat transfer coefficient corresponding to negligible molecular interactions may be regarded as an upper limit. It is designated here as $h_{i\max}$, and is expressible as (Ref. 5.21)

$$h_{i\,\max} = \left(\frac{2f_c}{2-f_c}\right)\left(\frac{\rho_v \kappa^2}{T_{im}}\right)\sqrt{\frac{M}{2\pi R_u T_{im}}} \tag{5.146}$$

where M is the molecular weight of the vapor molecules, R_u is the universal gas constant, and f_c is the condensation coefficient.

The physical properties in Eq. (5.146) are evaluated at the absolute mean temperature $T_{im} = 1/2(T_i + T_s)$ between the liquid-vapor interface and the adjacent saturated vapor. For liquid metals, f_c can be taken as unity.

In most heat pipe design situations, the vapor pressure is considerably greater than 0.1 psia (690 N/m^2), although vapor pressures of this magnitude may be encountered during the initial phases of startup in liquid metal heat pipes. The actual interface heat transfer coefficient may then be expected to be well below $h_{i\max}$.

Experimental data on h_i for sodium and potassium, based on both boiling and condensing, are quite imprecise and vary over a wide range. Nevertheless, these data indicate that the interface heat transfer coefficient at vapor pressures of several psia or higher can be as much as an order of magnitude or more below that predicted by Eq. (5.146) (Ref. 5.21).

The actual interface heat transfer coefficient h_i can be represented as

$$h_i = \theta_i h_{i\max} \tag{5.147}$$

where θ_i is the ratio of the actual interface heat transfer coefficient to the maximum value predicted from Eq. (5.146).

In view of the absence of firm data, engineering judgment may be necessary in the estimation of θ_i. If, for example, θ_i is taken to be 0.1 and the impact of h_i on the wick superheat for boiling and the boiling limit is found to be negligible, h_i can probably be ignored in the heat pipe design. On the other hand, if there is a significant reduction in the wick superheat for boiling and the boiling limit, an additional degree of conservatism in the heat pipe design may be warranted.

5.4.5. Solutions for Boiling Limit (r_n/r_c Is Negligible)

In most heat pipe design situations, the wick pore radius is a large multiple of the nucleation site radius. Since the radius of curvature can never be less than the wick pore radius, the radius of curvature will also be a large multiple of the nucleation site radius. The ratio r_n/r_c is then approximately zero, and the wick superheat at incipient boiling and the boiling limit are not dependent on the radius of curvature.

Interface Thermal Resistance Is Negligible. If the interface thermal resistance is also negligible, incipient boiling conditions may be calculated directly from Eqs. (5.143) and (5.144). The boiling parameter β varies significantly with the mean wick temperature, which is not known initially. Therefore, an iterative procedure will generally be required first to establish β and the mean wick temperature from Eq. (5.143). Then the boiling limit can be calculated from Eq. (5.144).

The likelihood of boiling at a given surface heat flux can also be predicted by solving Eqs. (5.138) and (5.140) for the radius of curvature at incipient boiling $r_c = r_{cb}$. The wick temperature drop is first found from Eq. (5.140). Since the interface thermal resistance is assumed to be negligible, $T_s = T_i$, and $T_m = T_s + \Delta T_w/2$. The boiling parameter β is then evaluated at the mean wick temperature.

If r_{cb} is less than the actual radius of curvature r_c (as determined from capillary pumping requirements), boiling will not occur. The boiling limit, even though not known, will then be greater than the actual surface heat flux. If r_{cb} should exceed r_c, the boiling limit will be less than the actual surface heat flux, and boiling can be expected.

This procedure can be used to establish the absence of boiling, even if the radius of curvature is not yet known. In this instance, r_{cb} is compared with the wick pore radius r_p. If r_{cb} is smaller than r_p, it must also be smaller than r_c, and hence boiling cannot occur. If r_{cb} is larger than r_p, no conclusions about the likelihood of boiling can be drawn until r_c has been determined.

Interface Thermal Resistance Is Significant. If the interface thermal resistance is significant, calculation of the boiling limit becomes more involved, even if it is not a function of the radius of curvature. For a given vapor temperature T_s, there are now two dependent variables—the mean interface temperature T_{im} as well as the mean wick temperature T_m. This situation requires the simultaneous solution of Eqs. (5.142) and (5.143) before the boiling limit can be calculated from Eq. (5.144).

An iterative procedure for obtaining a solution is now presented. The subscript n refers to the current value of a variable, and $n + 1$ refers to its subsequent value. The subscript m denotes the mean value.

1. Specify the following input parameters: T_s, θ_i, t_{wick}, r_o, $r_{ne} = r_n$, $\kappa(T_s)$. (Although the heat of vaporization κ is a function of temperature, the dependence is small enough that κ can be assumed constant and equal to its value at T_s.)
2. Assume initial values for $T_{nm} = T_m$ and $T_{inm} = T_{im}$.
3. Calculate C_{g1} and C_{g2} from Table 5.7.
4. Calculate $k_{wn} = k_w(T_{nm})$, using methods described in Section 6.5.1.
5. Calculate $P_v(T_{nm})$ from vapor pressure equations or tables.
6. Calculate $h_i(T_{inm})$ from Eqs. (5.146) and (5.147).
7. Calculate $\beta_n = \beta(T_{nm})$ from Eq. (5.137) or (5.139).
8. Calculate $T_{in} = T_s + 2(T_{inm} - T_s)$.
9. Calculate T_{wn} from Eq. (5.143).
10. Calculate $T_{(n+1)m} = (T_{wn} + T_{in})/2$.
11. Let y be some arbitrarily small value (such as 0.005).

12. Calculate $\delta_1 = (T_{(n+1)m} - T_{nm})/T_{nm}$.
13. If $\delta_1 \leq y$, proceed to step 14. If $\delta_1 > y$, let $T_{nm} = (T_{nm} + T_{(n+1)m})/2$ and return to step 4.
14. Calculate $T_{i(n+1)m} = (T_{in} + T_s)/2 = [(T_{wn} + T_s)/2] - (\beta_n/r_{ne})$ from Eq. (5.136).
15. Calculate $\delta_2 = (T_{i(n+1)m} - T_{inm})/T_{inm}$.
16. If $\delta_2 \leq y$, the solution has been obtained for T_{nm} and T_{inm}; proceed to step 17. If $\delta_2 > y$, let $T_{inm} = (T_{i(n+1)m} + T_{inm})/2$ and return to step 6.
17. Calculate the following incipient boiling conditions:

$$\Delta T_i = T_{in} - T_s = 2(T_{inm} - T_s) \qquad \text{(temperature drop across liquid-vapor interface)}$$

$$\Delta T_w = T_{wn} - T_{in} = 2(T_{nm} - T_{in}) \qquad \text{(temperature drop across wick)}$$

$$q_b = C_{g1}(k_{wn}/t_{\text{wick}})\Delta T_w \qquad \text{(surface heat flux = boiling limit)}$$

Effect of Interface Thermal Resistance on Incipient Boiling Conditions. The foregoing procedure was used to calculate the effect of the interface heat transfer coefficient h_i on the surface heat flux for incipient boiling q_b, and the associated interface and wick temperature drops ΔT_i and ΔT_w. Results are shown in Figures 5.19 and 5.20 for sodium and lithium heat pipes.

Both heat pipes are cylindrical with a wick thickness of 0.030 in. (7.62×10^{-4} m), a wick thermal conductivity assumed to be that of the heat pipe liquid, a nucleation site

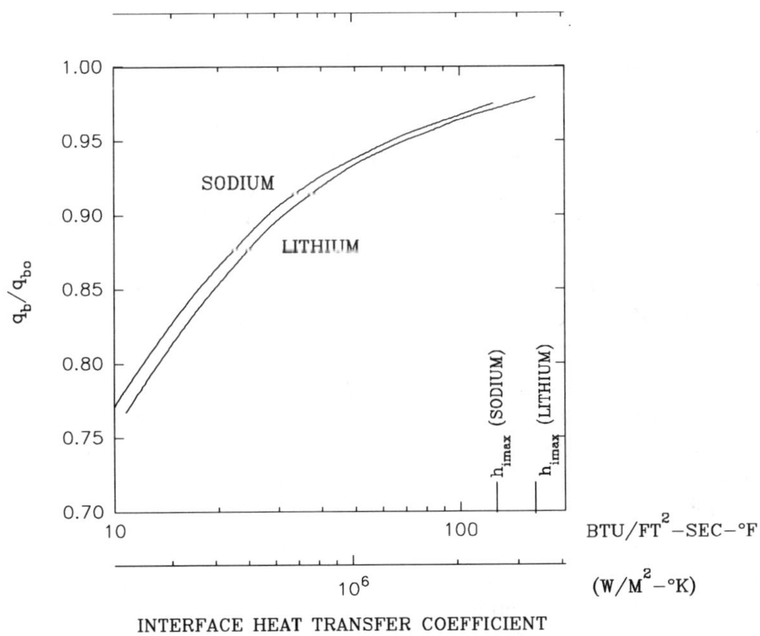

Figure 5.19. Effect of interface heat transfer coefficient on surface heat flux for incipient boiling.

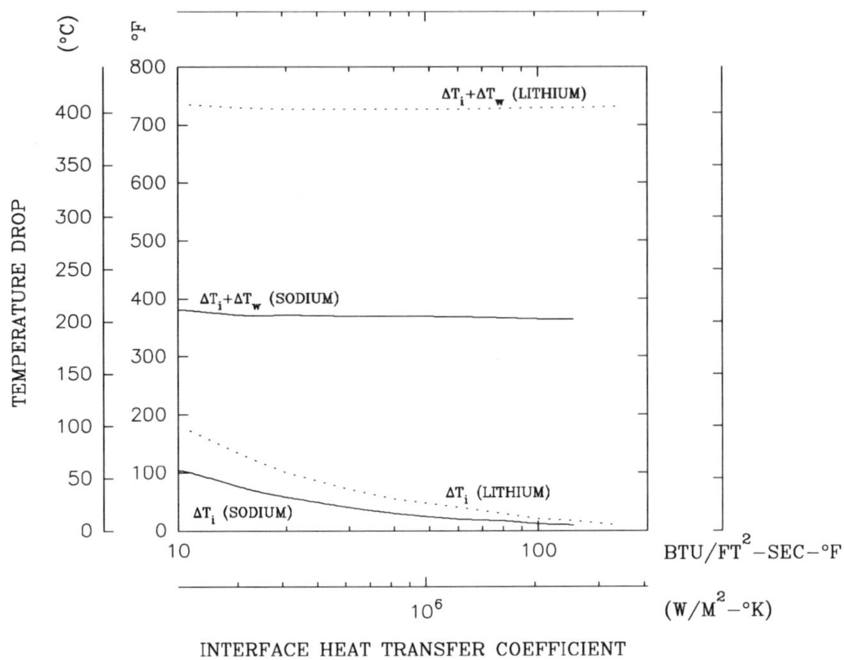

Figure 5.20. Effect of interface heat transfer coefficient on interface and wick temperature drops at incipient boiling.

radius of 1 micron, and a wick thickness/radius ratio much smaller than 1 (so that C_{g1} and C_{g2} are equal to 1).

The sodium heat pipe operates at a vapor temperature of 1500°F (816°C) and a vapor pressure of 7.2 psia (4.96×10^4 N/m^2). The surface heat flux for incipient boiling with negligible interface thermal resistance is 1295 Btu/ft^2-sec (1.47×10^7 W/m^2). Comparable figures for the lithium heat pipe are: 2200°F (1204°C), 5.5 psia (3.79×10^4 N/m^2), and 2450 Btu/ft^2-sec (2.78×10^7 W/m^2).

Values of the maximum interface heat transfer coefficient h_{imax} are: 125 Btu/ft^2-sec-°F (2.56×10^{10} W/m^2-°K) for sodium and 165 Btu/ft^2-sec-°F (3.37×10^{10} W/m^2-°K) for lithium.

In Figure 5.19, the ratio of surface heat fluxes for incipient boiling q_b/q_{bo} is shown as a function of the interface heat transfer coefficient h_i. The heat flux corresponding to negligible interface thermal resistance ($h_i = \infty$) is denoted by q_{bo}. The sodium and lithium curves are similar. When $h_i = h_{imax}$, q_b is about 95% of that when $h_i = \infty$. When h_i is an order of magnitude less than h_{imax}, q_b is 80–83% of q_{bo}.

These results show the extent to which the surface heat flux for incipient boiling is reduced by interface thermal resistance. Depending on the magnitude of h_i and specific design conditions, the reduction in q_b can be as much as 20% or more.

Curves of the temperature drop at incipient boiling are shown in Figure 5.20. For $h_i = \infty$, the interface temperature drop ΔT_i is zero. The wick temperature drop ΔT_w is

365°F (203°C) for sodium and 737°F (409°C) for lithium. As h_i decreases, the total temperature drop $\Delta T_i + \Delta T_w$ remains approximately constant while ΔT_i increases and ΔT_w becomes smaller. At $h_i = h_{imax}$, ΔT_i is around 10°F (5°C). At $h_i = 0.1 h_{imax}$, ΔT_i is 85°F (47°C) for sodium and 120°F (67°C) for lithium.

5.4.6. Solutions for Boiling Limit (r_n/r_c Is Significant)

The wick superheat and heat transfer rate at incipient boiling [(Eqs. (5.143) and (5.144)] are functions of r_{ne}, which in turn is dependent on the radius of curvature r_c at the liquid-vapor interface in the evaporator [Eq. (5.138)]. Another functional relation between the heat transport rate and the radius of curvature is given by the capillary pumping equation. If the ratio r_n/r_c is significant, a simultaneous solution of the boiling limit and capillary pumping equations is required in order to establish conditions at incipient boiling.

An explicit simultaneous solution for the boiling limit and capillary pumping equations will now be presented for a special case. The solution is applicable when the vapor pressure drop is negligible and the liquid pressure drop is solely due to friction during laminar flow.

Relation between Axial and Surface Heat Fluxes. When a simultaneous solution of the boiling limit and capillary pumping equations is called for, it is convenient to express the axial heat flux in terms of the associated surface heat flux. A simplified form of this relationship was presented in Section 5.2.5. A more general version will now be developed.

Consider the evaporator of a cylindrical heat pipe, as shown in Figure 5.21.

The evaporator length is L_e, the vapor space diameter is D_v, and the wick thickness is t_{wick}. The average surface heat flux at the outer diameter of the wick is q_m, while the maximum heat flux is q_{max}. The ratio q_m/q_{max} is ϕ_m, and the axial heat flux is ξ.

Equating the rate at which heat enters the evaporator to the rate at which it leaves,

$$\pi (D_v + 2t_{wick}) L_e \phi_m q_{max} = \frac{\pi}{4} D_v^2 \xi \tag{5.148}$$

From this equation, the relation between ξ and q_{max} can be expressed as

$$\xi = F_g \phi_m q_{max} \tag{5.149}$$

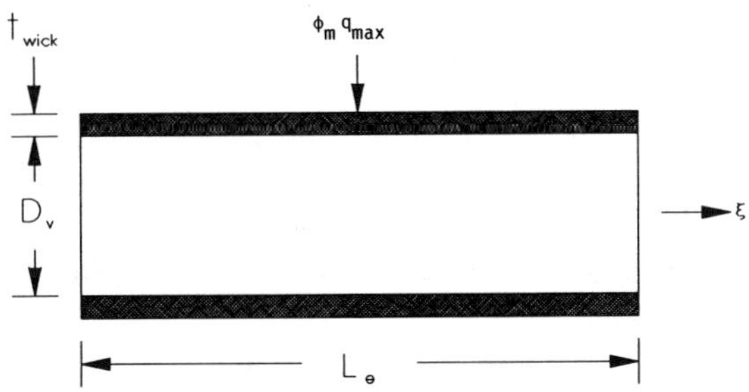

Figure 5.21. Axial heat flux–surface heat flux relationship in evaporator of a cylindrical heat pipe.

Equation (5.149) is a general relation between ξ and $q_{\max}$ in which F_g is a factor that depends on evaporator geometry.

For the cylindrical evaporator,

$$F_g = \frac{4\,L_e}{D_v}\left(1 + \frac{2t_{\text{wick}}}{D_v}\right) \tag{5.150}$$

Equation (5.150) also gives the geometric factor F_g for an evaporator consisting of a hemisphere combined with a cylinder having the same overall dimensions. This approximates the configuration of a heat pipe–cooled nose cone on a high-speed aircraft, although in actual situations the cross section behind the hemisphere-shaped nose usually diverges instead of remaining constant.

Capillary Pumping Limit Equation. The capillary pumping equation takes the form given by Eq. (5.64), which is repeated below:

$$Q_c = \left[\frac{D_{hl}^2\,A_l}{4\,\varepsilon_{po}\,D_p\,(L_p + L_a)}\right]\left(\frac{\sigma\,\kappa\,\rho_l}{\mu_l}\right) = G_p\,P_{ht} \tag{5.64}$$

In the equation, D_{hl} is the hydraulic diameter of the liquid flow passage, A_l is the cross-sectional area of the liquid flow passage, ε_{po} is the ratio of the pressure drop through porous and open liquid flow passages, D_p is the effective wick pore diameter for capillary pumping, Q_c is the capillary pumping limit, G_p is a geometry-dependent parameter, and P_{ht} is the liquid heat transport parameter. The other symbols refer to properties of the heat pipe liquid, and will not be reidentified here.

For a cylindrical heat pipe with a two-layer wick consisting of an outer liquid flow passage of thickness $t_{\text{wick}l}$, an inner capillary pumping layer of negligible thickness, and a vapor space diameter D_v much larger than $t_{\text{wick}l}$, the following relations can be written:

$$D_{hl} = 2t_{\text{wick}l} \qquad A_l = \pi D_v t_{\text{wick}l}$$

Equation (5.64) can be generalized to represent the heat transport rate Q due to capillary pumping at a given radius of curvature r_c by letting $Q_c = Q$ and $D_p = 2r_c$. Then, using the above relations for D_{hl} and A_l, Eq. (5.64) becomes

$$Q = \frac{\pi\,t_{\text{wick}l}^3\,D_v}{2\,\varepsilon_{po}\,r_c\,(L_p + L_a)}\,P_{ht} \tag{5.151}$$

Dividing both sides of Eq. (5.151) by $A_v = (\pi/4)D_v^2$, and noting that $\xi = Q/A_v$,

$$\xi = \frac{2\,t_{\text{wick}l}^3}{\varepsilon_{po}\,r_c\,(L_p + L_a)\,D_v}\,P_{ht} \tag{5.152}$$

Equation (5.152) is the desired form of the capillary pumping equation.

Radius of Curvature at Incipient Boiling. Upon substituting q_b from Eq. (5.140) and F_g from Eq. (5.150) into Eq. (5.149), equating Eqs. (5.149) and (5.152), substituting Eq. (5.138) for r_{ne}, and solving for $r_c/r_n = r_{cb}/r_n$, we have

$$r_{cb}/r_n = 1 + P_f P_g \tag{5.153}$$

where

$$P_f = \frac{P_{ht}}{k_w \, \beta} \tag{5.154}$$

and

$$P_g = \frac{t_{wickl}^4}{4\, C_{g1}\, L_e\, (L_p + L_a)\, \varepsilon_{po}\, \phi_m} \tag{5.155}$$

Equation (5.153) gives the radius of curvature r_{cb} at incipient boiling.

Note that P_f depends only on the mean wick temperature and the properties of the heat pipe fluid (with liquid properties evaluated at the mean wick temperature), whereas P_g is a function of heat pipe geometry and the average-to-peak ratio ϕ_m of the surface heat flux.

Surface Heat Flux at Incipient Boiling. From Eqs. (5.138) and (5.153), the value of the effective nucleation site radius r_{ne} at incipient boiling is expressible as

$$\frac{r_{ne}}{r_n} = \frac{1 + P_f P_g}{P_f P_g} \tag{5.156}$$

Substituting this equation for r_{ne} in Eq. (5.140), the surface heat flux at incipient boiling q_b (i.e., the boiling limit) becomes

$$q_b = 2C_{g1} \, \frac{k_w \, \beta}{t_{wickl} \; r_n} \left(\frac{P_f P_g}{1 + P_f P_g} \right) \tag{5.157}$$

Since in this case the thickness of the capillary pumping layer is negligible, $t_{wick} = t_{wickl}$.

If $\phi_m < 1$, q_b refers to the point on the evaporator surface where the incident heat flux is a maximum.

Equation (5.157) relates the surface heat flux at incipient boiling to the geometrical and fluid property parameters of the heat pipe. In order to evaluate the fluid property parameters, an iterative procedure similar to that described in Section 5.4.5 is required first to find the mean wick temperature. Again, if the interface thermal resistance is significant, the iterative procedure becomes quite involved.

Optimum Wick Thickness for Maximum Boiling Limit. There is an optimum wick thickness t_{wicko} for which the surface heat flux at incipient boiling takes on a maximum value q_{bmax}. This optimum wick thickness will now be found by letting $dq_b/dt_{wickl} = 0$. In the differentiation process, the geometry correction factor C_{g1} is assumed to be independent of t_{wickl}, implying either slab geometry or $t_{wickl}/r_o << 1$. If the liquid flow channel is open, $\varepsilon_{po} = 1$. If the liquid flow channel is porous, $e_{po} = KD_{hl}^2/32 = Kt_{wickl}^2/8$, where K is the friction factor of the porous channel. [See Eq. (4.35).]

Carrying out the differentiation on Eq. (5.157), the following results are obtained. The indicated expressions for r_{cb}/r_n, ΔT_{wb}, and q_{bmax} were derived from Eqs. (5.153), (5.156), (5.136), and (5.157).

Open liquid flow channel:

$$P_f P_g = 3$$

$$t_{wicko} = \left[\frac{12\, C_{g1}\, \phi_m\, L_e\, (L_p + L_a)}{P_f} \right]^{1/4}$$

$$r_{cb}/r_n = 4$$

$$\Delta T_{wb} = \frac{3\,\beta}{2 r_n}$$

$$q_{b\max} = \frac{3}{2}\, C_{g1}\, \frac{k_w\, \beta}{t_{wicko}\, r_n}$$

Porous liquid flow channel:

$$P_f P_g = 1$$

$$t_{wicko} = \left[\frac{C_{g1}\, \phi_m\, K L_e\, (L_p + L_a)}{2 P_f} \right]^{1/2}$$

$$r_{cb}/r_n = 2$$

$$\Delta T_{wb} = \frac{\beta}{r_n}$$

$$q_{b\max} = C_{g1}\, \frac{k_w\, \beta}{t_{wicko}\, r_n}$$

These results show that, when the wick thickness is optimized to maximize the boiling limit, the radius of curvature is between two and four times the nucleation site radius, depending on the configuration of the liquid flow channel. Since the wick pore radius is never larger than the radius of curvature, very fine-pored wicks would be required for the boiling limit to reach its maximum value.

Another interesting observation is that the wick temperature drop at the maximum boiling limit varies between one-half and three-quarters of that predicted from Eq. (5.136) if r_n/r_c is assumed to be negligible.

Incipient Boiling Conditions for Sodium Heat Pipes. The results derived previously for the boiling limit were used to establish boiling characteristics for sodium heat pipes, for the case in which the interface thermal resistance is negligible.

In Figure 5.22, the geometrical factor P_g for sodium heat pipes is shown as a function of mean wick temperature and the ratio of the radius of curvature at incipient boiling to the nucleation site radius r_{cb}/r_n. The curves of Figure 5.22 were obtained by calculating P_f for sodium as a function of mean wick temperature, and substituting these values for P_f in Eq. (5.153).

With the aid of Figure 5.22, incipient boiling conditions can be readily related to the heat pipe geometry. For example, consider a cylindrical sodium heat pipe with a single-layer wick, operating at a mean wick temperature of 2000°F (1093°C). The wick is composed of 100 mesh screen [K = 4.21×10^6 in.$^{-2}$ (6.53×10^9 m^{-2}), r_p = 127 microns], the wick thickness t_{wick} is 0.050 in. (1.27×10^{-3} m), the evaporator length L_e

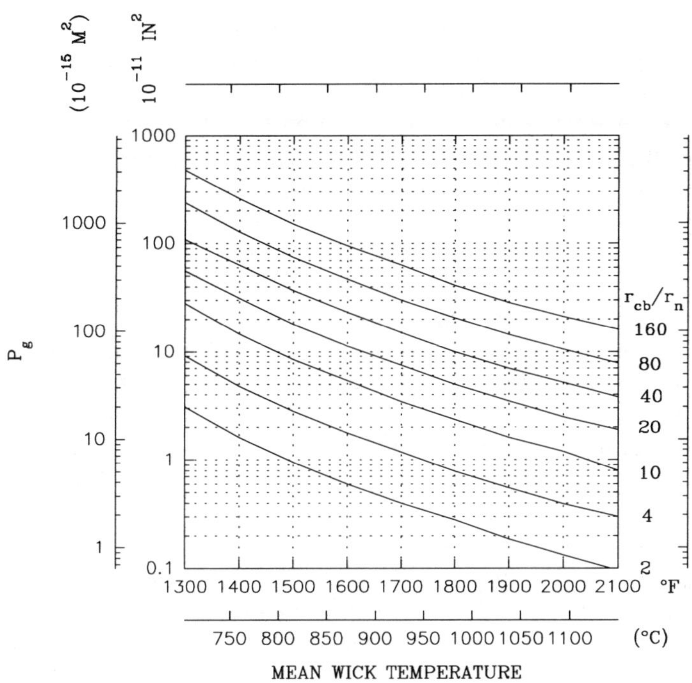

Figure 5.22. Geometrical factor P_g for incipient boiling in sodium.

is 3 in. (0.0762 m), the adiabatic length L_a is zero, and the heat pipe length L_p is 20 in. (0.508 m). The geometry factor C_{g1} is assumed to be 1, and the nucleation site radius r_n is assumed to be 2 microns.

The geometry parameter P_g is then 1.979×10^{-11} in.^{2} (1.277×10^{-14} m^2). From Figure 5.22, $r_{cb}/r_n = 17$. Then, $r_{cb} = 17r_n = 34$ microns. Since the radius of curvature r_{cb} that would be required for incipient boiling is less than r_p, boiling cannot occur in this heat pipe. If a second wick layer with an effective pore radius of 34 microns or less were to be placed inside the first wick layer, operation of the heat pipe at the heat transport rate corresponding to incipient boiling would be possible.

It has been shown that the wick thickness at which the boiling limit is a maximum occurs at $r_{cb}/r_n = 4$ for open liquid flow channels, and at $r_{cb}/r_n = 2$ for porous liquid flow channels. Therefore the r_{cb}/r_n curves in Figure 5.22 corresponding to these values can be used to establish the wick thickness t_{wicko} at the maximum boiling limit. This will now be illustrated for the porous (screen) wick under consideration.

For the porous wick case ($r_{cb}/r_n = 2$) and a mean wick temperature of 2000°F (1093°C), P_g from Figure 5.22 is 0.132×10^{-11} in.^{2} (8.52×10^{-16} m^2). Using the values for L_e, L_a, L_p, and K from the previous example, the wick thickness for the maximum boiling limit is 0.013 in. (3.30×10^{-4} m), about one-quarter the actual wick thickness. However, the achievement of incipient boiling at the optimum wick thickness would require a second wick of negligible thickness inside the first, with an effective wick pore radius no larger than 4 microns.

Concluding Remarks. It must be emphasized that the results obtained in this section apply only in cases in which the vapor pressure drop is negligible, and the liquid pressure drop is due solely to friction in laminar flow. Further, the conditions for which the boiling limit is a maximum apply only when C_{g1} is approximately equal to unity. It must also be assured that the axial heat flux associated with any boiling limit is not in excess of the sonic, entrainment, or capillary pumping limits, if the calculated boiling limit is to be achievable.

Nonetheless, these simplified results, including Figure 5.22, can serve as a useful guide to the prediction of incipient boiling during the initial design of heat pipes. Curves similar to those of Figure 5.22 can also be prepared for other heat pipe fluids.

The same approach to the prediction of incipient boiling also applies in situations in which the pressure drop accompanying capillary pumping is not simply that resulting from laminar flow of the heat pipe liquid. Again, the boiling limit and capillary pumping equations are solved simultaneously for the heat transport rate and radius of curvature at incipient boiling. An iterative solution procedure will generally be required when more general formulations of the capillary pumping limit equation are used (see Section 5.1.2).

It should also be pointed out that these results apply for a wick of uniform thickness over the entire heat pipe length. In design situations in which the evaporator is a small fraction of the total heat pipe length, it may be possible to increase the boiling limit by reducing the wick thickness in the evaporator section only, with relatively little effect on the liquid pressure drop and the radius of curvature. In that event, the boiling limit can be established independently of the radius of curvature.

5.5. HEAT PIPE OPERATIONAL BOUNDARIES

5.5.1. Sonic and Boiling Limit Curves for Sodium and Lithium Heat Pipes

The sonic and boiling limits define operational boundaries that are useful in the initial phase of heat pipe design. These boundaries are best illustrated by plotting the vapor temperature against the surface heat flux corresponding to the sonic and boiling limits. Two such curves are shown in Figure 5.23, one for a sodium heat pipe and the other for a lithium heat pipe.

Values for the boiling limit q_b were obtained by the method described in Section 5.4.5 for which r_n/r_c and the interface thermal resistance are negligible. Values for the surface heat flux q_{sl} corresponding to the axial heat flux ξ_s at the sonic limit were obtained using Eqs. (5.100), (5.116), (5.149), and (5.150).

The curves of Figure 5.23 apply to cylindrical heat pipes with the following characteristics:

Constant surface heat flux ($\phi_m = 1$)
Evaporator length–vapor space diameter ratio $L_e/D_v = 10$
Wick thickness $t_{\text{wick}} = 0.030$ in. (7.62×10^{-4} m)
Effective nucleation site radius $r_{ne} = 1$ micron
$t_{\text{wick}}/D_v << 1$

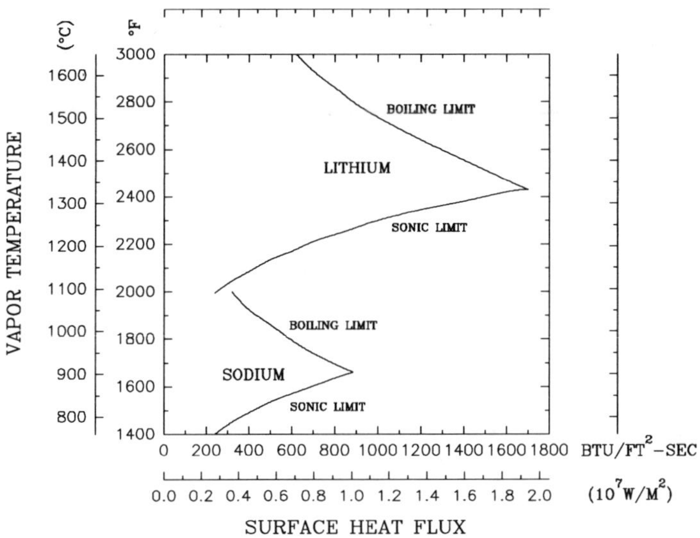

Figure 5.23. Operating temperature range versus surface heat flux for sodium and lithium cylindrical heat pipes.

The junction of the curves for each fluid represents the maximum possible surface heat flux q_{sb} for the particular heat pipe design. The surface heat flux curve corresponding to the sonic limit is to the left and below q_{sb}, and the curve corresponding to the boiling limit is to the left and above q_{sb}. The vapor temperature T_{sb} at which the maximum heat transport rate occurs is also the minimum temperature at which boiling is possible. At temperatures below T_{sb}, the sonic limit is always encountered before the boiling limit, and therefore boiling cannot occur.

For surface heat fluxes below q_{sb}, the sonic and boiling limit curves define operating temperature limits for the heat pipe. Below the lower temperature limit, the sonic limit would be exceeded. Above the upper temperature limit, the boiling limit would be exceeded. As the surface heat flux increases, the operating temperature range becomes smaller, finally shrinking to the single point T_{sb} at the maximum surface heat flux q_{sb}.

It must be emphasized that the sonic and boiling limit curves define upper limits on the maximum heat transport rate and operating temperature range. If capillary pumping or entrainment limits were to be encountered before the sonic limit was reached, the minimum boiling temperature and the lower limit on the operating temperature could be higher, and the maximum surface heat flux could be lower.

Nonetheless, the sonic-boiling limit curve can be quite useful in determining the adequacy of an initial heat pipe design. If the proposed heat transport rate exceeds the maximum heat transport rate, immediate remedial action on the heat pipe design is indicated. If the proposed vapor temperature is below the minimum temperature for boiling, heat pipe operation will not be disrupted by boiling.

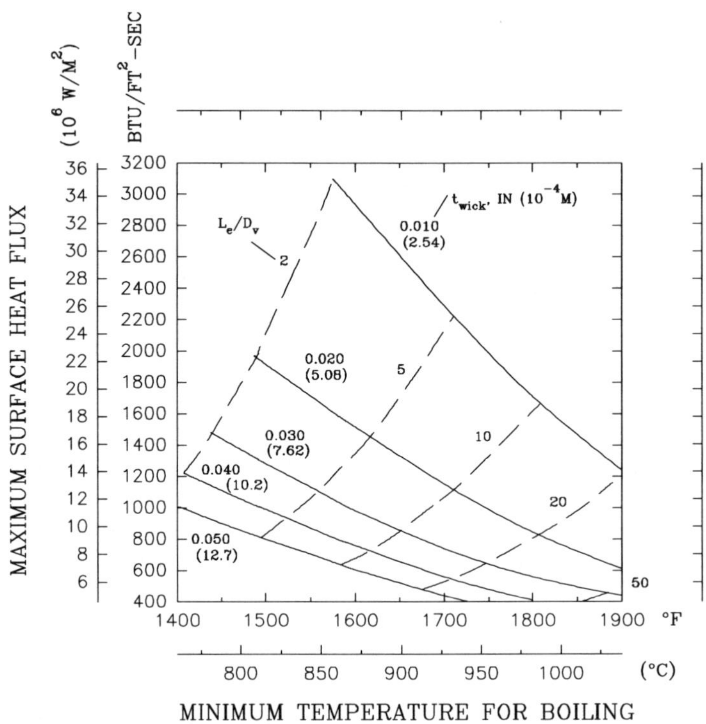

Figure 5.24. Maximum surface heat flux and minimum boiling temperature for a sodium heat pipe.

For the conditions of Figure 5.23, the maximum surface heat flux q_{sb} and associated vapor temperature T_{sb} are:

Heat pipe fluid	q_{sb}, Btu/ft^2-sec (W/m^2)	T_{sb}, °F (°C)
Sodium	860 (9.77×10^6)	1650 (899)
Lithium	1700 (1.93×10^7)	2440 (1338)

Boiling will therefore not occur at temperatures below 1650°F (899°C) in the sodium heat pipe, and at temperatures below 2440°F (1338°C) in the lithium heat pipe.

At a surface heat flux of 600 Btu/ft^2-sec (6.81×10^6 W/m^2), the maximum temperature range of the sodium heat pipe extends from 1580 to 1790°F (860 to 977°C). At a surface heat flux of 1200 Btu/ft^2-sec (1.36×10^7 W/m^2), the maximum temperature range of the lithium heat pipe extends from 2340 to 2640°F (1282 to 1449°C).

The variation of q_{sb} and T_{sb} with the heat pipe geometrical parameters L_e/D_v and t_{wick} is shown for the sodium and lithium heat pipes in Figures 5.24 and 5.25. At $L_e/D_v = 10$ and $t_{wick} = 0.030$ in. (7.62×10^{-4} m), the above cited values for q_{sb} and T_{sb} are obtained.

As L_e/D_v and t_{wick} decrease, the maximum surface heat flux becomes larger while the minimum vapor temperature for boiling becomes smaller. For example, in a lithium

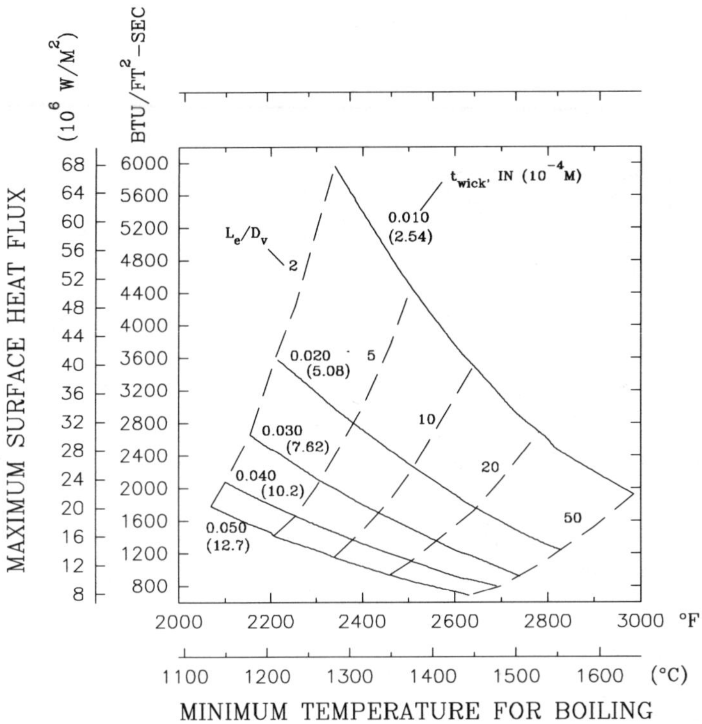

Figure 5.25. Maximum surface heat flux and minimum boiling temperature for a lithium heat pipe.

heat pipe for which $L_e/D_v = 2$ and $t_{wick} = 0.010$ in. (2.54×10^{-4} m), $q_{sb} = 5960$ Btu/ft^2-sec (6.77×10^7 W/m^2) and T_{sb} = 2340°F (1282°C). Thus, when a heat pipe is required to operate at a high surface heat flux, a small t_{wick} and L_e/D_v are desirable.

5.5.2. Conditions under Which Sonic and Boiling Limits Are Equal

A calculational procedure for explicit determination of T_{sb} and q_{sb} will now be presented.

When the sonic and boiling limits are equal, Eqs. (5.116), (5.139), (5.140), and (5.149) can be used to show that

$$A_s\,[P_v\,(T_{sb})]^{\alpha_s} = \frac{2\,\phi_m\,F_g\,C_{g1}\,k_w\,(T_m)}{r_{ne}\,t_{wick}}\,A_b\,[P_v(T_m)]^{\alpha_b} \tag{5.158}$$

Equation (5.158) is a function of the vapor temperature T_{sb} and the mean wick temperature T_m.

Another relation between T_{sb} and T_m can be obtained from Eq. (5.143) for the wick superheat at incipient boiling. (The boiling parameter β is a function of T_m.)

If $t_{wick}/D_v \ll 1$ and the interface thermal resistance is negligible, then $T_i = T_{sb}$, T_m

$= (T_w + T_{sb})/2$, and $T_w = 2T_m - T_{sb}$. Substituting this value of T_w in Eq. (5.143) and using Eq. (5.139), we obtain

$$T_m = T_{sb} + A_b \frac{[P_v(T_m)]^{\alpha_b}}{r_{ne}} \tag{5.159}$$

A simultaneous solution of Eqs. (5.158) and (5.159) gives the T_{sb} and T_m corresponding to the maximum heat transport rate.

The maximum surface heat flux q_{sb} is then found using Eq. (5.116) for the sonic limit and Eq. (5.149), which relates the surface and axial heat fluxes. The result is

$$q_{sb} = A_s \frac{[P_v(T_{sb})]^{\alpha_s}}{\phi_m F_g} \tag{5.160}$$

The associated wick temperature drop ΔT_w is obtained from the equation

$$\Delta T_w = T_w - T_{sb} = 2(T_m - T_{sb}) \tag{5.161}$$

The simultaneous solution of Eqs. (5.158) and (5.159) is facilitated by modifying Eq. (5.158) as follows:

First, Eq. (5.158) is rewritten in the form

$$[P_v(T_{sb})]^{\alpha_s} = \frac{2\,\phi_m F_g C_{g1} k_w(T_m)}{A_s t_{\text{wick}}} (T_m - T_{sb}) = I(T_m, T_{sb}) \tag{5.158a}$$

or

$$\log_{10} P_v(T_{sb}) = \frac{\log_{10} I(T_m, T_{sb})}{\alpha_s}$$

On substituting this equation into Eq. (5.100) for the saturation vapor pressure, the desired final form is obtained as

$$T_{sb} = \frac{B}{A - [\log_{10} I(T_m, T_{sb})]/\alpha_s} \tag{5.162}$$

where T_m and T_{sb} are absolute temperatures.

A solution for T_m and T_{sb} can be found from Eqs. (5.162) and (5.159) by employing the following iterative procedure:

1. Specify $T_{sb} = T_{sbn}$, the heat pipe fluid, the heat pipe geometry, and r_{ne}.
2. Assume an initial value of $T_m = T_{mn}$.
3. Calculate $T_m = T_{m(n+1)}$ from Eq. (5.159).
4. Let δ_1 be some arbitrarily small quantity (0.01, for example).
5. Calculate $t_1 = |(T_{m(n+1)} - T_{mn})/T_{mn}|$.
6. If $t_1 > \delta_1$, let $T_{mn} = (T_{m(n+1)} + T_{mn})/2$ and go back to step 3.
7. If $t_1 \le \delta_1$, let $T_{mn} = T_{m(n+1)}$ and go to step 8.
8. Calculate $I(T_{mn}, T_{sbn})$ from Eq. (5.158a).
9. Calculate $T_{sb} = T_{sb(n+1)}$ from Eq. (5.162).
10. Calculate $t_2 = |(T_{sb(n+1)} - T_{sbn})/T_{sbn}|$.
11. If $t_2 > \delta_1$, let $T_{sbn} = (T_{sb(n+1)} + T_{sbn})/2$ and go back to step 3.
12. If $t_2 \le \delta_1$, a solution has been obtained, with $T_{sb} = T_{sb(n+1)}$ and $T_m = T_{m(n+1)}$.
13. The maximum surface heat flux and the associated wick temperature drop are then calculated from Eqs. (5.160) and (5.161).

5.6. COMPARISON OF CALCULATED AND EXPERIMENTAL HEAT TRANSPORT LIMITS

In this chapter, methods for predicting capillary pumping, sonic, entrainment, and boiling heat transport limits in heat pipes have been presented. These methods are used in the design of heat pipes to insure that the actual heat transport rate will be well below the smallest of the heat transport limits.

5.6.1. Validity of Analytical Methods

There is no scarcity of experimental data on limiting heat pipe heat transport rates in the literature. Such data are frequently presented along with analytical predictions of heat transport limits. In principle, if the calculated and experimental limits are in reasonable agreement, the analytical methods may be considered to be valid. However, the validation is not so easily accomplished in practice, for the following reasons.

The value of key parameters needed to calculate heat transport limits may not be given, either because of an oversight or because they are not known. Calculational methods used may not be presented in sufficient detail for independent evaluation. Empirical adjustments that may have been introduced to improve the correlation between analysis and experiment may not be fully identified. The precision of key dimensional and temperature measurements may be limited. Finally, the experimental setup itself may not provide all the data needed to compare calculated and measured heat transport limits.

To illustrate the problems associated with correlating analytical predictions of heat transport limits with experimental data, calculations based on the methods presented in this chapter are compared with experimental heat transport limit data given in Ref. 5.28. The comparison is confined to a single test, identified as test #2, in which the heat transport limit was measured as a function of temperature for a specific heat pipe.

5.6.2. Experimental Heat Transport Limit Data

The heat pipe in question was cylindrical, about 0.75 in. (0.0191 m) in diameter and 12 in. (0.305 m) long. A two-layer wick structure was used, consisting of a series of equally spaced axial grooves in the heat pipe wall covered by a screen. The heat pipe was fabricated from nickel, and the heat pipe fluid was sodium.

An induction coil enclosing the evaporator section of the heat pipe provided the heat input. Heat was removed by an annular water jacket around the condenser section. The water jacket was separated from the heat pipe by a thin annular gap that could be filled with an arbitrary mixture of argon and helium to control the thermal resistance across the gap. The heat transport rate was determined by measuring the water inlet and outlet temperatures and the water flow rate.

The vapor temperature was measured with a thermocouple inside a well centered on the heat pipe axis, which extended over the entire heat pipe length. The diameter of the well was 0.125 in. (3.18×10^{-3} m).

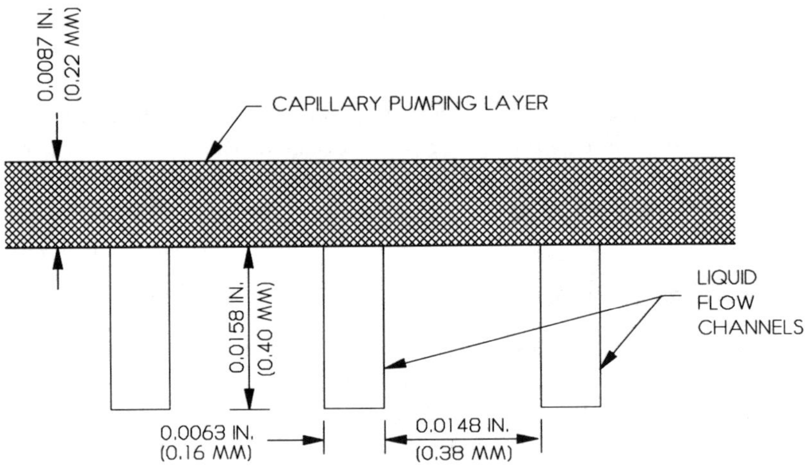

Figure 5.26. Configuration and dimensions of heat pipe wick for test #2 (Ref. 5.28).

The dimensions and configuration of the heat pipe wick are shown in Figure 5.26. Physical characteristics of the heat pipe, including wick dimensions, are given in Table 5.9.

Uncertainties in the Data. The various measurements were presented in Ref. 5.28 with an accuracy of one or two figures. The precision of the evaporator and condenser lengths was limited because of difficulty in determining the exact beginning and end of these regions. The precision of the other dimensions was limited because of the small dimensions involved. In a micrograph showing a cross section of the heat pipe with five

Table 5.9. Characteristics of heat pipe used for test #2 (Ref. 5.28)

Evaporator length	3.15 in. (0.08 m)
Condenser length	8.66 in. (0.22 m)
Heat pipe length	11.81 in. (0.30 m)
Heat pipe diameter	0.75 in. (0.0191 m)
Wall material	Nickel
Number of axial grooves	88
Groove depth	0.01575 in. (4×10^{-4} m)
Groove width	0.0063 in. (1.6×10^{-4} m)
Land (width between grooves)	0.0148 in. (3.8×10^{-4} m)
Screen thickness	0.00866 in. (2.2×10^{-4} m)
Screen void fraction	0.5
Screen material	?
Screen mesh	?
Vapor passage radius	0.286 in. (0.0073 m)

grooves visible, the width of the widest groove was about 25% greater than that of the narrowest groove.

The screen material and mesh were not specified, and hence had to be deduced from available evidence. The screen material was assumed to be the same as that for the heat pipe wall, i.e., nickel. Since the screen thickness corresponds approximately to that of a single layer of 100 mesh screen, it was assumed that the screen consisted of a single 100 mesh layer.

For calculational purposes, the following characteristics of 100 mesh screen, as obtained from Table C.8 of Appendix C, were used. Note that the void fraction does not correspond to that specified in Table 5.9, which was inferred from data for a similar type of screen that had been used in earlier experiments.

Void fraction	0.585
Friction factor	4.21×10^{6} in^{-2} (6.53×10^{9} m^{-2})

These properties were obtained from tests with multiple screen layers (Ref. 4.8), and may be expected to vary with the number of screen layers, the screen fabrication process, the dimensions of the screen wire, and the type of weave into which the screen has been formed. Since few if any of these factors are usually known, screen void fraction and friction factor data obtained from one set of tests may be subject to considerable uncertainty when applied in other circumstances.

Calculation of the boiling limit requires that the nucleation site radius and the radius of curvature in the screen pores at incipient boiling be known. Since such data were not available, a nucleation site radius of 3 microns and a radius of curvature of 80 microns were assumed.

The axial location where temperatures were measured in the thermocouple well was not identified. From the description of the observed drop in condenser temperature when a heat transport limit was reached, the temperature measurement point was assumed to be in the condenser section.

The heat transport limit calculations require that the vapor temperature at the beginning of the evaporator be specified. If the heat pipe is not operating isothermally when a heat transport limit is reached, the vapor temperature in the evaporator may differ significantly from that in the condenser. Therefore, if the temperature was not being measured at the beginning of the evaporator when the heat pipe was nonisothermal, another element of uncertainly is introduced into the heat transport limit calculations.

Liquid Flow through Capillary Pumping Layer. Normally, for a two-layer wick, the resistance to flow through the capillary pumping layer is much larger than that through the liquid flow channel. Parallel flow through the capillary pumping layer is then negligible, and may be neglected when calculating the capillary pumping limit. However, flow through the capillary pumping layer may be significant if its flow area is comparable to or greater than that of the liquid flow channel, and should then be taken into account. Such is the case for the heat pipe under consideration here. From Figure

5.26, it can be determined that the flow area of the capillary pumping layer is 1.837 times that of the liquid flow channel grooves.

As shown in Appendix E, the overall resistance to flow is $1/(1 + R_1/R_2)$ times that of the liquid flow channel alone, where R_1 is the flow resistance of the liquid flow channel and R_2 is the flow resistance of the capillary pumping layer. To account for the effect of flow through the capillary pumping layer, the frictional liquid pressure drop term in the capillary pumping equation is multiplied by this factor. The result is lower overall resistance to liquid flow and a higher capillary pumping limit.

Using Eq. (E.6) of Appendix *E*, $R_1/R_2 = 0.1724$ for the heat pipe under consideration here. Since this figure is significant, the correction for flow through the capillary pumping layer was included in the capillary pumping limit calculations.

5.6.3. Calculated versus Experimental Heat Transport Limits

In Figure 5.27, experimental and calculated heat transport limits are plotted as a function of vapor temperature, on the assumption that the temperatures measured in test #2 are a good representation of the vapor temperature at the start of the evaporator (i.e., the heat pipe is assumed to be isothermal). The experimental curve represents a good fit to individual data points (which are not shown in the figure). The temperature variation of vapor pressure and the axial temperature drop through the vapor are also shown.

Observations. The first observation to be made from Figure 5.27 is that the boiling limit is far above the measured heat transport limit for the indicated temperature range, and is likely to remain so until the vapor temperature has increased by at least another few hundred degrees. This is entirely consistent with more general observations that boiling is rarely observed in liquid metal heat pipes, except at temperatures at which the associated vapor pressure exceeds 20–40 psia ($1.38 \times 10^5 - 2.76 \times 10^5$ N/m^2).

The next important observation is that, as the vapor temperature falls below 1400°F (760°C) and the associated vapor pressure falls below about 4 psia (27,580 N/m^2), there is a rapid increase in the vapor temperature drop through the heat pipe, and the heat pipe becomes increasingly nonisothermal.

This means that calculations of the capillary pumping and entrainment limits, which are based on the assumption of isothermal heat transport, become increasingly inaccurate, predicting heat transport limits that are too high. If the effect of non-isothermality in the vapor were taken into account, the capillary pumping and entrainment limit curves in Figure 5.27 would fall closer to the experimental heat transport limit curve.

Third, at vapor temperatures below 1080°F (582°C), the sonic limit curve, which defines the highest possible heat transport rate, lies substantially below the experimental curve. This apparently contradictory situation can be explained if it is assumed that the vapor temperature was actually being measured in the condenser region.

A heat pipe is highly nonisothermal at the sonic limit, and the vapor temperature at the evaporator end may be significantly higher than that at the condenser end. Therefore,

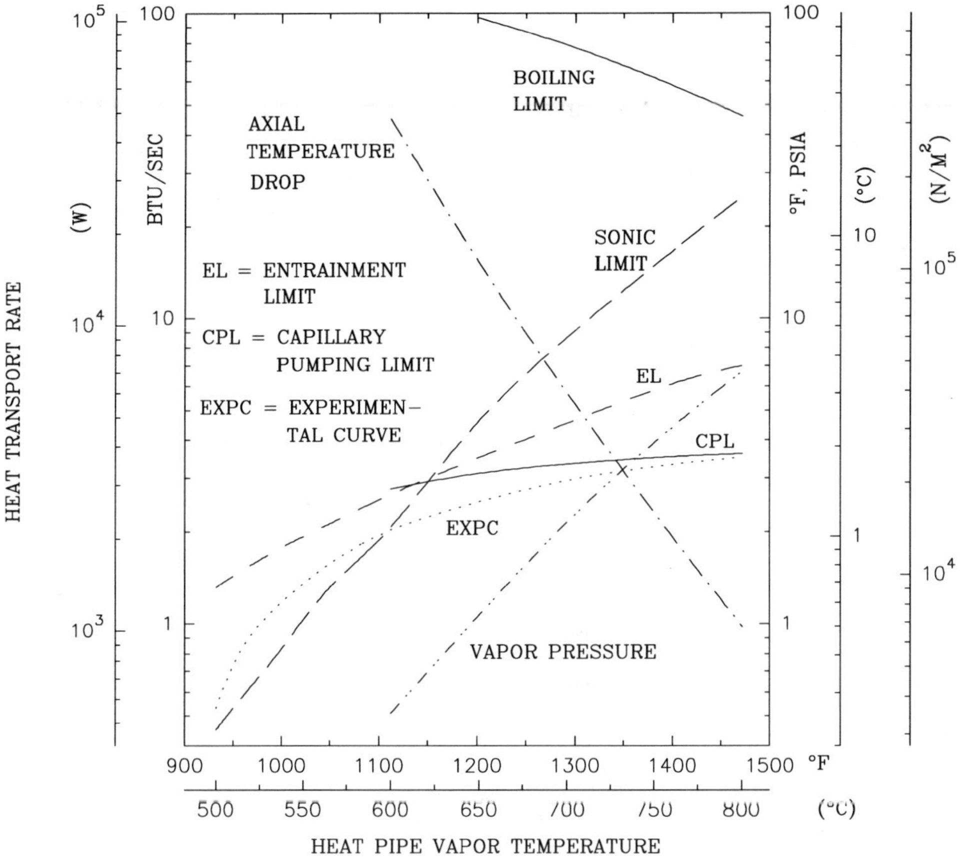

Figure 5.27. Comparison of experimental and calculated heat transport limits for test #2 (Ref. 5.28).

if the sonic limit were based on the vapor temperature at the beginning of the evaporator rather than in the condenser, it would be higher than is indicated in Figure 5.27.

Despite the uncertainties that have just been highlighted, it may be concluded from Figure 5.27 that heat transport limits are predicted with reasonable accuracy by the methods of this chapter.

The heat pipe appears to be sonic-limited at temperatures below about 1100°F (593°C), and capillary pumping–limited at temperatures above about 1150°F (621°C). In the temperature range between 1100 and 1150°F (593 and 621°C), it is not clear whether heat transport is limited by the onset of sonic velocity in the vapor, by entrainment of the heat pipe liquid, or by the cessation of capillary pumping.

Significance of Observations. The above observations serve to emphasize that the correlation of calculated heat transport limits with experimental data is often dependent on assumptions and data of uncertain validity. Thus, a cautious approach to the use of calculated heat transport limits in the design of heat pipes is advisable. It is suggested that a heat pipe should be designed initially so that the smallest heat transport limit is at least 1.5 times the actual heat transport rate. Higher safety factors may be warranted for specific design situations.

Once a prototype heat pipe has been fabricated, tested, and found to perform well below predicted heat transport limits, the safety factor may then be relaxed to permit a more refined design. The use of a thicker wick, a smaller vapor passage, or a larger pore size in the capillary pumping layer may then be feasible.

5.7. REFERENCES

5.1. Cotter, T. P., "Theory of Heat Pipes," LA-3246-MS, Los Alamos Scientific Laboratory, Los Alamos, New Mexico, March 1965.

5.2. Sears, F. W., *An Introduction to Thermodynamics, The Kinetic Theory of Gases, and Statistical Mechanics,* 2nd Ed., Addison-Wesley, Cambridge, Massachusetts, 1953, pp. 119–123, 248.

5.3. Levy, E. K., "Theoretical Investigation of Heat Pipes Operating at Low Vapor Pressures." Presented at the ASME Aviation and Space Conference, Los Angeles, California, June 1968.

5.4. Levy, E. K., and Chow, S. F., "Sonic Limit in Sodium Heat Pipes," Paper 71-WA/HT-11. Presented at the ASME Winter Annual Meeting, Washington, D.C., November 1971.

5.5. Deverall, J. E., Kemme, J. E., and Florschuetz, L. W., "Sonic Limitations and Startup Problems of Heat Pipes," LA-4518, Los Alamos Scientific Laboratory, Los Alamos, New Mexico, November 1970.

5.6. Kemme, J. E., "Ultimate Heat Pipe Performance," *IEEE Transactions on Electron Devices,* Vol. ED-16, No. 8, August 1969, pp. 717–723.

5.7. DeMichele, D. W., and Davis, M. V., "Vapor Transport Limits of Liquid Metal Heat Pipes," *Nuclear Technology,* Vol. 15, September 1972, pp. 366–383.

5.8. Busse, C. A., "Theory of the Ultimate Heat Transfer Limit of Cylindrical Heat Pipes," *International Journal of Heat and Mass Transfer,* Vol. 16, 1973, pp. 169–186.

5.9. Silverstein, C. C., "Correlation of Heat Pipe Heat Transport Limits with Vapor Pressure," AIAA-85-0939. Presented at the AIAA 20th Thermophysics Conference, Williamsburg, Virginia, June 1985.

5.10. Shapiro, A. H., *The Dynamics and Thermodynamics of Compressible Fluid Flow,* Vol. I, Ronald Press, New York, 1953, p. 166.

5.11. Davison, H. W., "Compilation of Thermophysical Properties of Liquid Lithium," TN D-4650, NASA, Washington, D.C., July 1968, pp. 12–13.

5.12. Jackson, C. B. (ed.), *Liquid-Metals Handbook: Sodium-NaK Supplement,* Atomic Energy Commission, Washington, D.C., July 1955, p. 31.

5.13. Hoffman, H. W., and Cox, B., "A Preliminary Collation of the Thermodynamic and Transport Properties of Potassium," ORNL-TM-2126, Oak Ridge National Laboratory, Oak Ridge, Tennessee, July 1968, p. 38.

5.14. Hoffman, H. W., and Robin, T. T., Jr., "A Preliminary Collation of the Thermodynamic and Transport Properties of Cesium," ORNL-TM-1755, Oak Ridge National Laboratory, Oak Ridge, Tennessee, June 1967, p. 16.

5.15. Cotter, T. P., "Heat Pipe Startup Dynamics." Presented at the 1967 Thermionic Conversion Specialist Conference, Palo Alto, California, October 30, 1967.

5.16. Ernst, D. M., "Evaluation of Theoretical Heat Pipe Performance." Presented at the 1967 Thermionic Conversion Specialist Conference, Palo Alto, California, October 30, 1967.

5.17. Kemme, J.E., "Vapor Flow Considerations in Conventional and Gravity-Assist Heat Pipes." Presented at the Second International Heat Pipe Conference, Bologna, Italy, March 1975.

5.18. Tien, C. L., and Chung, K. S., "Entrainment Limits in Heat Pipes," *AIAA Journal,* Vol. 17, No. 6, June 1979, pp. 643–646.

5.19. Kemme, J. E., "High Performance Heat Pipes." Presented at the 1967 Thermionic Conversion Specialist Conference, Palo Alto, California, October 30, 1967.
5.20. Busse, C. A., and Kemme, J. E., "The Dry-Out Limits of Gravity-Assist Heat Pipes with Capillary Flow." Presented at the Third International Heat Pipe Conference, Palo Alto, California, May 1978.
5.21. Silverstein, C. C., "Surface Heat Flux for Incipient Boiling in Liquid Metal Heat Pipes," *Nuclear Technology,* Vol. 12, September 1971, pp. 56–62.
5.22. Silverstein, C. C., "Boiling Limit in Alkali Liquid Metal Heat Pipes," Paper 71-WA/HT-10. Presented at the ASME Winter Annual Meeting, Washington, D.C., November 1971.
5.23. Kreith, F., *Principles of Heat Transfer,* 2nd Ed., International Textbook, Scranton, Pennsylvania, 1967, pp. 25–31.
5.24. Prenger, F. C., "Performance Predictions and Measurements for Space Power System Heat Pipes," LA-UR-81-909, Los Alamos Scientific Laboratory, Los Alamos, New Mexico, 1981.
5.25. Alario, J., Haslett, R., and Kosson, R., "The Monogroove High Performance Heat Pipe," AIAA 81-1156, American Institute of Aeronautics and Astronautics, New York, 1981.
5.26. Dwyer, O. E., *Boiling Liquid-Metal Heat Transfer,* American Nuclear Society, Hinsdale, Illinois, 1976, pp. 81–82.
5.27. Chen, J. C., "Incipient Boiling Superheats in Liquid Metals," *Journal of Heat Transfer,* August 1968, pp. 303–312.
5.28. Kemme, J. E., "Heat Pipe Capability Experiments," LA-3585-MS, Los Alamos Scientific Laboratory, Los Alamos, New Mexico, October 1966.

CHAPTER
SIX

HEAT PIPE DESIGN: STEADY STATE

Heat pipe design involves the determination of heat pipe characteristics under steady-state or transient operating conditions. This chapter is concerned with the steady-state aspects of heat pipe design. Calculation of transient heat pipe performance is treated in Chapter 7.

The heat pipe design process includes the following steps:

1. The criteria and constraints to which the design must conform, as established by requirements of the application, are defined. Such factors as the incident heat flux, geometrical constraints, and the type of heat sink are included in the design criteria.
2. The size and operating temperature of the heat pipe system are derived from heat transfer considerations, based on characteristics of the heat source and heat sink to which the heat pipe system is coupled. Material temperature limitations and the nature of the surrounding environment play an important role in the selection of heat pipe temperature.
3. Internal heat pipe dimensions (i.e., the vapor space size, wick thickness, and wick pore size) are calculated on the basis of heat pipe heat transport limit equations. In performing these calculations, it is necessary to specify first the heat pipe fluid, the wick type and material, and the wick frictional characteristics, and to establish the wick thermal conductivity.
4. Heat pipe wall (and, if required, rib) dimensions that will limit stress to tolerable levels and prevent buckling of the heat pipe structure are determined, using appropriate structural formulas. The heat pipe structural material and its characteristics must first be specified. The wall temperature drop needed to calculate thermal stress is found from the heat transfer calculations of step 2.

5. The amount of excess heat pipe liquid resulting from density changes with temperature is determined. Excess liquid is liquid that cannot be contained within the wick pores under some design conditions, and that consequently spills into the vapor space. If required by the application, a method for immobilizing the excess liquid is identified.
6. The weight of the heat pipe system (and overall volume, if required by the application) is calculated. This is determined by finding the volumes of the various heat pipe components and multiplying these volumes by the appropriate material densities.
7. Steps 2 through 6 are repeated as necessary to obtain a design that is in conformance with the design criteria.

The first section of this chapter deals with the types of heat pipe design criteria and constraints that may be encountered (step 1). Then, design procedures are presented for establishing the heat pipe size and temperature (step 2), the internal heat pipe dimensions (step 3), and the dimensions of structural components (step 4), and for accommodating excess heat pipe liquid (step 5). Weight and volume calculations (step 6) are straightforward, and are not discussed here.

The last two sections consider methods for calculating the thermal conductivity of composite wick materials, and design methods for heat pipe heat exchangers.

6.1. DESIGN CRITERIA AND CONSTRAINTS

Design criteria constitute externally imposed conditions to which the heat pipe design must conform. Design constraints represent more qualitative restrictions that are to be taken into account in the heat pipe design to the maximum practical extent. An upper temperature limit on a component being cooled by a heat pipe is an example of a design criterion. A statement that the weight of a heat pipe cooling system be held to a minimum is an example of a design constraint. Considerations involved in adhering to the following design criteria and constraints will now be discussed:

- Characteristics of the heat source
- Characteristics of the heat sink
- Heat pipe operating environment
- Size and weight constraints
- Temperature constraints
- Fabrication constraints
- Cost constraints

6.1.1. Heat Source Characteristics

Heat source design criteria may include the manner in which the heat source is thermally coupled to the heat pipes, the surface area over which heat is to be added, the rate of heat transfer, and the heat source temperature.

The heat source may be a solid or a fluid. Thermal coupling may be by direct solid-to-solid contact (conduction), via fluid-to-solid contact (convection), by radiation, or by electrical induction heating. The heat transfer rate and/or the heat source temperature may be given. Alternatively, information may be given only on the operational conditions under which the transfer of heat is to take place (for example, the conditions associated with high-speed flight). The heat transfer rate or heat source temperature would then have to be calculated.

An example of conductive coupling would be a radioisotope heat source that had been brazed to one end of a heat pipe in order to transfer heat at a lower heat flux to a thermoelectric generator at the other end. A representative convective heat source would be a hot gas flowing over a cooler heat pipe. An example of a radiative heat source would be a hot combustion liner radiating heat to a surrounding heat pipe cooling structure. Aerodynamic heating of a heat pipe–cooled leading edge on a high-speed aircraft represents an example of a gaseous heat source in which both convection and radiation play a role. An induction heating coil is commonly used as the heat source in laboratory measurements of heat pipe heat transport characteristics.

For the radioisotope heat source with conductive coupling, the heat transport rate would be specified. For a high-speed aircraft, the heat flux on the aerodynamically heated surface might be given, or alternatively information on speed, altitude, and vehicle geometry might be supplied from which the heat transfer rate would have to be calculated.

6.1.2. Heat Sink Characteristics

Like the heat source, the heat sink may be coupled to the heat pipe by conduction, convection, radiation, or some combination of these.

An example of conductive-convective coupling would be a heat pipe surrounded by an insulating layer and a liquid cooling jacket. The insulating layer separates the condenser section of the heat pipe from a convective heat sink. It provides sufficient thermal resistance to maintain the heat pipe at the temperature necessary for adequate heat transport capability and reasonably isothermal operation, while avoiding excessive coolant temperature.

The heat sink temperature would be specified, along with information needed to calculate the heat transfer rate between the heat pipe and the heat sink. For a convective heat sink, the heat transfer coefficient might be known, or would have to be calculated from fluid thermophysical and flow properties. For a radiative heat sink, the heat pipe and heat sink emissivities would be established, along with view factors for radiative heat transfer between the heat pipe and heat sink.

6.1.3. Heat Pipe Operating Environment

The operating environment includes the medium in which the heat pipe is immersed, the heat pipe orientation, the magnitude of the gravity field and external accelerations that may be present, and external pressures or structural loads to which the heat pipe is exposed.

If the heat pipe is to operate in the vacuum of space, or in an inert gas such as argon, virtually any containment material that is compatible with the heat pipe fluid may be used. If the heat pipe is to operate in air or in the presence of combustion products, the containment material must have adequate resistance to oxidation/corrosion at contemplated heat pipe operating temperatures. Alternatively, a protective coating may be required. Additional restrictions may be needed to prevent or minimize the diffusion of gases into the heat pipe interior, especially if hydrogen or hydrogen-bearing compounds such as water vapor are present.

The heat pipe orientation with respect to gravity or other acceleration forces, as well as the magnitude of these forces, will influence the liquid pressure drop, and must be taken into account in capillary pumping calculations. In a gravity-free and acceleration-free environment, heat pipe orientation is not a design consideration.

The heat pipe may be exposed to external pressure from the medium in which it is immersed. Additionally, portions of the heat pipe may be subjected to dynamic pressure due to motion of the heat pipe through the medium (i.e., stagnation pressure). Other external loads may be present if the heat pipe is acting as a structural member. These external pressures and loads must be taken into account in the structural aspects of the heat pipe design. It must be assured that resulting stresses are not excessive, and that buckling of the heat pipe containment wall will not occur.

6.1.4. Size/Weight Constraints

Constraints on heat pipe size and weight may be imposed by volume restrictions of the application and the need for minimal system weight. Limitations on heat pipe length may result from static liquid pressure drops produced by gravity and/or external acceleration.

In most applications, the area covered by the heat source, and hence the heat pipe evaporator surface area, will be well defined. On the other hand, the required condenser area for a given heat sink type will vary with the heat pipe temperature. Generally, the lower the operating temperature, the larger will be the condenser area required to transfer the heat load to the heat sink. The usually desirable objectives of minimal heat pipe size and weight will then be favored by elevated temperature operation.

If the objective is to cool a heated component, the size and weight of the heat pipe cooling system will be minimized by maintaining the temperature of the heated component as close as possible to the heat source temperature. In this way, the heat load to be transferred to the heat pipe cooling system is minimized. However, material strength and environmental compatibility considerations could limit the temperature of the component being cooled to a level well below that of the heat source.

If the objective is to transfer heat at a specified rate from a hotter to a colder fluid in a heat exchanger, size optimization becomes more involved. The heat exchanger will be composed of a relatively large number of individual heat pipes. The cumulative total area of the heat pipes will be determined by fluid properties and flow rates on the hot and cold sides of the heat exchanger, and by flow passage and heat pipe dimensions. System size is minimized by adjusting system variables to maximize hot and cold side heat transfer coefficients. However, a reduction in system size is usually accompanied by an increase in pressure drop of the heat exchanger fluids. The extent to which system size can be reduced is then determined by the allowable pressure drop for the application under consideration.

6.1.5. Temperature Constraints

If the design is not otherwise constrained by heat source or other temperature limitations, an upper limit on heat pipe temperature may be imposed by the strength of the heat pipe containment material, which generally decreases with an increase in temperature. An additional temperature limitation may arise from the decreasing effectiveness of oxidation- or corrosion-resistant coatings with an increase in temperature.

The materials temperature limitation may be overcome if another material with a higher temperature capability, which is also compatible with the heat pipe fluid being used, is available. For example. a sodium-superalloy heat pipe may be operated up to about 1800°F (982°C) in air. If higher-temperature operation is desirable, the refractory metal molybdenum may be substituted, but the requirement for a suitable oxidation-resistant coating is then introduced.

An upper limit on heat pipe operating temperature may also be imposed by the boiling limit. Should this occur, substitution of another heat pipe fluid could be considered. If the boiling limit were encountered with a sodium heat pipe at a given temperature, the substitution of lithium would almost surely solve that problem. However, it might be necessary to operate the lithium heat pipe at higher temperatures to insure that the vapor pressure is high enough to avoid capillary pumping, sonic, or entrainment limits.

6.1.6. Fabrication/Cost Constraints

The more difficult a heat pipe is to fabricate, the more costly it is likely to be. Some aspects of fabrication are under the control of the designer, whereas others may be inherent in the application. An example of the latter would be a heat pipe system that transports heat generated in a nuclear reactor to an associated energy conversion plant. The need to avoid direct nuclear radiation leakage paths may dictate that the heat pipe elements follow a tortuous path while passing through the radiation shield. Similarly, the very high incident heat flux in the stagnation regions of high-speed aircraft may dictate a very thin wick in the evaporator region of a heat pipe cooling system to avoid boiling.

Heat pipe fabrication will be simpler and less costly if readily available, easily machinable materials can be used. Thus, if the operating temperature of a sodium heat pipe can be held to 1700°F (927°C) or less, superalloy construction from an alloy such as Hastelloy X or Haynes 188 may be feasible. A more difficult to machine refractory alloy such as TZM molybdenum might be required at higher temperatures, with the additional requirement of an oxidation-resistant coating.

A single-layer wick, relatively thick with a relatively coarse pore structure, may be fabricated more easily and cheaply than a thin, two-layer wick of more complex design with a finer-pored wick structure.

Within constraints that may be imposed by the application, there may be considerable room for design tradeoffs. In such tradeoffs, the desire for a heat pipe system of minimal size and weight would be balanced against factors such as material strength and corrosion characteristics, fabricability, and cost. Final design decisions may well be based on the experience and judgment of the designer.

6.2. HEAT PIPE AREA-TEMPERATURE RELATIONS

Procedures for establishing heat pipe size and operating temperature will vary according to the manner in which the heat pipe is to be thermally coupled to the heat source and heat sink. A number of representative situations are schematically illustrated in Figure 6.1. The general procedure common to all cases is to equate the heat transfer rate between the heat source and the heat pipe to that between the heat pipe and the heat sink. Area-temperature relationships will now be considered for the cases shown in Figure 6.1.

6.2.1. Convective Heat Source and Heat Sink

Consider a heat pipe that is to be used to transfer heat from a hot gas at temperature T_h to a cooler gas at temperature T_c, as indicated in Figure 6.1A. This situation is representative of heat transfer in a single stage of a heat pipe heat exchanger. The heat transfer coefficient is h_h on the hot (evaporator) side and h_c on the cold (condenser) side. The surface area of the heat pipe is A_e on the hot side and A_c on the cold side. The heat transfer rate between the gases and the heat pipe is assumed to be small compared to their heat content, so that T_h and T_c remain approximately constant. The heat pipe temperature T_p is assumed to be constant. (Wall and wick ΔT's and the ΔT along the heat pipe length are assumed to be negligible.)

A heat balance between the two gases yields the equation

$$h_h A_e (T_h - T_p) = h_c A_c (T_p - T_c)$$

From this equation, the ratio of cold and hot side areas can be presented in the form

$$\frac{A_c}{A_e} = \frac{h_h}{h_c}\left[\frac{(T_h/T_c) - (T_p/T_c)}{(T_p/T_c) - 1}\right]$$

In case A of Figure 6.2, A_c/A_e is plotted as a function of T_p/T_c, with $T_h/T_c = 2$ and $h_h/h_c = 1$. Note that the required cold side area varies inversely with the heat pipe temperature, varying from zero at $T_p = T_h$ to infinity at $T_p = T_c$.

6.2.2. Constant Heat Source, Convective Heat Sink

In another example (Figure 6.1B), heat is added at a constant mean heat flux q_e over an area A_e to a heat pipe at temperature T_p. The heat is rejected by convective heat transfer

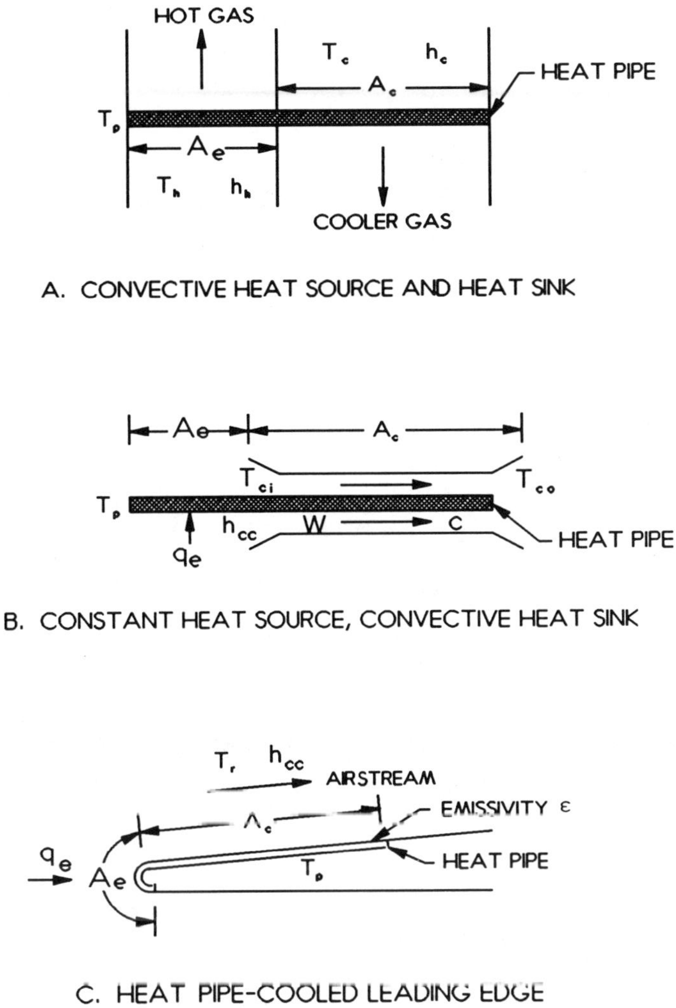

Figure 6.1. Heat source–heat sink configurations for heat pipes.

from area A_c to a surrounding liquid flow stream with inlet temperature T_{ci} and outlet temperature T_{co}. The specific heat of the liquid is c, its mass flow rate is W, and the convective heat transfer coefficient is h_{cc}. This situation is representative of a laboratory setup to measure heat pipe performance. The constant rate heat source might be an induction coil surrounding the evaporator section of the heat pipe.

The energy balance in this case is

$$q_e A_e = Wc(T_{co} - T_{ci})$$

Upon introducing the concepts of heat exchanger effectiveness ε and number of

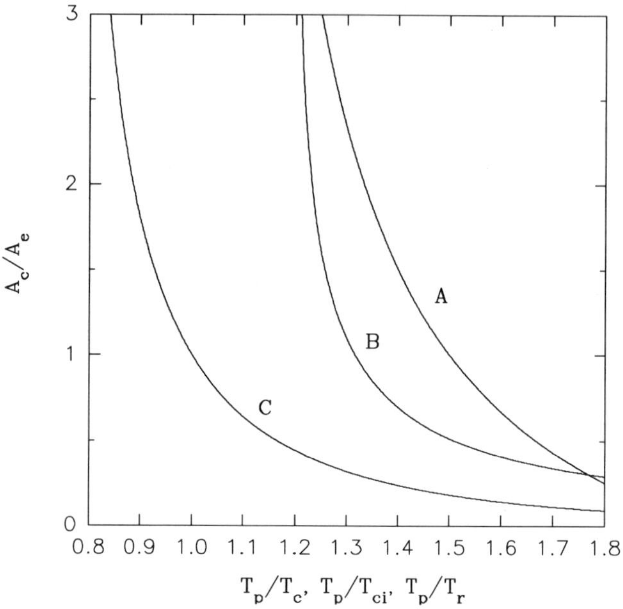

Figure 6.2. Variation of condenser heat transfer area with temperature. (A) Convective heat source, convective heat sink. (B) Constant heat source, convective heat sink. (C) Heat pipe–cooled leading edge.

heat transfer units NTU (see Section 6.6.2) for heat transfer between a condensing vapor (the heat pipe fluid) and a second fluid, where

$$\varepsilon = \frac{T_{co} - T_{ci}}{T_p - T_{ci}}$$

and

$$\text{NTU} = \frac{h_{cc} A_c}{Wc} = \ln \frac{1}{1 - \varepsilon}$$

the energy balance may be written (with some manipulation)

$$\frac{A_c}{A_e} = \frac{q_e}{h_{cc}\,(T_{co} - T_{ci})} \ln\left[\frac{(T_p/T_{ci}) - 1}{(T_p/T_{ci}) - (T_{co}/T_{ci})}\right]$$

The minimum heat pipe temperature in this case is T_{co}, when the denominator in the log term of the above equation is equal to zero and the condenser area is infinite. The ratio of condenser and evaporator areas is plotted against T_p/T_{ci} as case B of Figure 6.2, with $q_e/[h_{cc}(T_{co} - T_{ci})] = 1$ and $T_{co}/T_{ci} = 1.2$.

6.2.3. Heat Pipe–Cooled Leading Edge

In a final example (see Figure 6.1C), the outer surface on the the forward section of an aircraft wing is fabricated as a heat pipe structure. Aerodynamic heat incident on the area A_e of the leading edge at a constant mean heat flux q_e is to be dissipated over area A_c of the lateral surfaces by radiation. Variations of the aerodynamic heat flux with the leading edge temperature are neglected here. The lateral surfaces are also subjected to aerodynamic heating at a heat flux q_{ah} given by

$$q_{ah} = h_{cc}(T_r - T_p)$$

where h_{cc} is the convective heat transfer coefficient of the flowing air stream, T_r is the recovery temperature (temperature elevation due to deceleration of the air stream in the boundary layer), and T_p is the heat pipe temperature. Again, temperature drops through the heat pipe walls and along its length are neglected. A heat balance equating heat incident on the leading edge and lateral surfaces to heat radiated to the environment from the lateral surfaces (radiation from the leading edge is neglected) has the form

$$q_e A_e + h_{cc} A_c (T_r - T_p) = \varepsilon_s \sigma_r T_p^4 A_c$$

where ε_s is the surface emissivity and σ_r is the Stefan-Boltzmann radiation constant. With some manipulation, this equation may be recast into the form

$$\frac{A_c}{A_e} = \frac{q_e/(\varepsilon_s\, \sigma_r\, T_r^4)}{\left(\dfrac{T_p}{T_r}\right)^4 - \left(\dfrac{h_{cc}}{\varepsilon_s\, \sigma_r\, T_r^3}\right)\left(1 - \dfrac{T_p}{T_r}\right)}$$

When the denominator of the above equation is equal to zero, the lateral area needed to dissipate the heat incident on the leading edge becomes infinite. The temperature at which this occurs is called the radiation equilibrium temperature T_e, and represents the minimum possible heat pipe temperature. At T_e, only the aerodynamic heat incident on the lateral surfaces can be radiated away. When T_p is higher than T_e, both the leading edge heat load and that on the lateral surfaces can be dissipated radiatively.

The required radiator area decreases as T_p increases. When T_p exceeds the recovery temperature T_r, the lateral surfaces are subjected to aerodynamic cooling instead of aerodynamic heating.

The variation of A_c/A_e with T_p/T_r is shown as case C in Figure 6.2, with $q_e/(\varepsilon_s \sigma_r T_r^4) = 1$ and $h_{cc}/(\varepsilon_s \sigma_r T_r^3) = 1$.

As is evident from Figure 6.2, substantial reductions in heat pipe heat rejection area, and hence overall size, are associated with operation at the highest feasible heat pipe temperature.

6.3. HEAT PIPE INTERNAL DIMENSIONS

Once the heat pipe surface area and temperature have been determined from heat transfer considerations, the next step is to establish the internal heat pipe dimensions. These are determined from the requirement that the actual heat transport rate cannot exceed the heat pipe heat transport limits. The internal design of a heat pipe is most readily performed by starting with the vapor space, and then proceeding on to definition of the surrounding wick structure.

As was discussed in the introductory section of Chapter 5, there are two methods for using the various heat transport limits in the design process. In the first method, the heat transport limits are calculated, and the heat pipe design is modified as required to insure that all the limits are well above the actual heat transport rate.

In the second method, each heat transport limit (except for the sonic limit) is assumed to be the actual heat transport rate, and key design parameters corresponding to these limits are calculated. The sonic limit is assumed to be some multiple of the actual heat transport rate, such that relatively isothermal operation of the heat pipe can be expected. The design is then modified as required to insure that actual design parameters do not exceed those at which the heat transport limits are encountered.

As an example of the second method, consider a heat pipe designed to operate at 2 kW. Assuming this to be the capillary pumping limit, the diameter of curvature calculated from the capillary pumping limit equation might be 100 microns. Then, selection of a pore diameter of 50 microns for the capillary pumping layer of the wick will insure that the actual capillary pumping limit, although not known, is substantially larger than 2 kW. Hence, disruption of heat pipe operation due to loss of capillary pumping will not occur.

Although selection of either design method is a matter of individual choice, the second method is advantageous in the following respects:

1. By assuming that the heat transport limits (with the exception of the sonic limit) are equal to the actual heat transport rate, assumptions of constant vapor density used in deriving these limits are more likely to be valid than at some higher hypothetical heat transport rate corresponding to a heat transport limit.
2. The diameter of curvature corresponding to a specified capillary pumping limit can be calculated directly, whereas an iterative process is generally involved in calculating the capillary pumping limit corresponding to a specified diameter of curvature.

Method 2 is recommended for these reasons, and is generally used as the basis of the design procedures that are described here.

The following basic steps are involved in the calculation of internal heat pipe dimensions:

1. Establish dimensions of the vapor space from the sonic limit.
2. Select a wick type, and define wick characteristics.
3. Using the capillary pumping limit, determine the diameter of curvature D_c required for capillary pumping.
4. Using the entrainment limit, determine the maximum pore diameter D_{pe} at which entrainment can be avoided.
5. Using the boiling limit, determine the minimum diameter of curvature D_{cb} for which boiling can be avoided.
6. Select a wick pore diameter D_p for which all of the following conditions hold:

$$D_p < D_{pe}$$

$$D_p < D_c$$

$$D_{cb} < D_c \quad \text{or} \quad D_{cb} < D_p$$

7. If a pore diameter cannot be found for which all of the conditions in step 6 are satisfied, modify one or more design parameters and repeat the previous steps until the conditions of step 6 are met.

These steps will now be considered in more detail.

6.3.1. Vapor Space Dimensions

An initial criterion in designing vapor space dimensions is that the actual heat transport rate not exceed some specified percentage of the sonic limit. If this percentage is too high, the temperature drop through the heat pipe may be excessive. Also, the higher pressure drop associated with operation near the sonic limit could tax the capillary pumping capability of otherwise acceptable wick structures.

To avoid excessive temperature and pressure drops, the actual heat transport rate should generally not exceed 30–50% of the sonic limit. The lower figure is a nominal value that is suitable for initiating the internal heat pipe design. As the design process continues, other considerations may indicate the advisability of increasing or decreasing this percentage.

Using this criterion, the vapor space area A_v can then be determined from the equation

$$Q/A_v = 0.30\xi_s$$

or

$$A_v = Q/(0.30\xi_s) \tag{6.1}$$

where Q is the actual heat transport rate and ξ_s is the axial heat flux at the sonic limit (which is referred to here simply as the sonic limit). The sonic limit is calculated from Eq. (5.116).

If the vapor space area should be larger than desired, A_v can be reduced by allowing the axial heat flux to be a higher percentage of the sonic limit, at the expense of larger temperature and pressure drops through the heat pipe.

Alternatively, the vapor space area may be reduced by raising the operating temperature to increase the sonic limit. Since the sonic limit is a rapidly increasing function of temperature, only a modest temperature rise may be sufficient to satisfy vapor space area restrictions of a particular application.

6.3.2. Wick Types

If heat transport requirements are relatively modest, a *single-layer wick* fabricated from screen layers, fibers, or particles may suffice. However, these wick structures are inherently limited in heat transport capability because the same pore structure exists in the liquid flow channel and at the liquid-vapor interface. An increase in the coarseness of the pore structure to reduce the liquid pressure drop is accompanied by an unavoidable decrease in available capillary pressure.

A greater degree of control over the separate requirements of the liquid flow channel and capillary pumping is available with a *single-layer wick of the open groove type*. With this wick type, for example, the resistance to liquid flow can be reduced by increasing the groove depth, and hence the liquid flow area. The available capillary pressure, which is determined by the groove width, remains the same.

The *two-layer wick,* with independently selectable pore characteristics in the liquid flow channel and the capillary pumping layer, is the preferred choice when high heat transport capability is required. However, this wick type may be more sensitive to operational disruption from the presence of noncondensable gas than the single-layer wick. (See Section 3.5.2 for a discussion of the effect of noncondensable gas in heat pipes.)

A two-layer screen (or fiber or particle) wick, with relatively coarse pores in the liquid flow channel and relatively fine pores in the capillary pumping layer, is less sensitive to interference during startup or extended operation by any noncondensable gases that may be present. This type of wick may be the preferred choice in many situations.

In an *annular wick,* the internal pore structure of the liquid flow channel has been removed completely, and frictional resistance is at a minimum. A two-layer annular wick is one of the lightest, thinnest wick types. However, its sensitivity to operational disruption by noncondensable gas is fairly high, since a large fraction of the liquid flow passage could be blocked by a gas bubble.

The *covered grooved wick* is less sensitive to interruption by noncondensable gas than the annular wick, but will be thicker than an annular wick for the same liquid flow resistance. It is a good choice in situations in which a two-layer screened wick would be excessively thick and heavy.

With the *artery wick,* a thin distribution wick covers the inner surface of the heat pipe wall, and the liquid flow channel is confined to one or more arterial passages of low frictional resistance. The heat transport capability of the artery wick may be comparable to or greater than that of the annular wick. The artery wick is also subject to operational disruption by noncondensable gas. This susceptibility can be reduced by using multiple arteries.

The thin distribution wick lining the heat pipe wall is a major advantage when high surface heat fluxes are encountered. The thermal resistance to heat flow through the wick is then minimal, lessening the chance that the ΔT across the wick will exceed the superheat for boiling, even at high incident heat fluxes. However, at very high heat transport rates, the pressure drop of liquid flowing through the distribution wick could significantly reduce the capillary pumping capability of the artery wick.

If the heat input section covers only a small portion of the heat pipe length, a two-layer wick may achieve a resistance to boiling that is comparable to that of the artery wick. This would be accomplished by reducing the thickness of the liquid flow channel to a minimal value, or even zero, in the evaporator section. Flow would then be channeled through the capillary pumping layer in this region.

The foregoing considerations can be helpful in the preliminary selection of a wick type. In some cases, it may be advisable to choose two or more wick types during preliminary design studies in order to identify the preferred type for a particular application.

6.3.3. Wick Dimensions

Wick dimensions (liquid flow channel thickness and capillary pumping layer pore size) are determined from established relationships between these dimensions and the capillary pumping, entrainment, and boiling limits.

Wick Thickness and Pore Diameter for Capillary Pumping. The diameter of curvature D_c required for capillary pumping can be found from the capillary pumping equation, using the calculational procedure described in Section 5.1.2. The diameter of curvature is a directly varying function of liquid flow channel thickness. For minimum heat pipe size and weight, the smallest wick thickness should be used for which the associated diameter of curvature exceeds the smallest available wick pore size. The axial heat flux to be used is the value from which the vapor space dimensions were determined, as discussed in Section 6.3.1.

If the axial heat flux is a small fraction of the sonic limit, as may be the case for non–liquid metal heat pipes, then the simpler form of the capillary pumping equation given by Eq. (5.64) may be used. However, the validity of the assumptions on which this equation is based should be verified.

If the liquid flow channel is porous, its friction factor is needed to calculate the friction parameter ε_{po}. The friction factor depends on specific wick characteristics, such as fiber size, screen mesh, and porosity. Friction factor data for several screen mesh sizes are given in Table C.8 of Appendix C. If the liquid flow channel is open, ε_{po} is equal to unity.

For design purposes, the effect of axial liquid flow through the capillary pumping layer may conservatively be neglected in most cases. (The capillary pumping equation was derived on this basis.) Circumstances under which flow through the capillary pumping layer should not be neglected are discussed in Appendix F. The *thickness* of the capillary pumping layer should always be considered when calculating overall cross-sectional dimensions of the heat pipe, and when performing boiling limit calculations.

The pressure drop in the heat pipe vapor should be determined in the course of the capillary pumping calculation. The allowable vapor pressure drop is dependent on the permissible drop in the vapor temperature. If the vapor pressure drop is less that 3%, the vapor temperature drop is generally less than 10°F (6°C). (See Table 4.1.) In many cases, this is sufficiently close to the ideal of a completely isothermal vapor. Thus, an allowable vapor pressure drop of 3% is a reasonable design criterion. If the requirements of a particular application dictate a higher degree of isothermality in the vapor, the allowable vapor pressure drop should be reduced accordingly.

The calculation proceeds by selecting a thickness of the liquid flow channel and then calculating the corresponding value of the diameter of curvature D_c for capillary pumping and the percentage pressure drop in the vapor. If the vapor pressure drop should exceed 3%, the axial heat flux should be reduced, and the calculation repeated until the 3% vapor pressure drop criterion is satisfied.

Then, attention should be given to the calculated value of D_c. The calculations should be repeated with different thicknesses of the liquid flow channel in order to establish the relationship between D_c and the wick thickness. The smaller the wick thickness, the higher will be the liquid pressure drop and the smaller will be the diameter of curvature. In some cases, it may be desirable to establish the diameter of curvature/wick thickness relation for several wick types.

A lower limit on the wick thickness may be imposed by fabricability considerations. For example, if the liquid flow passage were to consist of 100 mesh screen, the minimum thickness would be the thickness of a single layer of screen (0.0088 in. = 2.24×10^{-4} m).

If the wick thickness is only available in finite increments (as is the case with screen layers), the actual wick thickness may have to be greater than that required by capillary pumping considerations. For example, if the minimum calculated thickness of a 100 mesh screen wick is 0.010 in. (2.54×10^{-4} m), two screen layers with a total thickness of $2 \times 0.0088 = 0.0176$ in. (4.47×10^{-4} m) will actually be required.

The capillary pumping layer of a two-layer wick is usually fabricated from one or more screen layers. Since the spacing between screen wires varies inversely with the screen mesh, the largest commercially available mesh should be considered for the capillary pumping layer when maximum capillary pumping capability is of primary concern. If cost is a principal design consideration, smaller mesh sizes may be preferable.

Because only a finite number of screen mesh sizes are available, the pore diameter of the capillary pumping layer may have to be appreciably smaller than the calculated diameter of curvature. For example, if the calculated D_c were 90 microns, 400 mesh screen with an effective pore diameter of about 60 microns would have to be used, since the pore diameter of the next smaller standard mesh size (200 mesh) is about 100 microns. Aside from such geometric limitations, the actual pore size should be well below the calculated diameter of curvature in any event, to account for uncertainties in such factors as wetting angle and friction factors.

There may be situations in which the required wick thickness is larger than can be accommodated by geometrical constraints of the application, or would result in a boiling limit that is too low. In that event, a different wick type with lower frictional resistance, or the development of a capillary pumping wick structure with pore sizes smaller than what is commercially available, could be considered.

Wick Pore Diameter for No Entrainment. Using the axial heat flux that was established in the previous step, the pore diameter D_{pe} above which entrainment of the wick liquid by the adjacent vapor will occur can be calculated from Eqs. (5.128) and (5.130) for the entrainment limit, with $D_{pe} \equiv \lambda_w$. In the absence of more definitive knowledge, the vapor flow can conservatively be assumed turbulent.

If the pore diameter D_p specified in the previous step is less than D_{pe}, entrainment will not occur. On the other hand, if $D_{pe} < D_p$, entrainment can be expected. In the latter event, the axial heat flux should be reduced and/or the operating temperature should be increased (to increase the vapor density) until $D_{pe} > D_p$.

Diameter of Curvature for No Boiling. From Eq. (5.144), the wick superheat for boiling $T_w - T_s$ corresponding to the maximum specified surface heat flux can be calculated. The diameter of curvature D_{cb} below which boiling will occur can then be found from Eqs. (5.138) and (5.143), letting $D_{cb} = 2r_c$. The boiling limit calculation requires that the nucleation site radius r_n and the interface heat transfer coefficient h_i be specified. Neither of these parameters is known with precision. For design purposes, a nominal value of 3 microns is recommended for r_n.

If the surface heat flux is on the order of several hundred Btu/ft^2-sec (or W/cm^2) or less, the influence of h_i on boiling conditions is small, and h_i can be neglected or assumed to be infinite. At surface heat fluxes on the order of 1000 Btu/ft^2-sec (or W/cm^2) or higher, the finite magnitude of h_i should be taken into account in the calculation of D_{cb}. Methods for estimating h_i, and for performing the boiling limit calculation when h_i is finite and significant, are described in Section 5.4.4.

Since the diameter of curvature cannot be less than the pore diameter, boiling will not occur if the calculated $D_{cb} < D_p$. Boiling can be expected if D_{cb} is greater than the actual diameter of curvature D_c required for capillary pumping.

If this is the case, the design should be modified until $D_{cb} < D_c$ (and preferably also $D_{cb} < D_p$ for greater design conservatism). This requires a thinner wick structure in the evaporator to reduce the wick superheat for boiling. Unless the evaporator section is just a small fraction of the heat pipe length, it may be necessary to increase the wick thickness in the remainder of the heat pipe to avoid a reduction in D_c as well. Alternately, a wick type that permits a thinner wick structure on the evaporator inner surface, such as an artery wick, could be considered.

6.3.4. Summary of Heat Pipe Internal Design Process

1. Select an axial heat flux that is well below the sonic limit and for which $\Delta P_v/P_v \leq 0.03$.
2. Select a wick type.
3. Select a thickness for the liquid flow channel, a thickness for the capillary pumping layer (if there is one), and a wick pore diameter D_p such that

$$D_p < D_{pe}$$

$$D_p < D_c$$

$$D_{cb} < D_c \qquad \text{or} \qquad D_{cb} < D_p$$

For a minimum-weight system, the axial heat flux should be as high as possible and the wick thickness as small as possible.

6.4. STRUCTURAL CONSIDERATIONS

Heat pipes may be subjected to structural loads of a type and magnitude that are dependent on the application under consideration. The heat pipe containment structure must be designed so that stresses resulting from these loads are compatible with material strength limitations, and so that excessive deformation or buckling resulting from structural instability does not occur. Stress in a heat pipe may develop from thermal or mechanical causes.

6.4.1. Thermal Stress

Thermal stress arises as the result of heat transfer through the containment wall, which produces a temperature difference across the wall thickness. (See Figure 6.3.) Full expansion of the hottest wall surface is constrained by adjacent cooler wall material. As a result, compressive stress develops in the circumferential direction, which is at a maximum at the hottest surface. Similarly, full contraction of the coolest surface is constrained by the adjacent hotter wall material. A circumferential tensile stress then results, which is at a maximum at the coolest surface.

In the heat pipe evaporator wall, where heat enters the heat pipe, the thermal stress is compressive at the outer (hottest) wall surface and tensile at the inner (coolest) surface. In the wall of the condenser section, where heat exits the heat pipe, the thermal stress is compressive at the inner (hottest) wall surface and tensile at the outer (coolest) surface.

Since the heat flux is usually (but not always) highest at the evaporator section, the wall temperature drop and thermal stress will also usually be greatest there. Therefore, only thermal stress at the heat input section will be considered in the succeeding analysis. Further, only the case of thin-walled containment (wall thickness small compared to the radius) will be considered, since this again is most characteristic of heat pipe structures with curvilinear geometry.

The maximum thermal stress s_{th} in a thin cylindrical or spherical wall occurs at the inner and outer radii in the tangential (circumferential) direction, and can be calculated from the equation (Ref. 6.1)

$$s_{th} = \frac{\alpha_{te}\, E\, \Delta T_{\text{wall}}}{2\,(1 - \upsilon)} \left[1 \pm \delta_g \frac{(r_o - r_i)}{3\, r_i} \right] \tag{6.2}$$

where α_{te} is the coefficient of thermal expansion at the mean wall temperature, E is the modulus of elasticity at the mean wall temperature, ΔT_{wall} is the wall temperature drop,

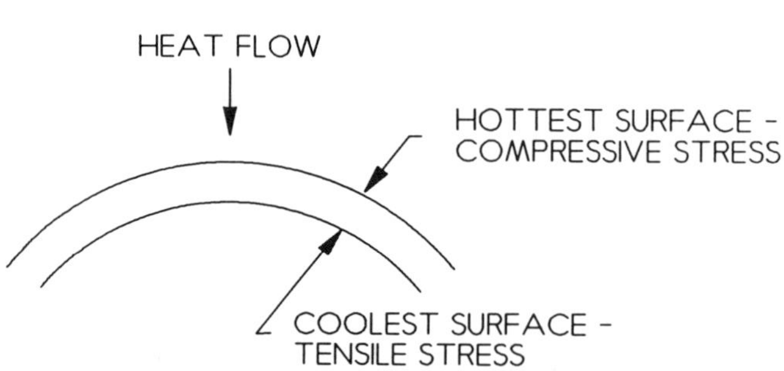

Figure 6.3. Thermal stress in a heat pipe containment wall.

υ is Poisson's ratio (with a value of 0.25–0.30 for most materials), r_o is the outer radius, r_i is the inner radius, and δ_g is a geometrical parameter. A representative value of 0.30 for υ is assumed here. The parameter δ_g is equal to 1 for a cylinder and 2 for a sphere. The one-dimensional wall temperature drop can be calculated for the various geometries using Eq. (5.140) and Table 5.7. In Eq. (5.140), the variables are reinterpreted to refer to the heat pipe wall instead of the wick.

The stress is compressive at the surface with the highest temperature, in which case the plus sign inside the brackets of Eq. (6.2) applies. At the surface with the lowest temperature, the stress is tensile, and the minus sign applies. If the ratio of wall thickness to radius is much less than 1, the maximum compressive and tensile stresses are equal, and also equal to the thermal stress in a flat plate of thickness equal to the wall thickness.

These results apply to the situation in which the cylinder is constrained at the ends so that its axial displacement is zero. Both tangential and axial stress components, of equal magnitude, exist at each surface and are calculated from Eq. (6.2). When the cylinder ends are unconstrained and free to expand axially, Ref 6.1 indicates that localized tangential stress at the free ends is 1.25 times the value predicted from Eq. (6.2).

Equation (6.2) can be used to estimate the thermal stress that develops in an evaporator or condenser wall of a heat pipe. If the surface heat flux is nonuniform, the one-dimensional wall temperature drop corresponding to the maximum heat flux can be conservatively used.

If the heat pipe is exposed to aerodynamic heating, as is the case for a heat pipe–cooled leading edge or nose tip, the thermal stress at the stagnation line or stagnation point can also be estimated from Eq. (6.2).

Equation (6.2) was derived on the assumption that thermal stresses lie in the elastic range. If the stress produced should be in excess of the yield strength, plastic flow and permanent deformation may occur. In general, heat pipes should be designed to limit stresses below the yield point in the elastic range. For extended operation at elevated temperature, most materials will undergo permanent deformation or creep. For such situations, stresses should not exceed levels corresponding to acceptable creep rates.

6.4.2. Mechanical Stress

In general, the heat pipe internal pressure (i.e., the vapor pressure) will differ from the pressure of the medium in which the heat pipe is located. The resulting pressure differential will produce mechanical stresses in the heat pipe walls. Such stresses will now be considered for cylindrical and flat plate geometries.

Cylindrical Geometry. The pressure differential across the wall of a typical thin-walled cylindrical heat pipe produces a fairly uniform circumferential stress, usually referred to as the hoop stress s_h. A force balance across the wall yields

$$s_h = \frac{r_m}{t_{\text{wall}}} \Delta P \tag{6.3}$$

where r_m is the mean wall radius, t_{wall} is the wall thickness, and ΔP is the pressure difference between the heat pipe interior and exterior. If the interior pressure is higher, a tensile hoop stress will be produced. If the external pressure is higher, the hoop stress will be compressive.

An axial stress will also be produced in the heat pipe wall as the result of the differential pressure across the end caps. The magnitude of this axial stress is one-half the circumferential stress given by Eq. (6.3).

An axial shear stress will be produced along the surface where the cylindrical wall and end cap of the heat pipe meet. The total force acting on an end cap is $\pi r_i^2 \Delta P$. This force acts over a shearing surface area equal to $2\pi r_i t_e$, where t_e is the end cap thickness. The shear stress is then

$$s_s = \frac{r_i \, \Delta P}{2t_e} \tag{6.4}$$

Since the end cap thickness is usually at least as great as the wall thickness, the shear stress is generally no more than half the hoop stress.

Flat Plate Geometry. If a heat pipe cooling structure is to extend over a relatively large area, it may be designed as a continuous shell rather than a series of individual contiguous cylindrical heat pipes. As shown in Figure 6.4, the heat pipe structure would then consist of an inner and an outer wall connected by uniformly spaced separating ribs. The ribs in effect divide the heat pipe structure into a series of individual heat pipe cells, each of which can function independently of the others. The ribs also provide an important structural function by limiting the wall stress that is produced by the pressure differential between the heat pipe interior and exterior.

The walls act in effect as a series of beams between fixed supports at the rib locations. The walls are then subjected to bending stress and deflection, which can be predicted from structural beam theory. The theory can be used to show that (Ref. 6.2)

$$s_b = \left(\frac{W_r}{t_{wall}}\right)^2 \frac{\Delta P}{2} \tag{6.5}$$

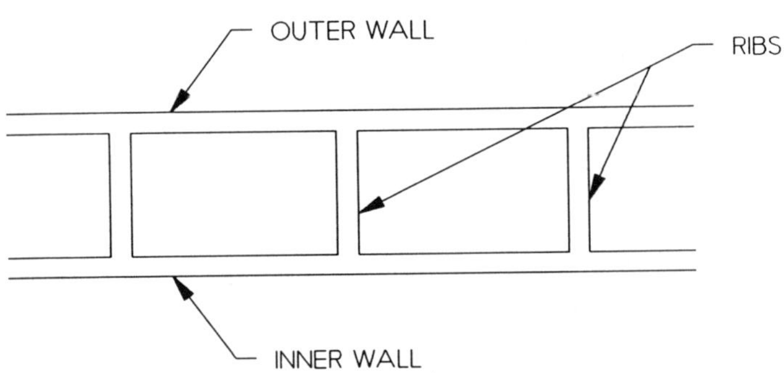

Figure 6.4. Heat pipe structural shell.

$$\frac{y_w}{W_r} = 0.0312\left(\frac{W_r}{t_{\text{wall}}}\right)^3 \frac{\Delta P}{E} \tag{6.6}$$

where W_r is the centerline distance between ribs, E is the modulus of elasticity, s_b is the maximum bending stress in the heat pipe walls, and y_w is the maximum wall deflection. The bending stress and deflection are at a maximum midway between the ribs, at the outer and inner wall surfaces. If the internal pressure is greater, the stress is tensile at the outer surface and compressive at the inner surface. If the external pressure is greater, the stresses at the outer and inner surfaces are reversed.

The ribs are subjected to tensile stress if the internal pressure is greater, and to compressive stress if the external pressure is greater. The magnitude of the rib stress s_r is equal to the total force acting on the ribs divided by the rib cross-sectional area, and is given by

$$s_r = \frac{W_r}{t_r}\Delta P - P_v \tag{6.7}$$

where t_r is the rib thickness and P_v is the vapor pressure. The second term is usually small compared to the first, and can frequently be neglected.

Depending on its type, thickness, and manner of construction, the wick structure may augment the strength/stiffness of an adjacent wall or rib. However, the extent of augmentation cannot be readily determined. Neglect of the wick contribution to wall strength/stiffness will yield conservative results for calculated stress (on the high side) and buckling loads (on the low side).

6.4.3. Buckling

Under certain conditions a structure under load may fail, not by tensile or compressive failure, but by sudden collapse into a highly deformed version of the original shape, a process that may or may not result in permanent distortion of the structure. Such failures develop when deflections caused by an imposed load become unstable and assume a new, usually unacceptable, shape. The onset of the unstable state is referred to as buckling.

For example, a relatively long rod under an axial compressive load may fail by buckling and bending, rather than by compression. Similarly, a thin spherical shell subjected to a radial load at a single point may fail by buckling and collapse of the wall, rather than by compression or shear.

Buckling of a heat pipe wall is of concern when the wall is subjected to large, relatively localized or axial forces. Such would be the case, for example, when heat pipe cooling is prescribed for the nose tip of a hypersonic vehicle that is subjected to high stagnation pressure. The hemispherical end of the heat pipe is then subject to buckling, as is the cylindrical main body of the heat pipe.

Nose Tip. Conservative criteria for designing a nose tip to prevent buckling and/or excessive deflection may be derived from Penning's study of clamped shallow shells

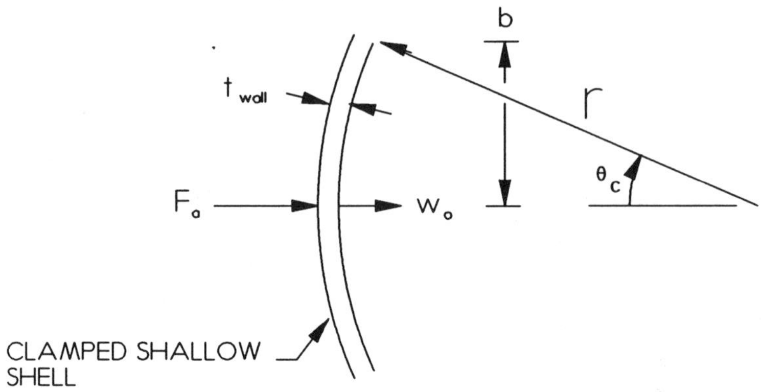

Figure 6.5. A clamped shallow shell subjected to buckling load.

under concentrated loads (Ref. 6.3). In Figure 6.5, a spherical shell of thickness t_{wall} and inner radius r is clamped at an angle θ_c from its axis. The axial force F_a produces the axial displacement w_o.

From Penning's data, it may be established that the axial displacement will be less than the shell thickness ($w_o/t_{wall} < 1$) and that buckling of the shell will not occur if

$$\frac{F_a r}{E\, t_{wall}^3} \leq 1$$

Hence, excessive deflection or buckling can be avoided if the imposed load F_a meets the requirement

$$F_a \leq \frac{E\, t_{wall}^3}{r} \tag{6.8}$$

Although Eq. (6.8) applies strictly to relatively shallow spherical shells, it is assumed to be approximately valid for a hemisphere ($\theta_c = 90$ deg).

The average pressure on a hemispherical nose tip at zero angle of attack is 0.351 times the stagnation pressure P_s, and occurs at an angle of 53.1 deg from the nose tip axis (Ref. 6.4). The total axial stagnation load on the hemisphere F_s is equal to the axial component of the average pressure times the nose tip area of $2\pi r^2$, or

$$F_s = 0.351(\cos 53.1\text{ deg})P_s(2\pi r^2) = 0.211 P_s r^2 \tag{6.9}$$

Assuming the distributed load to be approximately equivalent to a concentrated load at the stagnation point, the maximum allowable stagnation pressure may be obtained by equating F_s and F_a in the foregoing equations. Thus,

$$P_s = 4.74\, E\left(\frac{t_{wall}}{r}\right)^3 \tag{6.10}$$

If the actual stagnation pressure P_{sa} is comparable to or less than P_s, little or no structural deformation of the nose tip may be expected. If P_{sa} is appreciably larger than P_s, corrective action should be taken. If feasible, P_s should be increased, either by increasing the shell thickness or by reducing its radius.

Otherwise, stiffening ribs should be added to the internal wall surface. Rib requirements can be estimated as follows.

The rib plus wall thickness can be found by letting $P_s = P_{sa}$ in Eq. (6.10), solving for t_{wall}, and letting $t_{wall} \equiv t_{wall} + t_r$. Then

$$t_{wall} + t_r = 0.595\ r \left(\frac{P_{sa}}{E} \right)^{1/3} \tag{6.11}$$

The required rib thickness t_r is then $(t_{wall} + t_r) - t_{wall}$. Six or twelve stiffening ribs can be used, and the rib width can be taken equal to or less than the wall thickness.

To avoid interference with evaporative heat transfer from the heat pipe wick, the stiffening ribs should not extend all the way to the stagnation point. The unsupported distance b and associated angle θ_r (see Figure 6.5) can then be estimated by assuming that the unsupported load F_{su} can be approximated as

$$F_{su} \approx P_{sa}\,(\pi\, b^2) = F_a = \frac{E\, t_{wall}^3}{r}$$

On solving for b,

$$b = \left(\frac{E}{\pi\, P_{sa}\, r} \right)^{1/2} t_{wall}^{3/2} \tag{6.12}$$

From Figure 6.5, it then follows that

$$\theta_r = \sin^{-1}\,(b/r)$$

This calculational procedure is quite approximate, but is justifiable by the high degree of conservatism inherent in the buckling criterion of Eq. (6.8). Experimental validation of nose tip structural integrity would ultimately be required.

The shear stress s_s on the periphery of the unsupported nose tip wall is estimated by dividing the unsupported load $P_{sa}(\pi b^2)$ by the approximate shearing area $2\pi b t_{wall}$. Then

$$s_s = \left(\frac{b}{t_{wall}} \right) \frac{P_{sa}}{2} \tag{6.13}$$

Heat Pipe Wall. The total axial load F_s on the hemispherical nose tip is also transmitted to the cylindrical heat pipe section behind the nose tip. It is of interest to compare F_s to the axial load needed to buckle the heat pipe wall.

A hollow cylinder subjected to an axial load may buckle (collapse) in at least three ways (Ref. 6.5):

- Axisymmetric or nonsymmetric buckling (cylinder collapses accordion-style, with crinkles along the axis and around the periphery).
- Strut buckling (cylinder bends without crinkling).
- Wall buckling (wall bulges out symmetrically without crinkling).

The axial stress s_{ab} required for buckling in each of these modes is given by the following equations:

Axisymmetric or nonsymmetric buckling:

$$s_{ab} = \frac{E}{\sqrt{3}}\left(\frac{t_{\text{wall}}}{r_m}\right) \tag{6.14}$$

Strut buckling (hinged ends):

$$s_{ab} = \frac{\pi^2}{2} E \left(\frac{r_m}{L_{\text{cyl}}}\right)^2 \tag{6.15}$$

Wall buckling (hinged ends):

$$s_{ab} = \frac{\pi^2}{12} E \left(\frac{t_{\text{wall}}}{L_{\text{cyl}}}\right)^2 \tag{6.16}$$

where E is the modulus of elasticity, t_{wall} is the wall thickness, L_{cyl} is the cylinder length, and r_m is the mean wall radius.

The actual axial stress s_a is equal to the total axial load on the nose tip divided by the cross-sectional area of the cylindrical wall. Thus,

$$s_a = \frac{F_s}{2\,\pi\, r_m\, t_{\text{wall}}} \tag{6.17}$$

Since the axial stress required for axisymmetric or nonsymmetric buckling, and for strut buckling, is usually much greater than the actual axial stress, heat pipe buckling due to these modes is not likely. On the other hand, the axial stress for wall buckling may be substantially below the actual axial stress. If the heat pipe is surrounded by other structures, their additional stiffness will retard the outward bending of the heat pipe wall. The heat pipe wick will also add some stiffness to the heat pipe wall.

Nevertheless, a thicker wall, or the addition of uniformly spaced axial ribs to the wall, may be advisable. Ribs are generally preferable because of their lower weight and minimal impact on the wall ΔT. The required wall plus rib thickness may be estimated from Eq. (6.16) by letting $s_{ab} \approx 2s_a$ (the factor 2 is a safety factor), solving for t_{wall}, and assuming that t_{wall} is equal to the sum of the wall and rib thicknesses. The use of 12 uniformly spaced axial ribs is suggested, with the rib width equal to or less than the wall thickness.

A cylindrical heat pipe may also buckle under net external radial pressure. The radial pressure differential across the wall required for buckling is given by the equation (Ref. 6.2)

$$\Delta P_b = \frac{E}{4\,(1-\nu^2)}\left(\frac{t_{\text{wall}}}{r_m}\right)^3 \tag{6.18}$$

where ν is Poisson's ratio (usually 0.25–0.30).

6.5. ADDITIONAL DESIGN TOPICS

In this section, methods are presented for calculating the effective thermal conductivity of composite heat pipe wicks. The effect of temperature-dependent density changes in the heat pipe liquid on heat pipe design is also discussed.

6.5.1. Thermal Conductivity of the Heat Pipe Wick

Since the heat pipe wick is composed of both the solid wick material and the heat pipe liquid, the effective wick thermal conductivity will be an appropriately weighted average of the solid and liquid thermal conductivities. The proper weighting of the conductivities is strongly dependent on the wick geometry.

In the two simplest arrangements, the wick materials are arranged in series or parallel layers, as shown in Figure 6.6.

Each of these arrangements is a highly idealized representation of the heat flow path in actual wicks. Effective wick thermal conductivities obtained from the series and parallel models define lower and upper limits on the actual conductivity for any wick configuration (Ref. 6.6).

If the conductivities of the solid and liquid components of the wick are not too different, the lower and upper limits are fairly close to each other. On the other hand, if there is a large difference between the solid and liquid conductivities, the lower and upper limits may be so far apart as to provide little guidance in establishing the actual effective thermal conductivity of the wick.

Series Heat Flow. Let us first find the effective conductivity for the series arrangement of solid and liquid wick components shown in Figure 6.6. Heat flows at the same

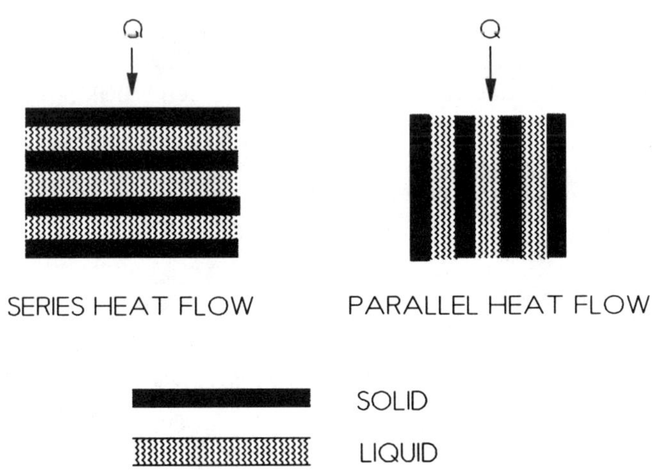

Figure 6.6. Series and parallel flow arrangements in wicks.

rate through each layer, and the total temperature drop across the wick thickness is the sum of the ΔTs in each layer. We may therefore write

$$q = k_s \frac{\Delta T_s}{t_s} = k_l \frac{\Delta T_l}{t_l} = k_{es} \frac{(\Delta T_s + \Delta T_l)}{(t_s + t_l)} \tag{6.19}$$

where q is the heat flux, k_s is the thermal conductivity of the solid component, k_l is the thermal conductivity of the liquid component, k_{es} is the effective thermal conductivity of the wick with series heat flow, ΔT_s is the cumulative temperature drop through the solid layers, ΔT_l is the cumulative temperature drop through the liquid layers, t_s is the cumulative thickness of the solid layers, and t_l is the cumulative thickness of the liquid layers.

Using Eq. (6.19), we may write

$$\Delta T_w = \Delta T_s + \Delta T_l = q \frac{t_s}{k_s} + q \frac{t_l}{k_l} = q \left(\frac{t_s + t_l}{k_{es}} \right)$$

where ΔT_w is the total temperature drop across the wick thickness.

Solving for $1/k_{es}$,

$$\frac{1}{k_{es}} = \left(\frac{1}{t_s + t_l} \right) \left(\frac{t_s}{k_s} + \frac{t_l}{k_l} \right)$$

Now, the liquid volume fraction in the wick $\phi = t_l/(t_s + t_l)$, and the solid fraction $1 - \phi = t_s/(t_s + t_l)$. Using these definitions, and solving the above equation for k_{es}, the effective thermal conductivity for series heat flow is

$$k_{es} = \frac{k_s \, k_l}{\phi \, k_s + (1 - \phi) \, k_l} \tag{6.20}$$

Letting $\kappa_{sl} = k_s/k_l$, Eq. (6.20) can also be written in the form

$$\frac{k_{es}}{k_l} = \frac{\kappa_{sl}}{1 - \phi \, (1 - \kappa_{sl})} \tag{6.21}$$

When $\kappa_{sl} >> 1$, as is the case for most heat pipes with non-liquid metal fluids, it may be shown from Eq. (6.21) that the effective thermal conductivity for series heat flow is equal to the liquid thermal conductivity divided by the liquid fraction.

Parallel Heat Flow. In the parallel arrangement of Figure 6.6, each component layer is equal to the wick thickness t_{wick}, and ΔT_w is the same for the layers and the wick. The heat flux for the solid and liquid layers is

$$q_s = k_s \frac{\Delta T_w}{t_{\text{wick}}} \qquad q_l = k_l \frac{\Delta T_w}{t_{\text{wick}}}$$

Assuming the wick to have unit length, the total heat flow rate Q is given by

$$Q = q_s \, t_s + q_l \, t_l = k_s \frac{\Delta T_w}{t_{\text{wick}}} + k_l \frac{\Delta T_w}{t_{\text{wick}}} = k_{ep} \frac{\Delta T_w}{t_{\text{wick}}} (t_s + t_l)$$

Solving for the effective thermal conductivity k_{ep} in parallel heat flow,

$$k_{ep} = \frac{k_s t_s + k_l t_l}{t_s + t_l}$$

Using the above definitions for ϕ and $\kappa_{sl,}$ the effective thermal conductivity for parallel heat flow can be written in the form

$$\frac{k_{ep}}{k_l} = \phi + (1 - \phi)\,\kappa_{sl} \qquad (6.22)$$

Evaluation of k_{es} and k_{ep}. Values of k_{es} and k_{ep} at limiting values of ϕ and κ_{sl} are given in Table 6.1.

The first two entries in Table 6.1 verify that the effective wick thermal conductivity is k_s when the liquid fraction is 0, and k_l when the liquid fraction is 1. The next entry shows that, when the solid conductivity is much smaller than the liquid conductivity, the effective conductivity approaches 0 for series heat flow and ϕk_l for parallel heat flow. The final entry shows that, when the solid conductivity is much larger than the liquid conductivity, the effective conductivity approaches k_l/ϕ for series heat flow and $(1 - \phi)k_s$ for parallel heat flow.

In Figure 6.7, the ratio of effective thermal conductivities for parallel and series flow is shown as a function of $\kappa_{sl} = k_s/k_l$ and ϕ. The ratio is seen to be symmetrical about $\kappa_{sl} = 1$.

When $0.5 \leq \kappa_{sl} \leq 2$, $k_{ep}/k_{es} \leq 1.125$. Therefore, for such heat pipes, letting $k_e = (k_{ep} + k_{es})/2$ will give the effective wick thermal conductivity with an error of no more than 6%, regardless of the actual wick configuration. This is a valid approach for evaluating the thermal conductivity of liquid metal heat pipe wicks such as sodium superalloy wicks ($\kappa_{sl} = 0.44$ at 1500°F [816°C]) and lithium-molybdenum wicks ($\kappa_{sl} = 1.94$ at 2200°F [1204°C]).

For larger or smaller values of $\kappa_{sl,}$ the upper limit on effective thermal conductivity k_{ep} exceeds the lower limit k_{es} by a large factor, and simple averaging of these limits is no longer adequate. Most of the non–liquid metal heat pipes fall into this category, such as water-copper ($\kappa_{sl} = 667$ at 68°F [20°C]) and alcohol-copper ($\kappa_{sl} = 1882$ at 68°F [20°C]) heat pipes. For these situations, thermal conductivity models are required that resemble actual wick geometries more accurately.

Table 6.1. Effective thermal conductivity for series and parallel heat flow at limiting values of ϕ and κ

Limiting value	k_{es}	k_{ep}
$\phi = 0$	k_s	k_s
$\phi = 1$	k_l	k_l
$\kappa_{sl} = 0$	0	ϕk_l
$\kappa_{sl} = \infty$	k_l/ϕ	$(1 - \phi)k_s$

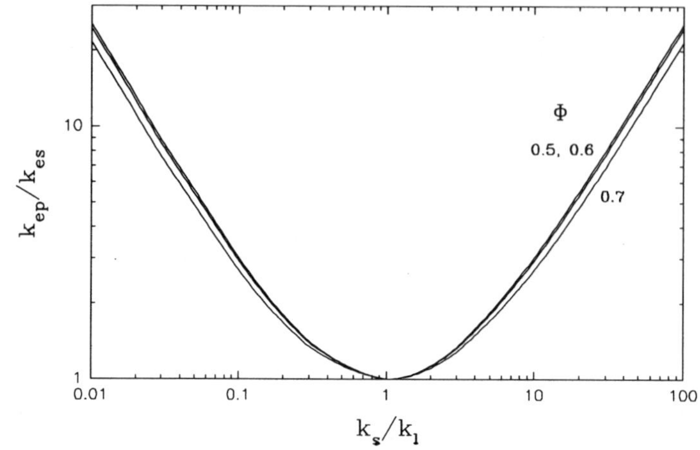

Figure 6.7. Ratio of effective thermal conductivities for parallel and series wick configurations.

Screened Wicks. In Refs. 6.6 and 6.7, an equation originally developed by Rayleigh is presented for the effective thermal conductivity k_e of a square array of cylinders dispersed in a fluid medium whose fluid volume fraction is relatively large. This equation is expressible in the form

$$\frac{k_e}{k_l} = \frac{2\,\kappa_{sl} + (1 - \kappa_{sl})\,\phi}{2 - (1 - \kappa_{sl})\,\phi} \tag{6.23}$$

The Rayleigh equation has been used to predict the effective thermal conductivity of screen-type wicks for which the screen layers are in mechanical contact with some accuracy.

With some manipulation and the use of Eqs. (6.21) and (6.22), Eq. (6.23) can also be expressed in terms of the upper and lower limits on the effective thermal conductivity as

$$\frac{k_e}{k_l} = \left(\frac{k_s + k_{ep}}{k_s + k_{es}}\right)\left(\frac{k_{es}}{k_l}\right) \tag{6.24}$$

For large values of κ_{sl}, Eqs. (6.21) and (6.22) can be used to show that k_e/k_l approaches $(2 - \phi)(k_{es}/k_l)$. The effective thermal conductivity is then at best twice the lower limit (i.e., that for series heat flow).

In Ref. 6.7, effective thermal conductivity data are given for 100 mesh stainless steel and copper screens whose pores were filled with various liquids. In Table 6.2, these data are used to present measured and calculated values of k_e/k_{es} as a function of κ_{sl}. The measured liquid volume fraction was 0.70 for the stainless steel screens and 0.69 for the copper screens.

Note that in all cases κ_{sl} is considerably greater than 1, and the measured k_e/k_{es} is less than 2. As is indicated in Table 6.2, two different equations were used to calculate k_e/k_{es}.

Table 6.2. Effective thermal conductivity ratios for stainless steel and copper screens

Screen	Liquid	$\kappa_{sl} = k_s/k_l$	k_e/k_{es} (measured)	k_e/k_{es}*	Deviation, %*	k_e/k_{es}**	Deviation, %**
Stainless	H_2O	25.20	1.285	1.341	+4.37	1.257	−2.19
	CH_3OH	71.09	1.329	1.348	+1.43	1.286	−3.27
	CCl_3F	158.65	1.404	1.362	−2.99	1.293	−7.93
Copper	H_2O	667	1.744	1.441	−17.37	1.309	−24.97
	CH_3OH	1882	1.693	1.631	−3.66	1.309	−22.70
	CCl_3F	4200	1.964	1.993	+1.48	1.308	−33.42

Note: *, From Eq. (6.25); **, from Eq. (6.24) (Rayleigh equation).

Equation (6.25),

$$k_e/k_{es} = 1.3372 + 0.0001561\kappa_{sl} \tag{6.25}$$

is based on correlation of the experimental data in Table 6.2. Equation (6.24) is the previously discussed form of the Rayleigh equation.

Equation (6.25) correlates the measured thermal conductivity within 5%, with the exception of data for the water-filled copper screen. The sharp deviation from this measurement compared with the other measurements suggests that the measured k_e for the copper-water screen may have been in error.

Except for the case when $\kappa_{sl} = 25.20$, the deviation from measured data is considerably greater for the Rayleigh equation than for Eq. (6.25). An equation derived for screens by Chang (Ref. 6.7) also provides better correlation of the measured data than the Rayleigh equation, but the correlation is not as good as with Eq. (6.25).

It is suggested that the Rayleigh equation be used to calculate the effective thermal conductivity of screen wicks for $\kappa_{sl} \leq 25$, and that Eq. (6.25) be used for $\kappa_{sl} > 25$. The correlations apply strictly for 100 mesh screen and a liquid volume fraction of 0.69–0.70, but may be used to approximate the effective thermal conductivity for other screen meshes and liquid volume fractions, in lieu of other data.

It is not clear from Ref. 6.7 whether the data presented in Table 6.2 are for single or multilayered screens or, if for multiple screens, whether the screen layers are in mechanical contact or have been sintered together.

If multiple layers of screens are sintered together, continuous flow paths will exist from layer to layer, enhancing the contribution of parallel heat flow through the solid component to the total heat transfer rate. If the screen layers are simply in mechanical contact, high thermal resistance at the contact points will greatly inhibit the contribution of parallel heat flow through the solid component. The fact that the measured values are within a factor of two of the lower limit shows that heat transfer through the screen is primarily by series heat flow, an indication that the contact is probably mechanical.

Therefore, the effective thermal conductivity of sintered screen wicks may be significantly higher than the values given in Table 6.2, and more representative of that for sintered fiber metal wicks (see the next section). The use of Table 6.2 data and the associated correlations would then give conservatively low values of the effective wick thermal conductivity.

Sintered Fiber Metal Wicks. In Ref. 6.8, the effective thermal conductivity of nickel-water and copper-water sintered fiber was measured in the direction normal to the plane in which the fibers lay, at an average temperature of 55°F (13°C). The average fiber diameter was 16 microns for the nickel fibers and 43 microns for the copper fibers. The limited data (three points for nickel-water and two points for copper-water) are well correlated by the equations

For nickel-water:

$$\frac{k_e}{k_l} = 3.204\ \phi^{-3.985} \qquad 0.664 \le \phi \le 0.893 \tag{6.26}$$

For copper-water:

$$\frac{k_e}{k_l} = 183.1(1 - \phi) \qquad 0.688 \le \phi \le 0.818 \tag{6.27}$$

For $\phi = 0.70$, k_e/k_l for a copper-water sintered fiber wick is, from Eq. (6.27), 54.93. Using Eq. (6.21) for the effective thermal conductivity of copper-water in series heat flow, we find that $k_{es}/k_l = 1.682$. Therefore,

$$\frac{k_e}{k_{es}} = \frac{k_e/k_l}{k_{es}/k_l} = \frac{54.93}{1.682} = 32.66$$

This figure may be compared with a value for k_e/k_{es} of 1.44 for a copper-water screen wick, as obtained from Eq. (6.24).

The effective thermal conductivity for the copper sintered fiber wick is thus almost 23 times greater than that for a copper screen wick. The higher conductivity of the fiber wick may be attributed to the continuous (although devious) conduction path through the wick thickness via the sintered zones connecting adjacent fibers. The creation of similar conduction paths in a screen wick via sintering can be expected to improve significantly the thermal conductivity of screen wicks.

A more involved method for estimating the thermal conductivity of sintered materials is described in Ref. 6.6.

Grooved Wicks. Grooved wicks, while simple in concept, are more complex than other wick types. The reason is that, with screen, fiber, or particle wicks, the wick structure may be regarded as a relatively homogeneous mixture of solid and liquid components of fixed geometry. Heat flow at a constant heat flux across these wick structures may be regarded as one-dimensional (at least on a macroscopic scale), with the heat pipe liquid vaporizing at a constant rate from the liquid-vapor interface in the evaporator.

With an open grooved wick (i.e., one in which the grooves are not covered by an additional wick layer), there is a high degree of inhomogeneity between the liquid in the groove and the adjacent solid "land." This inhomogeneity is due in part to the extension of alternate liquid and solid legs with different thermal conductivities over the entire wick thickness. Additionally, there is some question as to whether the inner surface of the land (the solid leg) remains wetted during heat pipe operation.

If the rate of evaporation from the wetted land surface exceeds the rate at which the liquid film is supplied from liquid in the grooves, the land surface will dry out. Heat transfer from the land inner surface will then be by convection into the adjacent heat pipe vapor, at a much lower rate than by evaporative heat transfer. As a result, the land inner surface becomes, for practical purposes, thermally insulated.

The presence of these inhomogeneous effects causes the flow of heat through the wick to be two-dimensional, with heat incident on an individual groove-land pair flowing laterally as well as normal to the externally imposed heat flux. More heat will flow along the paths of lower thermal resistance (higher conductivity) within groove-land pairs.

The situation is further complicated in an open grooved wick because the radius of curvature of the meniscus (the curved liquid-vapor interface in the grooves) is not constant, but decreases as the heat load increases, causing the meniscus to recede into the groove. The extent of meniscus recession varies along the heat pipe length, and is generally at a maximum at the beginning of the evaporator section.

Under these circumstances, precise determination of the effective thermal conductivity requires a two-dimensional calculation of heat flow through the heat pipe wall, the liquid-filled grooves, and the adjacent land. A two-dimensional analysis for circumferential grooves is described in Ref. 6.9. The thermal resistance of the liquid legs varies with the extent to which the meniscus recedes into the grooves, which in turn depends on the heat load and the axial position along the heat pipe. The effective thermal conductivity will then also vary with the heat load and axial position.

For design purposes, and particularly in the preliminary design phase, a simpler, more approximate approach is preferable, although the options in this regard are limited by the complexity of the problem. A semiempirical approach is proposed here, based on experimental heat transfer data for a specific open grooved wick configuration.

Overall effective thermal conductivity of wall and grooved wick. Consider the typical grooved wick configuration shown in Figure 6.8.

Heat is added at the rate Q to the evaporator section of a cylindrical heat pipe of outer radius r_o and vapor space radius r_v. The width of the liquid-filled grooves is w_g, and their depth is d_1. The total (wall plus wick) thickness is l. The grooves are separated by solid land of width w_l. The mean temperature at the outer evaporator surface is T_{om}, and the vapor temperature (assumed equal to the liquid-vapor interface temperature) is T_v. The temperature on the inner surface of the land is undetermined.

If the land surface is dry (i.e., there is no liquid film present), its temperature may be substantially higher than T_v, and virtually all of the heat flowing into the land leg will enter the liquid-filled groove and vaporize liquid at the meniscus. This heat flow pattern is shown schematically in Figure 6.8 for one groove.

If the land surface should remain covered with a liquid film, heat entering the land leg will then flow primarily to the land surface, vaporizing the liquid film on that surface instead of at the meniscus.

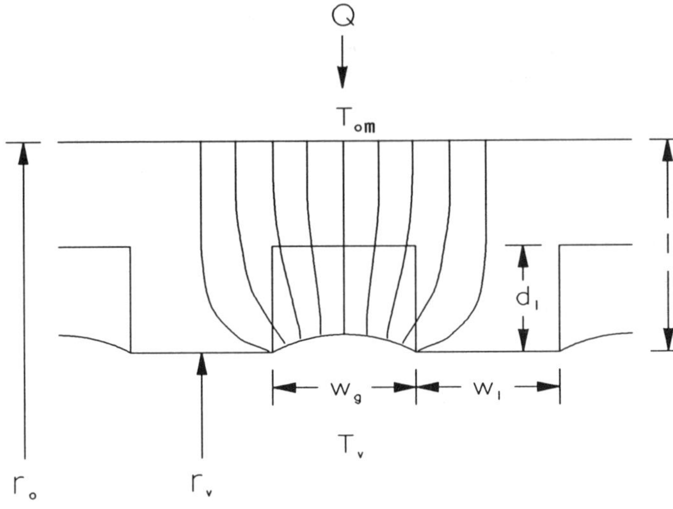

Figure 6.8. Configuration of a heat pipe open grooved wick.

From the heat transfer standpoint, the actual grooved configuration of Figure 6.8 may be approximated as shown in Figure 6.9.

Section 1 includes the heat pipe wall and the liquid-filled groove. Section 2 includes the heat pipe wall and its extension into the land. A small portion of the land is assumed to consist of the heat pipe liquid, and is included so that the increased thermal resistance when heat flows from the land through a portion of the liquid in the groove to the meniscus can be taken into account. The thickness d_p of the "liquid land" is to be established from experimental data.

Heat is assumed to flow in parallel through sections 1 and 2. Within each section, the heat flows in series through the solid and liquid zones.

From Figure 6.9, the liquid volume fraction in each section is

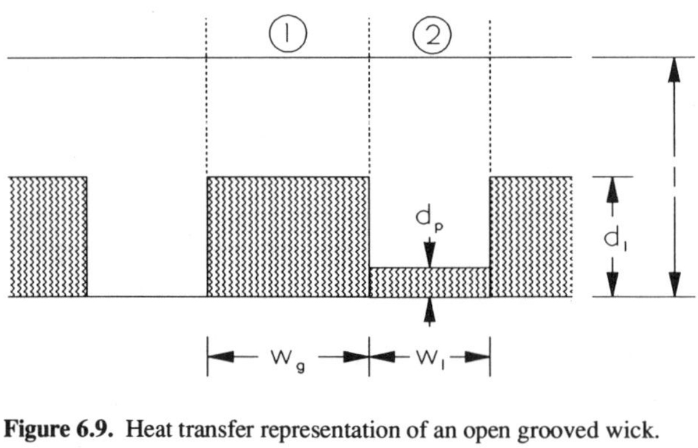

Figure 6.9. Heat transfer representation of an open grooved wick.

$$\phi_{l1} = d_1/l \qquad \phi_{l2} = d_p/l$$

and the fraction of the total volume occupied by section 1 is

$$\phi_{1v} = \frac{w_g}{w_g + w_l}$$

From Eq. (6.21) for series flow, the following relationships may be written for the effective thermal conductivity in each section:

$$\frac{k_{e1}}{k_l} = \frac{\kappa_{sl}}{1 - \phi_{l1}\,(1 - \kappa_{sl})} \tag{6.28}$$

$$\frac{k_{e2}}{k_l} = \frac{\kappa_{sl}}{1 - \phi_{l2}\,(1 - \kappa_{sl})} \tag{6.29}$$

Using Eq. (6.22) for parallel flow, the overall effective thermal conductivity k_{eo} of sections 1 and 2 is then given by

$$\frac{k_{eo}}{k_l} = \phi_{1v}\left(\frac{k_{e1}}{k_l}\right) + (1 - \phi_{1v})\left(\frac{k_{e2}}{k_l}\right) \tag{6.30}$$

In terms of k_{eo}, the heat transfer rate Q into a cylindrical grooved heat pipe with evaporator length L_e is given by

$$Q = \frac{(T_{om} - T_v)\,2\,\pi\,k_{eo}\,L_e}{\ln\,(r_o/r_v)}$$

The heat flux q_v, based on the inner surface area, is then

$$q_v = \frac{Q}{2\,\pi\,r_v\,L_e} = \frac{(T_{om} - T_v)\,k_{eo}}{r_v\,\ln\,(r_o/r_v)} \tag{6.31}$$

The ratio of this heat flux to the temperature drop across the wall-wick has been referred to in the literature as a film coefficient, in analogy to the convective heat transfer coefficient. A more appropriate term, the unit thermal conductance U, will be used here.

From Eq. (6.31), the unit thermal conductance for the wall-wick combination is then

$$U = \frac{q_v}{T_{om} - T_v} = \frac{k_{eo}}{r_v\,\ln\,(r_o/r_v)} \tag{6.32}$$

The unit thermal conductance has been measured for specific wall-wick geometries and solid-liquid combinations. The effective overall thermal conductivity for such cases can then be determined by solving Eq. (6.32) for k_{eo}. Thus,

$$k_{eo} = U\,r_v\,\ln\,(r_o/r_v) \tag{6.33}$$

Effective thermal conductivity of open grooved wick. Now, an expression for the effective thermal conductivity of the wick (grooved) section will be derived.

Equation (6.28) can be used to calculate k_{e1}/k_l. From Eq. (6.30), k_{e2}/k_l can then be determined in the form

$$\frac{k_{e2}}{k_l} = \frac{(k_{eo}/k_l) - (\phi_{1v}\, k_{e1}/k_l)}{1 - \phi_{1v}} \tag{6.34}$$

From Eq. (6.29), the liquid volume fraction in section 2 is calculable as

$$\phi_{l2} = \frac{1 - [\kappa_{sl}/(k_{e2}/k_l)]}{1 - \kappa_{sl}} \tag{6.35}$$

It follows from Figure 6.9 that $d_p = l\phi_{l2}$.

Once d_p has been determined, the effective thermal conductivity k_{ew} for the wick section alone can be estimated by letting $\phi_{l2w} = d_p/d_1$.

Then, for section 2 of the wick, using Eq. (6.29) with the subscript w added to denote the wick, the effective thermal conductivity is

$$\frac{k_{e2w}}{k_l} = \frac{\kappa_{sl}}{1 - \phi_{l2w}\,(1 - \kappa_{sl})} \tag{6.36}$$

Here, $\phi_{l2w} = d_p/d_1$.

For section 1 of the wick, which is all liquid, the effective thermal conductivity is found from the equation

$$k_{e1w}/k_l = 1 \tag{6.37}$$

The effective thermal conductivity of the wick is then obtained from Eq. (6.30), with the subscript w added to denote the wick.

$$\frac{k_{ew}}{k_l} = \phi\left(\frac{k_{e1w}}{k_l}\right) + (1 - \phi)\left(\frac{k_{e2w}}{k_l}\right) \tag{6.38}$$

Here, ϕ denotes the volume fraction of the wick that is occupied by the liquid in the grooves.

Using Eq. (6.32), replacing $T_{om} - T_v$ by the wick temperature drop ΔT_w, and letting $r_o = r_v + d_1$, the wick temperature drop is expressible as

$$\Delta T_w = q_v \frac{r_v}{k_{ew}} \ln\left(1 + \frac{d_1}{r_v}\right) \tag{6.39}$$

The wick temperature drop ΔT_w can then be compared with that at incipient boiling to evaluate the likelihood of boiling in an open grooved wick.

Representative data for open grooved wicks. Representative data from which the effective thermal conductivity can be calculated for high-κ_{sl} heat pipes with open rectangular grooved wicks are given in Tables 6.3A and 6.3B. The data apply for the evaporator section of aluminum heat pipes designed for use in the ATS-F satellite program (Ref. 6.10).

Using the data of Tables 6.3A and 6.3B, along with the calculational procedure described earlier, the overall (wall + wick) effective thermal conductivity k_{eo} and the wick effective thermal conductivity k_{ew} were calculated. Results are presented in Table 6.4.

It must be emphasized that the data on effective thermal conductivities apply only for the specific heat pipe geometry and materials for which heat transfer data were

Table 6.3A. Heat pipe dimensions, in. (m)

Outer radius	0.3071 (0.00780)
Inner (vapor space) radius	0.1835 (0.00466)
Groove depth	0.039 (0.00099)
Groove width	0.026 (0.00066)
Land width (between grooves)	0.017 (0.00043)

Table 6.3B. Data for aluminum heat pipes with rectangular, open-grooved wicks

Fluid	Temperature, °F (°C)	k_s/k_l	Unit thermal conductance, Btu/hr-ft^2-°F (W/m^2-°K)	k_l, Btu/hr-ft-°F (W/m-°K)
Ammonia	99 (37)	407	1000 (5678)	0.29 (0.502)
Methane	–189 (–123)	1137	2240 (12,720)	0.102 (0.177)
Nitrogen	–315 (–193)	1782	555 (3151)	0.078 (0.135)

obtained. The data may also be useful for estimating the thermal conductivity of geometrically similar wall-wick configurations (i.e., configurations whose dimensions are a constant multiple of those for which the data were obtained).

For situations in which $0.5 < \kappa_{sl} < 2$, a condition that is representative of liquid metal heat pipes, it has been shown that the upper and lower limits on the effective wick thermal conductivity differ by 12% at most. Therefore, the effective wick thermal conductivity for a liquid metal heat pipe with a grooved wick may be taken as the average of the upper and lower limits with an error of no more than 6%.

Effective Thermal Conductivity of Wall with Grooved Wick. As is evident from Figure 6.8, the flow of heat through the outer wall into the wick is two-dimensional,

Table 6.4. Thermal conductivity for open grooved wicks

Fluid	k_{eo}, Btu/hr-ft-°F (W/m-°K)	k_{ew}, Btu/hr-ft-°F (W/m-°K)
Ammonia	7.88 (13.64)	2.77 (4.79)
Methane	1.891 (3.27)	0.610 (1.056)
Nitrogen	0.433 (0.749)	0.1369 (0.237)

even if the surface heat flux is constant. A larger fraction of the incident heat flows into the higher-thermal-conductivity leg of the grooved wick layer. The flow path is then longer for the heat entering the wall that is farthest from the higher-conductivity leg, and the effective wall thermal conductivity is then lower than that of the wall material.

It is suggested that the effective wall thermal conductivity for a grooved wick $k_{e\text{wall}}$ can be conservatively approximated with the equation

$$k_{e\text{wall}} = k_s \left[\frac{t_{\text{wall}}}{t_{\text{wall}} + \dfrac{w_g}{4}(1 - k_l/k_s)} \right] \tag{6.40}$$

This equation satisfies limiting conditions on the relative magnitudes of the liquid and solid thermal conductivities. When $k_l = k_s$, heat flow across the wick thickness is one-dimensional, and so $k_{e\text{wall}} = k_s$. When $k_l << k_s$, essentially no heat flows through the liquid leg, and two-dimensional effects are maximized. The mean heat flow path through the wall is then approximately the mean of the flow paths t_{wall} and $t_{\text{wall}} + w_g/2$ at the edges of a representative wick cell. It then follows that

$$k_{e\text{wall}} = k_s \left(\frac{t_{\text{wall}}}{t_{\text{wall}} + w_g/4} \right)$$

Covered Grooved Wicks. A grooved wick may be covered with another wick layer, such as one or more layers of screen, as shown in Figure 6.10. The second wick layer would be added in situations in which a greater capillary pumping capability is desired than is available with the grooved wick alone.

With a covered grooved wick, the entire groove volume, as well as the pores of the second wick layer, will be filled with the heat pipe liquid. The liquid-vapor interface will then lie within the pores of the second wick layer. Heat flow into the land leg of the grooved wick can then move directly to the liquid-vapor interface through the second wick layer.

Heat flow is then approximately one-dimensional. The effective thermal conductivity k_{l1} for layer 1 can be found from Eq. (6.22) for parallel flow. The effective thermal conductivity k_{l2} for layer 2 corresponds to that of screens or sintered fibers. The overall

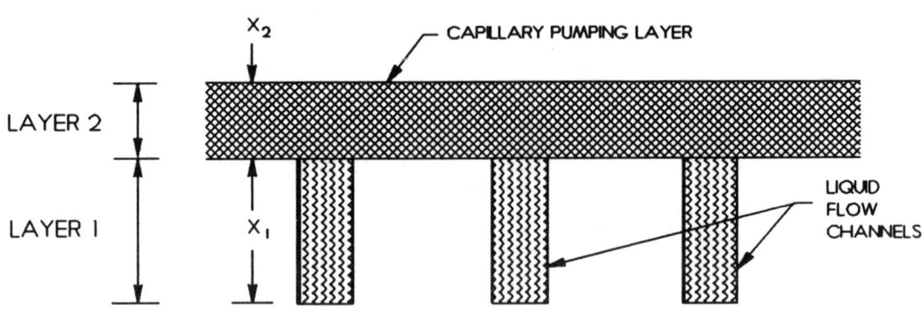

Figure 6.10. Covered grooved (two-layer) wick.

wick thermal conductivity k_{e2l} can then be found from Eq. (6.41) or (6.42) for two-layer wicks (see the next section).

Effective Thermal Conductivity of Two-Layer Wicks. The solution already obtained for the effective thermal conductivity for series heat flow through a liquid-solid wick is also applicable to series heat flow through a two-layer wick. For use with a two-layer wick, it is convenient to modify the terms in Eq. (6.21) as follows:

$$\kappa_{21} = \frac{k_2}{k_1} \qquad \phi = \frac{x_1}{x_1 + x_2} \qquad 1 - \phi = \frac{x_2}{x_1 + x_2}$$

where x is the layer thickness, subscript 1 refers to the first layer, and subscript 2 refers to the second layer.

We may then write

$$\frac{k_{e2l}}{k_1} = \frac{\kappa_{21}}{1 - \phi\,(1 - \kappa_{21})} \tag{6.41}$$

where k_{e2l} is the effective thermal conductivity of the two-layer wick.

The above equation was derived for rectangular geometry. For cylindrical geometry, the same procedure used to derive Eqs. (6.21) and (6.41) may be used to show that

$$\frac{k_{e2l}}{k_1} = \frac{\kappa_{21} \ln\,(1 + \gamma_1 + \gamma_2)}{\kappa_{21} \ln\,(1 + \gamma_1 + \gamma_2) - (\kappa_{21} - 1)\ln\,(1 + \gamma_2)} \tag{6.42}$$

where $\gamma_1 = x_1/r$, $\gamma_2 = x_2/r$, and r is the inner wick radius (also the radius of the heat pipe vapor space).

When γ_1 and γ_2 are both $<<1$, Eq. (6.42) reduces to Eq. (6.41) for rectangular geometry, as would be expected. When γ_1 and $\gamma_2 \le 0.1$ and $0.5 \le \kappa_{21} \le 2$, the effective thermal conductivity may be calculated from Eq. (6.41) with little error (<2%). If $\kappa_{21} <<$ 1 or >> 1, Eq. (6.42) for cylindrical geometry should be used for best accuracy.

6.5.2. Effect of Liquid Density Changes

Since the density of heat pipe liquids usually decreases as the temperature rises, the volume occupied by the liquids increases with temperature. If there is a large difference between the heat pipe operating temperature and its nonoperational temperature (usually ambient temperature), the liquid density and volume changes can be significant. For example, when lithium is heated from its melting point of 367°F (186°C) to 2200°F (1204°C), its density decreases by 19.7% (Refs. 6.11 and 6.12).

If the lithium charge were such that all the wick pores were filled with lithium at its melting point, then 24.6% of the lithium liquid would have spilled out of the wick by the time an operating temperature of 2200°F (1204°C) had been reached. It is evident that the consequences of liquid density variations with temperature must be taken into account in the heat pipe design.

One possibility would simply be to allow the spillage into the vapor space, with the liquid accumulating as a slug (in zero gravity), a puddle (in an acceleration field), or as a liquid film along a portion of the heat pipe length. If the volume of excess liquid were small compared to the vapor space volume, its presence might be tolerable. This might be particularly true if vapor drag forces and/or gravity/acceleration acted to confine the excess liquid to a small volume at the condenser end of the heat pipe. If the excess liquid were to accumulate in the evaporator section, as might occur during the deceleration of a reentry vehicle, increased thermal resistance would reduce the boiling limit, increasing the probability of boiling.

Another possibility would be to provide a separate storage reservoir (probably a porous body) at some point within the vapor space or external to the vapor space but connected to it. Vapor drag or gravity/acceleration force would be relied on to move the excess liquid to the reservoir for collection. If the porous reservoir structure were physically connected to the wick, the heat pipe liquid would expand into the reservoir automatically.

A third approach would be to accommodate the change in liquid volume with temperature in the wick structure. With this approach, the liquid charge would be such that the wick pores would be full at the design operating temperature. At lower temperatures, the liquid-vapor interface would recede into the wick pores, as indicated in Figure 6.11.

With a two-layer wick, the liquid-vapor interface should always lie within the capillary pumping layer to insure that the capillary pumping capability of this layer, rather than that of the liquid flow channel, is available over the entire operating temperature range. This condition can be assured by using a capillary pumping layer of sufficient thickness. The required minimum thickness can be determined by reference to Figure 6.11.

Let the liquid density at the operating temperature be ρ_{lo} and the density (liquid or solid) at the initial (or ambient) temperature be ρ_{la}. The fractional change in the liquid volume v_f as the heat pipe temperature drops from the operating temperature T_p to the ambient temperature T_a is then

$$v_f = 1 - \frac{\rho_{lo}}{\rho_{la}} \tag{6.43}$$

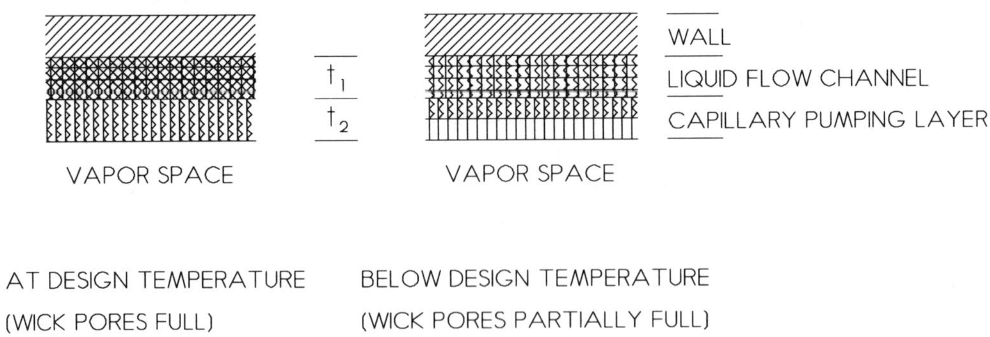

Figure 6.11. Variation of liquid-vapor interface position with temperature.

In order for the liquid-vapor interface to remain within the capillary pumping layer, the following condition must be valid:

$$\frac{\phi_2 t_2}{\phi_1 t_1 + \phi_2 t_2} > v_f$$

where ϕ_1 is the porosity of the liquid flow channel, t_1 is its thickness, ϕ_2 is the porosity of the capillary pumping layer, and t_2 is its thickness. The equation can be expressed in the form

$$\frac{t_2}{t_1} > \frac{\phi_1 v_f}{\phi_2 (1 - v_f)} \tag{6.44}$$

This equation defines the minimum ratio of the capillary pumping and liquid flow channel thicknesses, if there is to be no spillage of excess liquid into the vapor space, the liquid-vapor interface is to remain within the capillary pumping layer, and there is no external reservoir. The initial charge of liquid would be such that the liquid volume is just equal to the total wick pore volume at the heat pipe operating temperature.

For the previously considered case of lithium operating between 367°F and 2200°F (186°C and 1204°C), $v_f = 0.1971$. Then, if $\phi_1 = \phi_2$, $t_2/t_1 > 0.246$.

With an open grooved wick, even slight underfilling of the wick grooves can effectively destroy the capillary pumping capability of the wick (Refs. 6.10 and 6.13). Therefore, allowing the liquid-vapor interface to recede into the wick grooves as the temperature drops below its design value is not feasible. If the heat pipe is to remain functional during the startup process, the only alternatives are either to allow spillage of the excess liquid as the temperature increases, or to provide for a separate reservoir to hold the excess liquid.

If the heat pipe is not required to be operational before its design operating temperature is reached, another option is to raise the heat pipe temperature to its operating point by uniformly heating the entire heat pipe with an external heat source. Liquid in the underfilled wick could then expand to the required "just full" volume prior to the start of operation.

6.6. HEAT PIPE HEAT EXCHANGERS

In this section, the characteristics of conventional and heat pipe heat exchangers are reviewed. Then the basic relationships between the heat transfer rate and heat transfer area are presented. Expressions for the pressure drop in a heat exchanger are given next. Temperature-area relationships for single-stage and multiple-stage heat pipe heat exchangers are then developed. Finally, a direct comparison is made between the heat transfer area requirements of heat pipe and conventional heat exchangers.

There is more to the design of heat exchangers than determining the heat transfer area. Overall dimensions must also be determined such that pressure drop as well as heat transfer requirements are satisfied. This aspect of heat pipe heat exchanger design is treated in the design example of Chapter 8 that deals with a heat pipe recuperator.

6.6.1. Introduction

Conventional Heat Exchangers. A heat exchanger is a device in which heat is transferred from a flowing hotter fluid to a flowing colder fluid. Heat transfer takes place through the surface that separates the fluids. In order to fit the required heat transfer area into a heat exchanger of reasonable size, the heat transfer area is generally divided into a large number of small flow passages. The flow passages that contain the hotter fluid are referred to as the *hot side,* and those flow passages that contain the cooler fluid are referred to as the *cold side*.

The common wall between the hot and cold side flow passages is called the *primary heat transfer area*. A *secondary heat transfer area* may be present in the form of fins, which extend from the common wall into one or both of the fluids. The secondary area does not transfer heat as effectively as the primary area. The inefficiency of the secondary heat transfer area is accounted for by introducing the *fin effectiveness*. The total effective heat transfer area is then equal to the sum of the primary area and the secondary area multiplied by the fin effectiveness.

Flow passage cross sections in a heat exchanger can have a variety of shapes. The hot and cold side fluids may flow in the same direction (parallel flow), in opposite directions (counterflow), or at right angles to each other (cross flow). Inlet and outlet plenums are required to direct the hot and cold side streams into the multiple flow passages of the heat exchanger, and then to reassemble the fluids leaving the multiple flow passages into single streams. The portion of the heat exchanger containing the heat transfer surface (as opposed to the plenums) is called the *core*.

The heat exchanger surface inside the heat exchanger may be stationary, or it may rotate between the hot and cold streams. Generally, a heat exchanger with stationary heat transfer surfaces is called a *recuperator.* A heat exchanger with rotating heat transfer surfaces is referred to as a *rotary regenerator.*

Heat Pipe Heat Exchangers. Heat pipes may also be used to transfer heat from the hot side fluid to the cold side fluid in a heat exchanger. Since heat pipes have the capability to transport heat over appreciable distances virtually isothermally, it is not necessary to subdivide the original flow streams into interspersed multiple flow passages in a heat pipe heat exchanger. Instead, the original flow streams remain intact and separated. Heat transfer is accomplished via multiple small heat pipes extending through the common wall of the hot and cold side flow streams. Such an arrangement for a counterflow heat pipe heat exchanger is shown in Figure 6.12.

As in conventional heat exchangers, fins may be added to the external surface of the heat pipes to increase the effective surface area. However, the contribution of fins in augmenting heat transfer area must be balanced against their increased weight.

Since it is not necessary to intersperse the hot and cold side flow streams in a heat pipe heat exchanger, heat transfer is accomplished with little or no disruption or diversion of the flow streams. Inlet and outlet flow distribution plenums will be considerably simplified, or may not be needed at all. Complete separation of the flow streams

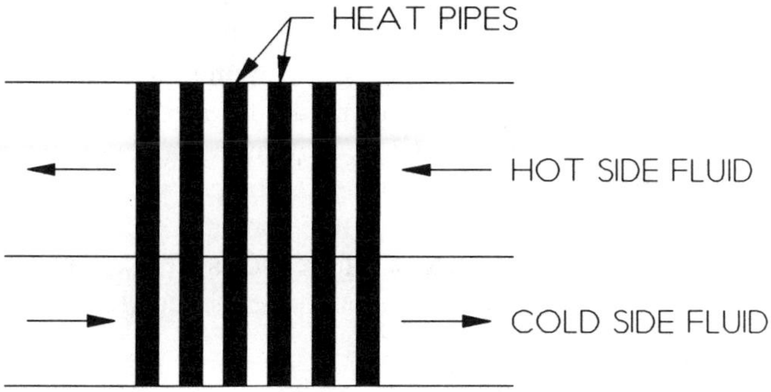

Figure 6.12. Heat pipe heat exchanger—counterflow configuration.

can be particularly advantageous when the hot and cold side fluids are chemically reactive, and mixing of the streams in the event of a leak must be avoided.

The heat pipes in a heat pipe heat exchanger are arranged in *stages,* each stage consisting of a single row of heat pipes, all of which are at the same temperature. The heat pipe temperature changes from stage to stage, varying from a minimum value at one end of the heat exchanger to a maximum value at the other end. If the temperature range is large enough, more than one heat pipe fluid could be required to assure adequate heat transport capability in all stages.

The required heat transfer area for a given heat transfer rate decreases as the number of stages is increased, but the pressure drop of the hot and cold side fluids also increases with the number of stages. Therefore, pressure drop limitations may establish the maximum number of stages and the minimum heat transfer area.

In a heat pipe heat exchanger, the core volume and weight are proportional to the sum of the hot side (evaporator section) area and the cold side (condenser section) area. In an unfinned conventional heat exchanger in which one fluid flows inside tubular flow passages and the other fluid flows outside these passages, the hot and cold side areas are virtually equal and extend over the same length. The core volume and weight are then proportional to either the hot or the cold side areas, rather than their sum.

Therefore, in comparing the heat transfer area of unfinned heat pipe and conventional heat exchangers, the outside surface area of the heat pipes should be compared to the hot or cold side area of the conventional heat exchanger flow passages. Depending on specific design circumstances, a heat pipe heat exchanger may be larger or smaller than a conventional heat exchanger with the same heat transfer rate.

6.6.2. Relation between Heat Transfer Rate and Heat Transfer Area

The heat transfer rate Q to or from a fluid is equal to its rate of change of enthalpy, and is given by

$$Q = Wc\Delta T \tag{6.45}$$

where W is the fluid mass flow rate, ΔT is the fluid temperature change during heating or cooling, and c is the fluid specific heat. If the fluid is a gas, the specific heat at constant pressure should be used. If the fluid is an ordinary incompressible liquid, the normally designated specific heat for that liquid should be used.

The product Wc is equal to the rate of heat transfer to or from the fluid per unit of temperature change. It is referred to here as the *unit thermal capacity rate*. The unit thermal capacity rate of the cold side fluid is designated as $W_c c_c$, and that of the hot side fluid as $W_h c_h$. The smaller unit thermal capacity rate is referred to as $(Wc)_{low}$, and the larger unit thermal capacity rate as $(Wc)_{high}$. It is convenient to define the *unit thermal capacity ratio* as

$$\psi = (Wc)_{low}/(Wc)_{high} \tag{6.46}$$

Values of ψ lie in the range between 0 and 1.

The maximum possible heat transfer rate Q_{max} occurs when the temperature of the fluid with $(Wc)_{low}$ changes from its inlet value to the inlet value of the other fluid. If that fluid were the cold side fluid, it would be heated from T_{ci} to T_{hi}, where T_{ci} is the inlet temperature of the cold side fluid and T_{hi} is the inlet temperature of the hot side fluid. If the fluid with $(Wc)_{low}$ were the hot side fluid, it would be cooled from T_{hi} to T_{ci}. Thus,

$$Q_{max} = (Wc)_{low}(T_{hi} - T_{ci}) \tag{6.47}$$

Since the heat transfer rate is the same for the fluid with $(Wc)_{high}$, its temperature change at Q_{max} is less than $T_{hi} - T_{ci}$.

Temperature Distributions in Heat Exchanger. The temperature distribution of the fluids in a heat exchanger is dependent on their relative unit thermal capacity rates and their relative flow directions. The temperature distribution for two common flow arrangements is shown in Figure 6.13.

With parallel flow, both the hot side and cold side fluids flow in the same direction.

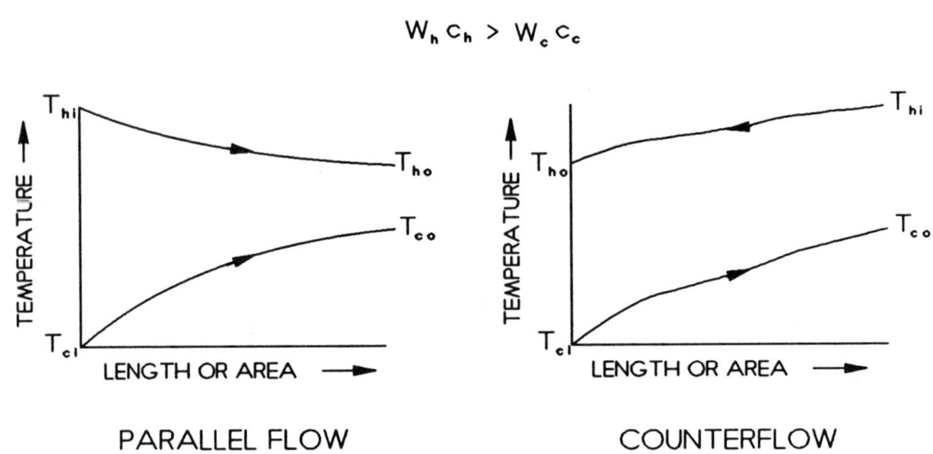

Figure 6.13. Temperature distributions in heat exchangers.

With counterflow, the two fluid streams flow in opposing directions. In Figure 6.13, $W_h c_h$ is greater than $W_c c_c$. Therefore, the hot side temperature drop is less than the cold side temperature rise.

For a given heat transfer area, the heat transfer rate is higher with the counterflow arrangement. However, the difference decreases as $\psi = (Wc)_{low}/(Wc)_{high}$ becomes smaller, and disappears as ψ goes to zero. The temperature distribution for the latter case is then the same for parallel flow and counterflow. This situation is closely approximated if one of the fluids is a constant-temperature vaporizing liquid or condensing vapor. Its specific heat is then effectively infinite, as is $(Wc)_{high}$, and $\psi \approx 0$.

Calculation of the relationship between the heat transfer rate and the heat transfer area is simplified considerably by the introduction of dimensionless forms of these variables. The dimensionless heat transfer rate is referred to as the *heat exchanger effectiveness,* and the dimensionless heat transfer area as the *number of heat transfer units.*

Heat Exchanger Effectiveness. The heat exchanger effectiveness ε is defined as the ratio of the actual heat transfer rate Q to the maximum possible heat transfer rate Q_{max}. From Eqs. (6.45) and (6.47), the heat exchanger effectiveness is then

$$\varepsilon = \frac{Wc\,\Delta T}{(Wc)_{low}\,(T_{hi} - T_{ci})} \tag{6.48}$$

where the terms in the numerator refer to either the hot side or the cold side.

If $W_c c_c = (Wc)_{low}$, it follows that

$$\varepsilon = \frac{T_{co} - T_{ci}}{T_{hi} - T_{ci}} \tag{6.49}$$

where T_{co} is the outlet temperature of the cold side fluid.

If $W_h c_h = (Wc)_{low}$, it follows that

$$\varepsilon = \frac{T_{hi} - T_{ho}}{T_{hi} - T_{ci}} \tag{6.50}$$

where T_{ho} is the outlet temperature of the hot side fluid.

If $W_c c_c = W_h c_h$, it follows that

$$\varepsilon = \frac{T_{co} - T_{ci}}{T_{hi} - T_{ci}} = \frac{T_{hi} - T_{ho}}{T_{hi} - T_{ci}} \tag{6.51}$$

The latter condition is approximated reasonably well in a gas turbine recuperator.

Number of Heat Transfer Units. Before defining the number of heat transfer units, additional concepts involved in the definition must be introduced.

The total resistance to the flow of heat between a hot fluid and a cooler fluid is equal to the sum of the resistances in the hot and cold fluids plus the resistance of the heat transfer path (usually a wall) between the fluids. The resistance of the heat transfer path between the fluids is usually small compared to the fluid thermal resistance, and will be neglected here. We may then write for the total thermal resistance R_t

$$R_t = \frac{1}{UA} = \frac{1}{U_h A_h} = \frac{1}{U_c A_c} = \frac{1}{h_h A_h} + \frac{1}{h_c A_c} \tag{6.52}$$

where h_c is the heat transfer coefficient of the cold side fluid, h_h is the heat transfer coefficient of the hot side fluid, A_c is the cold side heat transfer area, A_h is the hot side heat transfer area, U_c is the unit thermal conductance based on the cold side area, U_h is the unit thermal conductance based on the hot side area, and $UA = U_hA_h = U_cA_c$ is the overall thermal conductance.

Note that the overall thermal conductance can be defined in terms of the hot side or cold side parameters, and A is the heat transfer area associated with U. If the heat transfer surfaces are finned, A_c and A_h represent the total effective heat transfer areas.

The heat transfer rate between the fluids Q can be expressed in terms of UA as

$$Q = UA\,\Delta T_m \tag{6.53}$$

where ΔT_m is the mean temperature difference between the hot side and cold side fluids.

The heat transfer rate given by Eq. (6.53) is also equal to the heat transfer rate to or from either fluid, as given by Eq. (6.45).

A reference heat transfer area A_{ref} will now be defined as

$$A_{ref} = Q_{max}/(U\Delta T_{max}) \tag{6.54}$$

where $\Delta T_{max} = T_{hi} - T_{ci}$.

As is evident from Eq. (6.53), the reference area is equal to the area that would be needed to transfer heat between the hot and cold side fluids at the maximum rate if the mean temperature difference were equal to the maximum temperature difference between the fluids.

With the aid of Eq. (6.54), the number of heat transfer units NTU can now be defined as

$$\text{NTU} = \frac{A}{A_{ref}} = UA\,\frac{\Delta T_{max}}{Q_{max}} \tag{6.55}$$

where $UA = U_cA_c = U_hA_h$ is the overall thermal conductance.

Substituting for Q_{max} from Eq. (6.47), NTU assumes the final form

$$\text{NTU} = \frac{UA}{(Wc)_{low}} \tag{6.56}$$

The magnitude of NTU can be less than, equal to, or greater than 1.

Relation between Heat Exchanger Effectiveness and NTU. It can be shown that the heat exchanger effectiveness ε is a function of the number of heat transfer units NTU, the unit thermal capacity ratio ψ, and the relative flow direction of the two fluids (Ref. 6.14). For counterflow and parallel flow, the relationship takes the forms

Counterflow:

$$\varepsilon = \varepsilon_{cf} = \frac{1 - e^{-(1-\Psi)\,\text{NTU}}}{1 - \Psi\, e^{-(1-\Psi)\,\text{NTU}}} \tag{6.57}$$

When NTU << 1, $\varepsilon_{cf} \approx$ NTU. When NTU >> 1, $\varepsilon_{cf} \approx 1$.

When $\psi = 0$, which corresponds to constant temperature operation of one of the fluids,

$$\varepsilon_{cf} = 1 - e^{-\mathrm{NTU}} \qquad \mathrm{NTU} = \ln \frac{1}{1 - \varepsilon_{cf}} \tag{6.58}$$

When $\psi = 1$, which corresponds to equal unit thermal capacity rates in the two fluids,

$$\varepsilon_{cf} = \frac{\mathrm{NTU}}{1 + \mathrm{NTU}} \qquad \mathrm{NTU} = \frac{\varepsilon_{cf}}{1 - \varepsilon_{cf}} \tag{6.59}$$

Parallel flow:

$$\varepsilon = \varepsilon_{pf} = \frac{1 - e^{-(1+\Psi)\,\mathrm{NTU}}}{1 + \Psi} \tag{6.60}$$

When NTU << 1, $\varepsilon_{pf} \approx$ NTU. When NTU >> 1, $\varepsilon_{pf} \approx 1/(1 + \psi)$.
When $\psi = 0$,

$$\varepsilon_{pf} = 1 - e^{-\mathrm{NTU}} \qquad \mathrm{NTU} = \ln \frac{1}{1 - \varepsilon_{pf}} \tag{6.61}$$

When $\psi = 1$,

$$\varepsilon_{pf} = \frac{1 - e^{-2\,\mathrm{NTU}}}{2} \qquad \mathrm{NTU} = -\frac{\ln (1 - 2\,\varepsilon_{pf})}{2} \tag{6.62}$$

In general, the heat exchanger effectiveness is smaller with parallel flow. However, the counterflow and parallel flow results are identical when $\psi = 0$. Also, when $\psi = 1$, the maximum value of the heat exchanger effectiveness is 0.5 with parallel flow (as NTU becomes very large), compared to a maximum value of 1 with counterflow.

Derivations of the foregoing equations are given in Appendix D.

Mean Temperature Difference. The mean temperature difference between the hot side and cold side fluids is, from Eq. (6.53),

$$\Delta T_m = \frac{Q}{U A}$$

Introducing Eqs. (6.45), (6.48), and (6.56), and simplifying,

$$\frac{\Delta T_m}{\Delta T_{\max}} = \frac{\varepsilon}{\mathrm{NTU}} \tag{6.63}$$

Thus, the mean temperature difference in a heat exchanger is readily calculated from ε and NTU.

When $\psi = 1$, substitution of the appropriate value for NTU yields

Counterflow:

$$\frac{\Delta T_m}{\Delta T_{\max}} = 1 - \varepsilon_{cf} \tag{6.64}$$

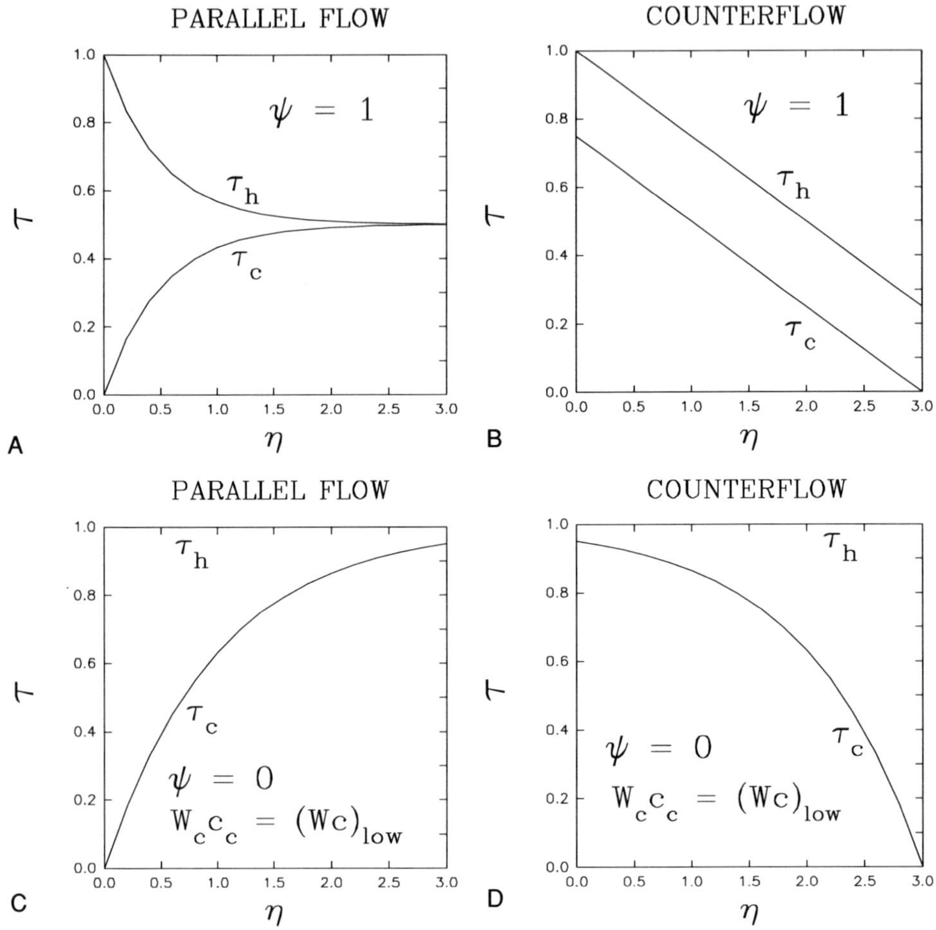

Figure 6.14. Heat exchanger temperature distributions.

Parallel flow:

$$\frac{\Delta T_m}{\Delta T_{\max}} = -\frac{2\,\varepsilon_{pf}}{\ln\,(1 - 2\,\varepsilon_{pf})} \tag{6.65}$$

Dimensionless Temperature Distributions. In Figure 6.14, the temperature distribution in a heat exchanger is shown as a function of heat transfer area along the heat exchanger length for parallel flow and counterflow, and for $\psi = 1$ and $\psi = 0$. The parallel flow and counterflow distributions are mirror images for the latter case.

The plots are presented in terms of dimensionless temperatures and areas. Derivations of the dimensionless temperature distributions are given in Appendix D.

The cold side dimensionless temperature is $\tau_c = (T_c - T_{ci})/\Delta T_{\max}$. The hot side

Table 6.5. Characteristics of Figure 6.14 heat exchangers

Figure	Flow direction	ψ	NTU	ε	$\Delta T_m/\Delta T_{max}$
6.14A	Parallel	1	3	0.499	0.166
6.14B	Counter	1	3	0.750	0.250
6.14C	Parallel	0	3	0.950	0.317
6.14D	Counter	0	3	0.950	0.317

dimensionless temperature is $\tau_h = (T_h - T_{ci})/\Delta T_{max}$. In these expressions, T_h and T_c are the hot and cold side temperatures at some point along the heat exchanger. The dimensionless heat transfer area is $\eta = Ua/(Wc)_{low}$, where a represents the cumulative heat transfer area at a point along the heat exchanger length. At the end of the heat exchanger, η = NTU.

The total heat exchanger size in all cases is NTU = 3. Overall heat exchanger characteristics are summarized in Table 6.5.

The heat exchanger effectiveness and mean temperature difference are largest for Figures 6.14C and 6.14D, in which the hot side temperature is constant. The next largest values occur in Figure 6.14B, with counterflow and equal hot and cold side unit thermal capacity rates. The lowest values occur in Figure 6.14A, in which the unit thermal capacity rates are also equal but the flows are in parallel.

The above results apply to a conventional heat exchanger, in which fluid temperatures change continuously along the heat exchanger length. A heat pipe heat exchanger is divided into discrete stages along the heat exchanger length, in each of which there is a discrete heat pipe temperature and fluid temperature distribution. Therefore, the heat exchanger equations that have been presented thus far require modification when applied to heat pipe heat exchangers. These modifications are described in Sections 6.6.4 and 6.6.5.

6.6.3. Pressure Drop through a Heat Exchanger

The hot and cold side fluids undergo a drop in pressure while flowing through the heat exchanger. Since a decrease in fluid pressure can adversely affect the performance of the system of which the flowing fluids are a part, there is generally a limit on the allowable hot and cold side fluid pressure drops. In a gas turbine recuperator, for example, a representative requirement might be that the sum of the hot and cold side pressure drops not exceed 6%.

The same factors that act to increase heat transfer coefficients and reduce heat exchanger size, such as an increase in the flow velocity or a decrease in the hydraulic diameter of flow passages, also act to increase the pressure drop. Therefore, the minimum size of a heat exchanger with a specified effectiveness will generally be determined by the allowable overall (hot side plus cold side) pressure drop.

The fractional pressure drop of a fluid flowing through one side of a heat exchanger can be expressed as (Ref. 6.14)

$$\frac{\Delta P}{P_1} = \frac{G^2 v_1}{2 P_1} \left[(K_c + 1 - \sigma_a^2) + 2\left(\frac{v_2}{v_1} - 1\right) + 4f\frac{L}{D_h}\frac{v_m}{v_1} - (1 - \sigma_a^2 - K_e)\frac{v_2}{v_1} \right] \quad (6.66)$$

The symbols in Eq. (6.66) have the following definitions. P_1 is the inlet pressure, ΔP is the net pressure drop, v_1 is the specific volume at the core inlet, v_2 is the specific volume at the core exit, v_m is the mean specific volume in the core, G is the mass flow rate per unit of minimum free flow area in the core, K_c is the entrance loss coefficient, K_e is the exit loss coefficient, σ_a is the ratio of minimum free-flow area in the core to the core frontal area, f is the friction coefficient, L is the core length, and D_h is the hydraulic diameter of the flow passages (equals four times the hydraulic radius).

The variables f, K_c, and K_e depend on the core geometry and the Reynolds number. Data on these variables may be found in Ref. 6.14.

The first term in the brackets of Eq. (6.66) accounts for entrance losses due to both the momentum increase associated with the increase of velocity as the fluid enters the core $(1 - \sigma_a^2)$ and the irreversible losses associated with the abrupt contraction of flow area (the entrance loss coefficient K_c). The second term accounts for the pressure change due to the momentum change in the core as the fluid density changes during heating or cooling. The third term accounts for the core friction pressure loss. The last term accounts for the increase in exit pressure due to the momentum loss associated with the deceleration of gas leaving the core $[-(1 - \sigma_a^2)]$, and the irreversible pressure loss associated with the abrupt expansion (the exit loss coefficient K_e).

For the case of *flow normal to banks of tubes,* the effects of entrance and exit losses are incorporated into the friction coefficient. Therefore, for this case K_c and K_e are equal to zero. The pressure drop equation then takes the simpler form

$$\frac{\Delta P}{P_1} = \frac{G^2 v_1}{2 P_1} \left[\left(\frac{v_2}{v_1} - 1\right)(1 + \sigma_a^2) + 4f\frac{L}{D_h}\frac{v_m}{v_1} \right] \quad (6.67)$$

In a heat pipe heat exchanger in which the hot and cold side fluids flow over and normal to banks of heat pipes, Eq. (6.67) for the pressure drop would apply.

The mass flow rate per unit of minimum free flow area in the core G is expressible as

$$G = \frac{W}{\sigma_a A_f} \quad (6.68)$$

where A_f is the core frontal area.

The calculation of $\Delta P/P_1$ from Eq. (6.66) or (6.67) requires that v_2 and v_m be known. For gaseous hot and cold fluids, the perfect gas law can be used to obtain the following relationships:

$$\frac{v_2}{v_1} = \frac{T_2}{T_1}\frac{P_1}{P_2} \qquad \frac{v_m}{v_1} = \frac{T_m}{T_1}\frac{P_1}{P_m} \quad (6.69)$$

Since the allowable percentage pressure drop is small in most cases,

$$\frac{v_2}{v_1} \approx \frac{T_2}{T_1} \qquad \frac{v_m}{v_1} \approx \frac{T_m}{T_1} \quad (6.70)$$

The inlet and outlet temperatures T_1 and T_2 are known, or are calculable from the heat exchanger design criteria. The mean temperature T_m can then be determined from the temperature distribution in the heat exchanger fluid under consideration, according to the equation

$$T_m = \frac{1}{L}\int_0^L T\,dL \tag{6.71}$$

For counterflow heat exchange with equal hot and cold side unit thermal capacity rates, the temperature distribution will be linear, and T_m is simply the arithmetic average of T_1 and T_2.

If greater accuracy is desired, v_2 and v_m should be calculated from Eq. (6.69). An iterative solution for $\Delta P/P_1$ in Eqs. (6.66) and (6.67) will then be required. A value of P_2 is assumed initially, P_m is calculated as the arithmetic average of P_1 and P_2, v_2/v_1 and v_m/v_1 are calculated from Eq. (6.69), and $\Delta P/P_1$ is calculated from Eq. (6.66) or (6.67). If $P_2 = P_1 - \Delta P$ differs from the assumed value of P_2 by more than some specified fraction δ_1, a new value is assumed for P_2. The calculations are repeated until the P_2 obtained differs from the prior value of P_2 by less than δ_1.

6.6.4. Single-Stage Heat Pipe Heat Exchangers

The transfer of heat from a hot fluid to a cooler fluid in a heat exchanger can be accomplished with the use of heat pipes. The basic concept involved for counterflow is illustrated in Figure 6.15.

In general, a heat exchanger will consist of multiple rows or stages of heat pipes, with multiple heat pipes in each stage. In Figure 6.15, a single heat pipe is shown for simplicity, and is taken to be representative of the heat pipes in a single stage. The multiple-stage heat pipe heat exchanger is considered in Section 6.6.5.

In Figure 6.15, a heat pipe of length L_p and diameter D_{hp} extends across the total

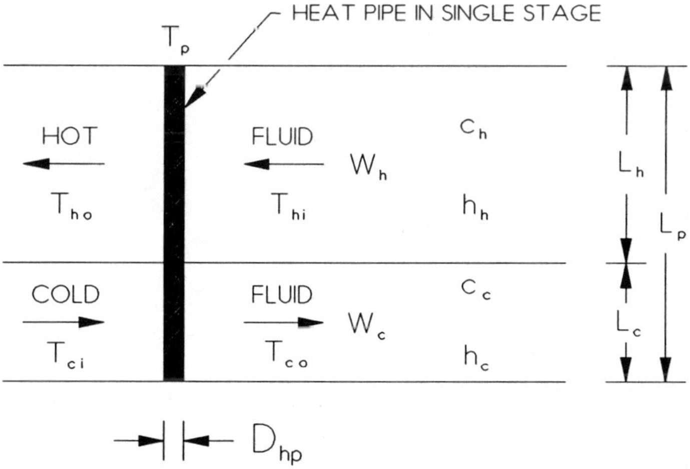

Figure 6.15. Heat pipe heat exchanger concept—single stage.

width of two adjacent flow ducts. The nominally constant temperature of the heat pipe is T_p. (Variations in the surface temperature due to the transfer of heat into and out of the heat pipe are usually small in a heat pipe exchanger, and are neglected here.)

A hot fluid flows at the rate W_h through a duct of width L_h. The fluid temperature is T_{hi}, its specific heat at constant pressure is c_h, and its heat transfer coefficient corresponding to heat transfer between the fluid and the cooler heat pipe, is h_h.

A cold fluid flows at the rate W_c through an adjacent duct of width L_c in a direction opposite to that of the hot fluid. The fluid temperature is T_{ci}, its specific heat is c_c, and its heat transfer coefficient for heat transfer between the hotter heat pipe and the fluid is h_c.

As the hot fluid flows over the heat pipe, the fluid temperature drops to T_{ho}. In the process, heat is transferred to the heat pipe at the rate $Q = W_h c_h (T_{hi} - T_{ho})$. The heat pipe liquid evaporates from the wick at the rate required to transport heat through the heat pipe at the rate Q. The vapor condenses on the wick along the width of the cold flow duct, raising the temperature of the cold fluid to T_{co} as it flows over the heat pipe. Heat is transferred out of the heat pipe at the same rate $Q = W_c c_c (T_{co} - T_{ci})$.

Heat Transfer Relationships. It is convenient to consider the process by which heat is transferred from the hot fluid to the cold fluid as two separate processes. In the first process, heat is transferred from the hot fluid to the evaporating liquid in the constant-temperature heat pipe. In the second, heat is transferred from condensing vapor in the constant-temperature heat pipe to the cooler fluid. The hot and cold sides of the heat exchanger may then be considered as separate heat exchangers, which are coupled together by a common heat transfer rate and a common heat pipe temperature T_p.

In each case, the unit thermal capacity ratio $\psi = 0$. Then, using the definitions of NTU and ε that were presented in Section 6.6.2, we may write for the hot side

$$\mathrm{NTU}_h = \frac{h_h A_h}{W_h c_h} \tag{6.72}$$

$$\varepsilon_h = \frac{T_{hi} - T_{ho}}{T_{hi} - T_p} = 1 - e^{-\mathrm{NTU}_h} \tag{6.73}$$

$$Q = W_h c_h (T_{hi} - T_{ho}) = W_h c_h \varepsilon_h (T_{hi} - T_p) \tag{6.74}$$

and for the cold side

$$\mathrm{NTU}_c = \frac{h_c A_c}{W_c c_c} \tag{6.75}$$

$$\varepsilon_c = \frac{T_{co} - T_{ci}}{T_p - T_{ci}} = 1 - e^{-\mathrm{NTU}_c} \tag{6.76}$$

$$Q = W_c c_c (T_{co} - T_{ci}) = W_c c_c \varepsilon_c (T_p - T_{ci}) \tag{6.77}$$

Heat Pipe Temperature. Upon equating Q from Eqs. (6.74) and (6.77), and then solving for T_p,

$$T_p = \frac{T_{hi} + \Omega_{tc} E_h T_{ci}}{1 + \Omega_{tc} E_h} \tag{6.78}$$

where

$$\Omega_{tc} = \frac{W_c c_c}{W_h c_h} \qquad E_h = \frac{\varepsilon_c}{\varepsilon_h}$$

Equation (6.78) gives the heat pipe temperature for given hot and cold side heat exchanger conditions.

In some cooling situations, the heat transfer rate may be small compared to the rate at which heat is being transported by the fluid heat source. The associated temperature change in the fluid is then small, in which case ε and NTU will also be small. Then ε_h and ε_c may be approximated as

$$\varepsilon_h \approx \text{NTU}_h = \frac{h_h A_h}{W_h c_h} \tag{6.79}$$

$$\varepsilon_c \approx \text{NTU}_c = \frac{h_c A_c}{W_c c_c} \tag{6.80}$$

Then, Eq. (6.78) for T_p takes the form

$$T_p = \frac{T_{hi} + (h_c A_c / h_h A_h)\ T_{ci}}{1 + (h_c A_c / h_h A_h)} \tag{6.81}$$

From Figure 6.15, it is evident that the heat pipe temperature T_p must lie between the hot and cold side fluid outlet temperatures.

If Eq. (6.78) is solved for E_h,

$$E_h = \frac{T_{hi} - T_p}{\Omega_{tc} (T_p - T_{ci})} \tag{6.82}$$

If the heat pipe temperature T_p is specified, the associated ratio E_h of cold and hot side effectivenesses can be found from Eq. (6.82).

For a given overall heat transfer rate, the variation of area with T_p can be found by selecting a value of T_p, and calculating ε_h, ε_c, NTU_h, and NTU_c from Eqs. (6.73) and (6.76). A_h and A_c can then be found from Eqs. (6.72) and (6.75). The calculations are repeated for various values of T_p. In general, there will be an optimum value of T_p for which the total heat exchanger area $A_c + A_h$ is minimized.

Overall Heat Exchanger Effectiveness. In many situations, the heat transfer area is to be determined for a specified overall heat exchanger effectiveness. The overall effectiveness ε is defined as

$$\varepsilon = \frac{W_c c_c (T_{co} - T_{ci})}{(Wc)_{\text{low}} (T_{hi} - T_{ci})} = \frac{W_h c_h (T_{hi} - T_{ho})}{(Wc)_{\text{low}} (T_{hi} - T_{ci})} \tag{6.83}$$

Upon substituting $T_{hi} - T_{ho}$ and $T_{co} - T_{ci}$ from Eq. (6.83) into Eqs. (6.73) and (6.76) for ε_h and ε_c,

$$\varepsilon_h = \left(\frac{T_{hi} - T_{ci}}{T_{hi} - T_p}\right) \varepsilon \qquad \Omega_{tc} \geq 1$$
$$\varepsilon_h = \left(\frac{T_{hi} - T_{ci}}{T_{hi} - T_p}\right) \Omega_{tc}\, \varepsilon \qquad \Omega_{tc} \leq 1 \tag{6.84}$$

$$\varepsilon_c = \left(\frac{T_{hi} - T_{ci}}{T_p - T_{ci}}\right) \frac{\varepsilon}{\Omega_{tc}} \qquad \Omega_{tc} \geq 1$$
$$\varepsilon_c = \left(\frac{T_{hi} - T_{ci}}{T_p - T_{ci}}\right) \varepsilon \qquad \Omega_{tc} \leq 1 \tag{6.85}$$

From the above equations, the hot and cold side effectivenesses necessary for an overall single-stage effectiveness of ε can be determined. The associated hot and cold side NTUs can then be found, and then the required hot and cold side heat transfer areas, from Eqs. (6.73), (6.76), (6.72), and (6.75).

For a heat exchanger application, ε is generally made as large as possible to maximize heat transfer between the hot and cold fluids. Since NTU and the heat transfer area also increase with ε, restrictions on overall heat exchanger size generally limit the maximum feasible value of ε to 0.95 or less. In weight-sensitive applications such as aircraft gas turbine recuperators, ε may be limited to a practical upper limit of 0.80–0.85.

For a single-stage (one-row) heat pipe heat exchanger, ε may be restricted to substantially lower values. From Figure 6.15, the hot side fluid can be cooled at most to $T_{ho} = T_p$, and the cold side fluid can be heated at most to $T_{co} = T_p$. Therefore, the largest possible value of $\varepsilon = \varepsilon_{max}$ occurs when $T_{co} = T_{ho} = T_p$. Upon solving for T_{co} and T_{ho} from Eq. (6.83), equating these variables, and solving for $\varepsilon = \varepsilon_{max}$, it may be shown that

$$\varepsilon_{max} = \frac{1}{1 + \Omega_{tc}} \qquad \Omega_{tc} \leq 1$$
$$\varepsilon_{max} = \frac{\Omega_{tc}}{1 + \Omega_{tc}} \qquad \Omega_{tc} \geq 1 \tag{6.8}$$

If $\Omega_{tc} = 1$, $\varepsilon_{max} = 0.5$. Thus, for a single-stage gas turbine recuperator with $\Omega_{tc} \approx 1$, the maximum value of the overall heat exchanger effectiveness would be limited to 0.5. In Section 6.6.5, it is shown that ε_{max} increases with the number of stages.

There are also upper and lower limits on the heat pipe temperature. The lower limit $T_{pl} = T_{co}$, and the upper limit $T_{pu} = T_{ho}$. Again, Eq. (6.83) can be used to establish that

$$\frac{T_{pl} - T_{ci}}{T_{hi} - T_{ci}} = \varepsilon \qquad \Omega_{tc} \leq 1$$
$$\frac{T_{pl} - T_{ci}}{T_{hi} - T_{ci}} = \frac{\varepsilon}{\Omega_{tc}} \qquad \Omega_{tc} \geq 1 \tag{6.87}$$

$$\frac{T_{hi} - T_{pu}}{T_{hi} - T_{ci}} = \Omega_{tc}\,\varepsilon \qquad \Omega_{tc} \le 1$$

$$\frac{T_{hi} - T_{pu}}{T_{hi} - T_{ci}} = \varepsilon \qquad \Omega_{tc} \ge 1 \tag{6.88}$$

Equations (6.87) and (6.88) are useful for finding T_{pl} and T_{pu} when T_{co} and T_{ho} have not been determined.

The use of these equations is illustrated in the following example. Consider a single-stage heat pipe heat exchanger for which $\Omega_{tc} = 1$, $\varepsilon = 0.3$, $T_{hi} = 1200$°F (649°C), $T_{ci} = 800$°F (427°C), and $T_p = 940$°F (504°C). Equations (6.84) through (6.88) may be used to show that

$$\varepsilon_{max} = 0.5$$
$$T_{pl} = 920\text{°F (493°C)}$$
$$T_{pu} = 1080\text{°F (582°C)}$$
$$\varepsilon_c = 0.857$$
$$\varepsilon_h = 0.462$$

6.6.5. Multiple-Stage Heat Pipe Heat Exchangers

Instead of just a single stage, a heat pipe heat exchanger may incorporate two or more stages. All the heat pipes in each stage operate at the same temperature, but the heat pipe temperature varies from stage to stage. With multiple stages, more heat transfer area can be incorporated into the heat exchanger without excessively large dimensions in the direction normal to the flow direction. Also, when $\psi > 0$, the maximum heat exchanger effectiveness increases with the number of stages.

The same equations that were given for a single-stage heat pipe heat exchanger in Section 6.6.4 also apply for each stage of a multiple-stage heat exchanger. However, in Eq. (6.83) for the overall heat exchanger effectiveness, the inlet and outlet temperatures now refer to the complete heat exchanger rather than an individual stage. Also, the heat pipe temperature now ranges from a minimum value at the cold side inlet to a maximum value at the hot side inlet.

Heat Pipe Temperature. Equation (6.78) for the heat pipe temperature in single-stage heat exchangers can also be used for individual stages of a multistage heat exchanger, provided that appropriate parameters for the stage under consideration are used in the equation. Generally, the heat pipe temperatures at the cold side inlet and the hot side inlet are of most interest.

At the stage on the cold side inlet, it is usually more convenient to express T_p in terms of the hot side outlet temperature T_{ho} instead of the hot side inlet temperature T_{hi}. Similarly, it is more convenient to express T_p at the stage on the hot side inlet in terms of the cold side outlet temperature T_{co} instead of the cold side inlet temperature T_{ci}. Expressions for T_p incorporating these changes will now be derived.

Stage at cold side inlet. T_{hi} can be obtained from Eq. (6.73) as

$$T_{hi} = \frac{T_{ho} - \varepsilon_h T_p}{1 - \varepsilon_h}$$

Upon substituting this equation for T_{hi} in Eq. (6.78), replacing E_h with $\varepsilon_c/\varepsilon_h$, and solving for T_p,

$$T_p = \frac{\varepsilon_h T_{ho} + \Omega_{tc} (1 - \varepsilon_h) \varepsilon_c T_{ci}}{\varepsilon_h^2 + (1 - \varepsilon_h)(\varepsilon_h + \Omega_{tc} \varepsilon_c)} \tag{6.89}$$

If $\Omega_{tc} = 1$, Eq. (6.89) reduces to

$$T_p = \frac{\varepsilon_h T_{ho} + (1 - \varepsilon_h) \varepsilon_c T_{ci}}{\varepsilon_h^2 + (1 - \varepsilon_h)(\varepsilon_h + \varepsilon_c)} \tag{6.90}$$

Stage at hot side inlet. From Eq. (6.76), T_{ci} can be expressed as

$$T_{ci} = \frac{T_{co} - \varepsilon_c T_p}{1 - \varepsilon_c}$$

Upon substituting T_{ci} from this equation into Eq. (6.78), replacing E_h with $\varepsilon_c/\varepsilon_h$, and solving for T_p,

$$T_p = \frac{\varepsilon_h (1 - \varepsilon_c) T_{hi} + \Omega_{tc} \varepsilon_c T_{co}}{\Omega_{tc} \varepsilon_c^2 + (1 - \varepsilon_c)(\varepsilon_h + \Omega_{tc} \varepsilon_c)} \tag{6.91}$$

If $\Omega_{tc} = 1$, Eq. (6.91) reduces to

$$T_p = \frac{\varepsilon_h (1 - \varepsilon_c) T_{hi} + \varepsilon_c T_{co}}{\varepsilon_c^2 + (1 - \varepsilon_c)(\varepsilon_h + \varepsilon_c)} \tag{6.92}$$

Heat Exchanger Effectiveness for Multistage Counterflow Heat Exchanger. Conditions in a three-stage counterflow heat exchanger will now be considered, and then extended to any number of stages by induction. It is assumed that $\Omega_{tc} = 1$. In each stage, the heat pipe temperature assumes a value such that the heat transfer rate from the hot side fluid to the heat pipes is equal to the heat transfer rate from the heat pipes to the cold side fluid. The hot and cold side stage effectivenesses ε_h and ε_c are assumed to be constant for all stages, but not necessarily equal to each other. It follows that the hot and cold side stage numbers of heat transfer units NTU_h and NTU_c are also constant for all stages, but not necessarily equal to each other.

The temperature distribution is then as shown in Figure 6.16. The temperature change ΔT is the same for the hot and cold sides in all stages. The maximum possible temperature change per stage is ΔT_{hp} on the hot side and ΔT_{pc} on the cold side. Both ΔT_{hp} and ΔT_{pc} are constant for all stages.

From the basic definition of ε, it follows that

$$\Delta T = \varepsilon_h \Delta T_{hp} = \varepsilon_c \Delta T_{pc} \tag{6.93}$$

and

$$\varepsilon = \frac{T_{h4} - T_{h1}}{T_{h4} - T_{c1}} = \frac{T_{c4} - T_{c1}}{T_{h4} - T_{c1}} \tag{6.94}$$

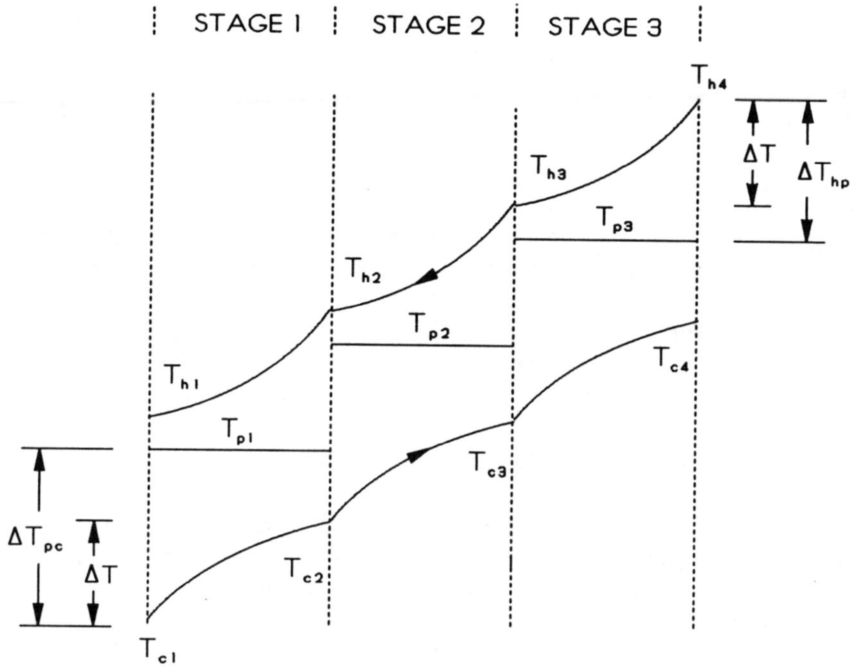

Figure 6.16. Temperature distribution in three-stage counterflow heat pipe heat exchanger.

where the subscript 1 denotes conditions at the cold side inlet and the subscript 4 denotes conditions at the hot side inlet. Also, for simplicity, the subscript *cf* (denoting counterflow) has not been added to ε.

From the relationships of Figure 6.16, and using Eq. (6.93), ε can be expressed as

$$\varepsilon = \frac{3\,\Delta T}{\Delta T_{hp} + 2\,\Delta T + \Delta T_{pc}} = \frac{3\,\varepsilon_h\,\varepsilon_c}{(1 + 2\,\varepsilon_h)\,\varepsilon_c + \varepsilon_h}$$

By induction, the preceding equation for ε becomes, for *n* stages,

$$\varepsilon = \frac{n\,\varepsilon_h\,\varepsilon_c}{[1 + (n-1)\,\varepsilon_h]\,\varepsilon_c + \varepsilon_h} \tag{6.95}$$

Upon solving Eq. (6.95) first for ε_c and then for ε_h,

$$\varepsilon_c = \frac{\varepsilon\,\varepsilon_h}{n\,\varepsilon_h - \varepsilon\,[1 + (n-1)\,\varepsilon_h]} \tag{6.96}$$

$$\varepsilon_h = \frac{\varepsilon\,\varepsilon_c}{n\,\varepsilon_c - \varepsilon\,[1 + (n-1)\,\varepsilon_c]} \tag{6.97}$$

Equation (6.95) gives the overall heat exchanger effectiveness as a function of the hot and cold side stage effectivenesses and the number of stages. Equations (6.96) and (6.97) give the cold and hot side stage effectivenesses associated with a given overall effectiveness, hot or cold side effectiveness, and number of stages. In applying the above equations, it should be noted that none of the effectivenesses can exceed unity, and that the NTU and the heat transfer area become infinite if any of the effectivenesses should be equal to 1.

The maximum value of $\varepsilon = \varepsilon_{max}$ occurs when $\varepsilon_h = \varepsilon_c = 1$. Then, from Eq. (6.95),

$$\varepsilon_{max} = \frac{n}{n+1}$$

Therefore, the maximum attainable heat exchanger effectiveness when the hot and cold side unit thermal capacity rates are equal increases with the number of stages, and approaches 1 as the number of stages becomes large. For a single-stage heat pipe heat exchanger, $\varepsilon_{max} = 0.5$. For a nine-stage heat exchanger, $\varepsilon_{max} = 0.9$.

When the number of stages is large and $\varepsilon_{max} \to 1$, heat exchanger performance approaches that of conventional heat exchangers. Then ε-NTU relations for heat pipe heat exchangers will approximate those of conventional heat exchangers.

The relationship between ε_c and ε_h is shown in Figure 6.17 for a ten-stage heat exchanger and several values of ε. The dashed line is the locus of the path for which $\varepsilon_c = \varepsilon_h$.

Consider the curve for an overall effectiveness of 0.8. The stage effectivenesses when equal have a value of 0.571. As ε_h falls below this figure, ε_c becomes larger. The smallest possible value of ε_h is 0.4, occurring when ε_c reaches its maximum value of 1. As ε_h becomes larger than 0.571, ε_c becomes smaller. As ε_h approaches its maximum value of 1, ε_c approaches its minimum value of 0.4.

The heat pipe temperature corresponding to each stage can be found from Eq. (6.78), with all the temperatures in the equation applying to the stage under consideration.

Heat Transfer Area. From Eqs (6.72) and (6.75), the hot and cold side heat transfer areas per stage Ah and A_c may be expressed as

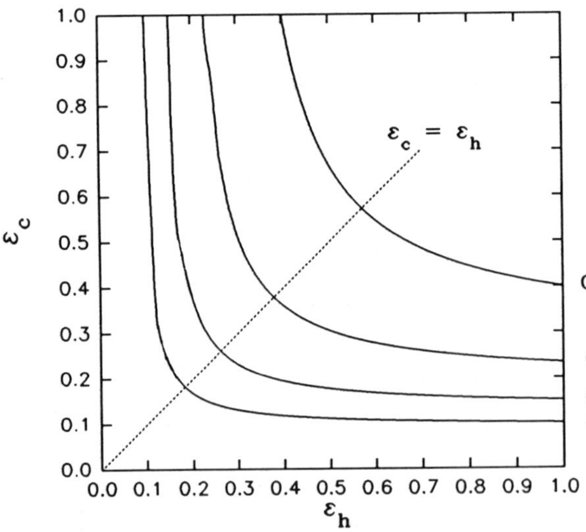

Figure 6.17. ε_c versus ε_h for a ten-stage heat pipe heat exchanger.

$$A_h = \frac{W\,c\,\mathrm{NTU}_h}{h_h} \tag{6.98}$$

$$A_c = \frac{W\,c\,\mathrm{NTU}_c}{h_c} \tag{6.99}$$

In the preceding equations, since $\Omega_{tc} = 1$, $W_h c_h = W_c c_c = Wc$. Also, NTU_h and NTU_c are obtained from Eqs. (6.73) and (6.76).

The total heat transfer area A_t for n stages is then expressible as

$$A_t = n\,(A_h + A_c) = \frac{n\,W\,c\,\mathrm{NTU}_h}{h_h}\left(1 + \frac{\mathrm{NTU}_c}{\mathrm{NTU}_h}\,\frac{h_h}{h_c}\right) \tag{6.100}$$

If the hot and cold side stage effectivenesses are equal, then $\varepsilon_h = \varepsilon_c = \varepsilon_{st}$, $\mathrm{NTU}_h = \mathrm{NTU}_c = \mathrm{NTU}_s$, and Eq. (6.96) takes the form

$$\varepsilon_{st} = \frac{2\,\varepsilon}{n - (n-1)\,\varepsilon} \tag{6.101}$$

Letting A_{ts} be the total heat transfer area for this case, we can write, using Eq. (6.100),

$$A_{ts} = \frac{n\,W\,c\,\mathrm{NTU}_s}{h_h}\left(1 + \frac{h_h}{h_c}\right) \tag{6.102}$$

Dividing Eq. (6.100) by Eq. (6.102),

$$\frac{A_t}{A_{ts}} = \frac{\mathrm{NTU}_h}{\mathrm{NTU}_s}\left(\frac{1 + (\mathrm{NTU}_c/\mathrm{NTU}_h)(h_h/h_c)}{1 + (h_h/h_c)}\right) \tag{6.103}$$

Equation (6.103) gives the ratio of total heat transfer area when the hot and cold side stage effectivenesses are unequal to the area when they are equal.

A plot of A_t/A_{ts} as a function of ε_h and h_h/h_c is shown in Figure 6.18 for a ten-stage heat exchanger with an overall effectiveness of 0.8. The plot was developed by specifying a value of ε_h and h_h/h_c, calculating ε_c from Eq. (6.96); calculating ε_s from Eq. (6.101); calculating NTU_h, NTU_c, and NTU_s from Eqs (6.73) and (6.76); and finally calculating A_t/A_{ts} from Eq. (6.103).

All the curves for the various values of h_h/h_c pass through the point $A_t/A_{ts} = 1$, $\varepsilon_h = 0.571$, which is the symmetrical condition ($\varepsilon_h = \varepsilon_c$). All the curves also pass through a minimum. At $\varepsilon_h = 0.4$, for which $\varepsilon_c \to 1$, A_t/A_{ts} approaches infinity for all the curves.

From Figure 6.18, it is evident that if $h_h/h_c \ll 1$ or $\gg 1$, the heat transfer area of a ten-stage heat exchanger with an overall effectiveness of 0.8 may be reduced by 10–20% if optimum values for ε_h and ε_c are used, instead of letting $\varepsilon_h = \varepsilon_c$.

The condition for which $h_h/h_c \ll 1$ can be approximated in aircraft gas turbine recuperators. Because of large density differences, the hot side (turbine exhaust stream) heat transfer coefficient may be 0.1–0.2 times that on the cold side (compressor discharge stream).

The ultimate reduction in heat transfer area occurs when $h_h/h_c = 0$ or $\to \infty$. Using Eq. (6.103), we may then write

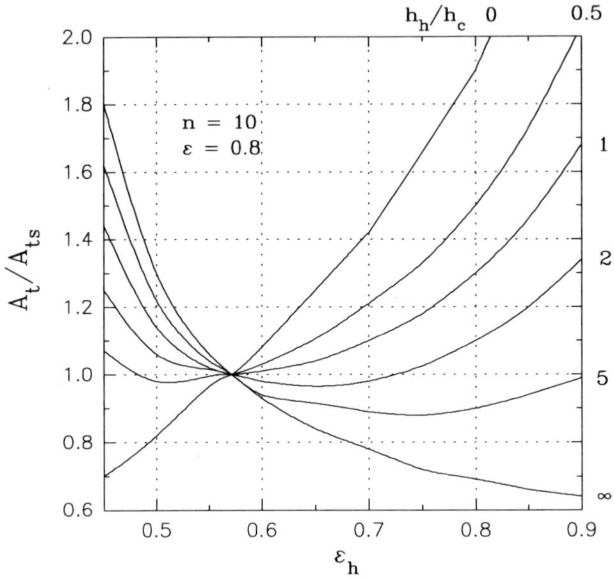

Figure 6.18. Ratio of heat transfer area for unequal hot and cold side stage effectivenesses to area when effectivenesses are equal.

$$\frac{A_t}{A_{ts}} = \frac{\text{NTU}_h}{\text{NTU}_s} \qquad \frac{h_h}{h_c} = 0$$

$$\frac{A_t}{A_{ts}} = \frac{\text{NTU}_c}{\text{NTU}_s} \qquad \frac{h_h}{h_c} \to \infty$$

The area ratios for these limiting cases are shown as a function of ε_h and ε in Figure 6.19, for a ten-stage heat exchanger. The lower ends of the curves correspond to $\varepsilon_c = 1$ when $h_h/h_c = 0$, and to $\varepsilon_h = 1$ when $h_h/h_c \to \infty$.

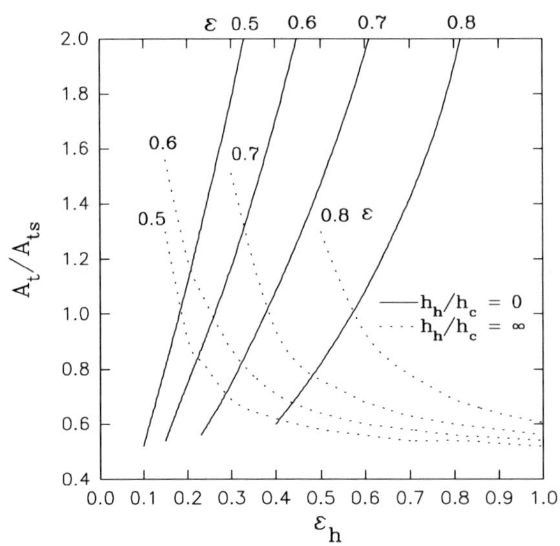

Figure 6.19. Ultimate limits on heat transfer area for a ten-stage heat pipe heat exchanger.

From Figure 6.19, it is clear that reductions in heat transfer area on the order of 40–50% are possible compared to the area when $\varepsilon_h = \varepsilon_c$. Such reductions can be expected when $h_h/h_c \ll 1$ or $\gg 1$, and the stage effectiveness on the low-resistance (high-h) side is close to unity. The heat transfer area will be reduced further as the number of stages increases beyond ten.

Generally, as the number of stages increases, there will be less surface area per stage. The lateral heat exchanger dimensions may become smaller, resulting in an increase in the fluid flow velocity. The combination of higher velocity and a longer flow path will produce a higher pressure drop through the heat exchanger. The number of allowable stages could then be limited by the allowable pressure drop.

The analysis presented in this section has been restricted to counterflow heat exchangers. Similar relations for parallel flow heat exchangers can be developed following the same general procedure. However, parallel flow heat exchangers are usually of less interest because of their lower effectiveness.

6.6.6. Comparison of Heat Pipe and Conventional Heat Exchangers

Typical elements of unfinned heat pipe and conventional heat exchangers are shown in Figure 6.20. The hot and cold side fluid directions are assumed to be in counterflow, and the unit thermal capacity ratio $\Omega_{tc} = W_c c_c / W_h c_h$ is assumed to be 1.

The unit thermal capacity rate is designated as Wc for both fluids, and is assumed to be the same for both heat exchanger types. The hot and cold side heat transfer coefficients are h_h and h_c for the heat pipe heat exchanger, and h_h' and h_c' for the conventional heat exchanger. The hot and cold side heat transfer areas in one stage of the heat pipe heat exchanger are A_h and A_c, and are the same in all stages. For n stages, the total heat transfer area A_t is $n(A_h + A_c)$. In the conventional heat exchanger, the total hot and cold side areas A_h' and A_c' are approximately equal, and are designated as A_t'.

In the heat pipe heat exchanger, both the hot and cold side areas are on the external surface of the heat pipes. In the conventional heat exchanger, the hot side area is on the

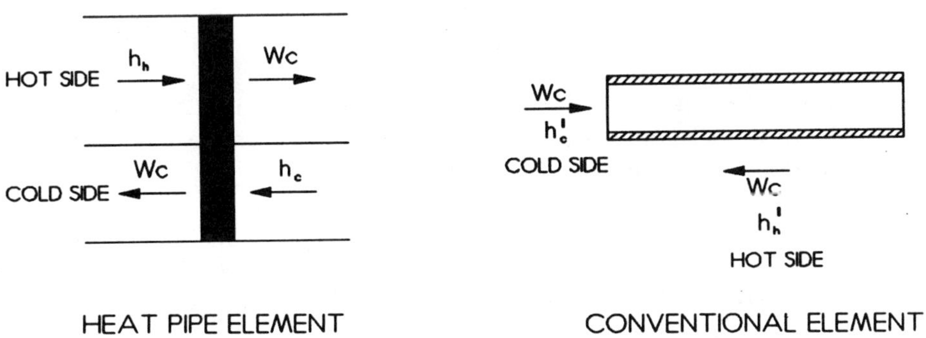

Figure 6.20. Typical heat pipe and conventional heat exchanger elements.

external surface of the flow elements, and the cold side area is on the internal surface. For heat exchanger elements of a specified diameter and wall thickness, the volume and weight of the elements are proportional to the external surface area.

Then, if $A_t > A_t'$, the core (region where heat transfer occurs) will be larger and heavier for the heat pipe heat exchanger. In addition to the core, a complete determination of heat exchanger size and weight will include the inlet and outlet plenums and any additional fluid containment structure. We will be concerned here only with the heat exchanger core.

Expressions for A_t and A_t' will now be derived and compared for various design conditions. As will be shown, $A_t > A_t'$ for some conditions, and $A_t < A_t'$ for others.

Heat Pipe Heat Exchanger. Heat transfer occurs first between the hot side heat exchanger fluid and vaporizing heat pipe liquid, and then between condensing heat pipe vapor and the cold side fluid. The thermal resistance of the intervening heat pipe wall and the vaporizing and condensing heat pipe fluid is generally small compared to that of the hot and cold side fluids, particularly if these fluids are gases. It is assumed here that only the thermal resistance of the hot and cold side fluids is significant.

Conventional Heat Exchanger. As with the heat pipe heat exchanger, the thermal resistance of the intervening wall between the hot and cold sides is usually small, and will be neglected here. Then, from Eq. (6.52), using primes to denote the conventional heat exchanger, it may be established that

$$U' = \frac{h_h' \, h_c'}{h_h' + h_c'}$$

where U′ denotes the unit thermal conductance for the conventional heat exchanger.

Also, from the basic definition of NTU,

$$\text{NTU}' = \frac{U' \, A_t'}{W \, c}$$

Substituting the previous equation for U′ in this equation and solving for the external heat transfer area At′,

$$A_t' = \frac{W \, c \, \text{NTU}'}{h_h'} \left(1 + \frac{h_h'}{h_c'} \right) \tag{6.104}$$

Heat Pipe and Conventional Heat Exchanger Comparison. The ratio of external heat transfer surface area for heat pipe and conventional heat exchangers is obtained by dividing Eq. (6.100) by Eq. (6.104), with the result

$$\frac{A_t}{A_t'} = \frac{n \, \text{NTU}_h \, h_h'}{\text{NTU}' \, h_h} \left[\frac{1 + (\text{NTU}_c/\text{NTU}_h)\,(h_h/h_c)}{1 + (h_h'/h_c')} \right] \tag{6.105}$$

The following procedure is used to evaluate Eq. (6.105):

1. Select values for n and ε_h.
2. Select or calculate values of h_h, h_c, h_h', and h_c'.

3. Calculate $\varepsilon_{max} = n/(n + 1)$.
4. Select a value of $\varepsilon < \varepsilon_{max}$.
5. Calculate ε_c from Eq. (6.96).
6. Calculate: $NTU_h = \ln [1/(1 - \varepsilon_h)]$; $NTU_c = \ln [1/(1 - \varepsilon_c)]$; $NTU' = \varepsilon/(1 - \varepsilon)$.
7. Calculate A_t/A_t' from Eq. (6.105).

A_t/A_t' will now be determined for three cases.

Case 1. Consider the situation when the hot and cold side stage effectivenesses of the heat pipe heat exchanger are equal, and when the hot and cold side heat transfer coefficients are the same for the heat pipe and conventional heat exchangers. Then,

$$\varepsilon_h = \varepsilon_c = \varepsilon_{st}$$

$$h_h = h_h' \qquad h_c = h_c'$$

It follows that

$$NTU_h = NTU_c = NTU_s$$

Equation (6.105) then reduces to

$$\frac{A_t}{A_t'} = n \frac{NTU_s}{NTU'}$$

The preceding equation can be presented in more convenient form by expressing the NTU terms as functions of n and ε.

Upon solving Eq. (6.73) for NTU_h, letting $NTU_h = NTU_s$ and $\varepsilon_h = \varepsilon_{st}$,

$$NTU_s = \ln \frac{1}{1 - \varepsilon_{st}}$$

Letting $\varepsilon_{cf} = \varepsilon$ and $NTU = NTU'$ in Eq. (6.59),

$$NTU' = \frac{\varepsilon}{1 - \varepsilon}$$

Using these equations for NTU_s and NTU', and Eq. (6.101) for ε_{st}, the equation for A_t/A_t' becomes

$$\frac{A_t}{A_t'} = n \frac{(1 - \varepsilon)}{\varepsilon} \ln \left[\frac{n - (n - 1)\,\varepsilon}{n - (n + 1)\,\varepsilon} \right] \tag{6.106}$$

In general, A_t/A_t' from Eq. (6.106) is greater than 2, and approaches 2 as the number of stages becomes very large. For the case when $n = 10$ and $\varepsilon = 0.8$, $A_t/A_t' = 2.12$.

Case 2. In this case, the hot and cold side heat transfer coefficients remain the same as in case 1 ($h_h = h_h'$, $h_c = h_c'$), but ε_h and ε_c may differ in magnitude. Equation (6.105) then assumes the form

$$\frac{A_t}{A_t'} = n \frac{NTU_h}{NTU'} \left[\frac{1 + (NTU_c/NTU_h)\,(h_h/h_c)}{1 + (h_h/h_c)} \right] \tag{6.107}$$

In Figure 6.21, A_t/A_t' from Eq. (6.107) is plotted against ε_h and h_h/h_c for a ten-stage heat exchanger and an overall heat exchanger effectiveness of 0.8. The ε_c corresponding to ε_h is

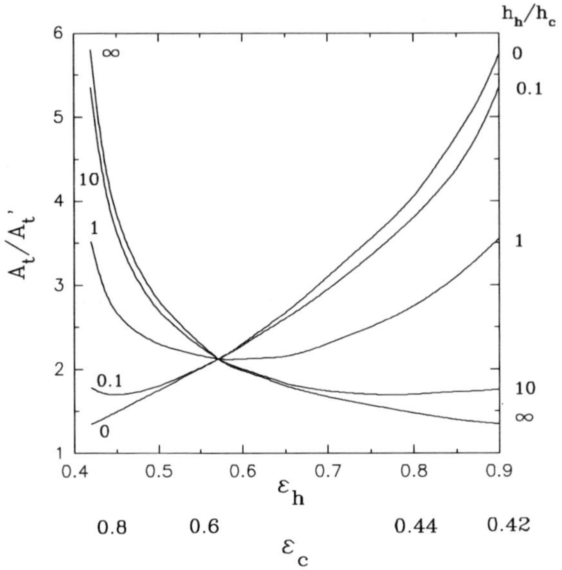

Figure 6.21. Ratio of external heat transfer area for heat pipe and conventional heat exchangers.

also shown. The curve for $h_h/h_c = 0.1$ roughly approximates the situation in a gas turbine recuperator, for which h_h would be the heat transfer coefficient in the turbine exhaust stream and h_c would be the heat transfer coefficient in the compressor discharge stream.

All of the curves pass through the point $A_t/A_t' = 2.12$, $\varepsilon_h = \varepsilon_c = 0.571$. Thus, when the hot and cold side stage effectivenesses are equal, $A_t/A_t' = 2.12$ regardless of the value of h_h/h_c. This point is also the minimum of the $h_h/h_c = 1$ curve, indicating that, when $h_h/h_c = 1$, the area of the heat pipe heat exchanger is at a minimum when the hot and cold side stage effectivenesses are equal.

When $h_h/h_c \neq 1$, the minimum A_t/A_t' is less than 2.12, and occurs when $\varepsilon_h \neq \varepsilon_c$. The lowest value of the minimum A_t/A_t' is 1.277. It occurs when $h_h/h_c = 0$, $\varepsilon_h = 0.4$, and $\varepsilon_c = 1$, and also when $h_h/h_c = \infty$, $\varepsilon_h = 1$, and $\varepsilon_c = 0.4$. (These limiting values are not shown in Figure 6.21.) As the number of stages increases, the minimum value of A_t/A_t' moves closer to 1. For example, when n increases from 10 to 20, the lowest value of the minimum A_t/A_t' decreases from 1.277 to 1.116.

It is clear that the discrepancy in core external surface area between heat pipe and conventional heat exchangers is small when the number of stages is large, the hot side heat transfer coefficient is either much smaller or much larger than the cold side heat transfer coefficient, and the stage effectivenesses have been optimized.

Case 3. For this case, different hot side and cold side heat transfer coefficients for the heat pipe and conventional heat exchangers are allowed.

The assumption made in cases 1 and 2 that the hot and cold side heat transfer coefficients are the same for heat pipe and conventional heat exchangers is not very realistic in general. Even if key design variables such as the hydraulic diameter and fluid flow velocity are the same for both heat exchanger types, the heat transfer coefficients may still differ if the core configurations are different.

In the examples that have been considered here, flow is parallel to the axes of the tubular heat transfer surfaces in the conventional heat exchanger, and perpendicular to the axes of the heat pipes in the heat pipe heat exchanger. The greater turbulence and interruption of the boundary layer as fluid flows across the heat pipes results in higher heat transfer coefficients (as well as higher friction factors) than in the conventional heat exchanger.

An accurate comparison of the relative size and weight of heat pipe and conventional heat exchangers requires detailed design studies for each contemplated application, taking into account design criteria such as allowable pressure drops, heat exchanger length, and heat exchanger lateral dimensions. Additionally, the size and weight of inlet and outlet plenums and other structural elements such as containment walls must be considered.

For illustrative purposes, the external surface area ratio A_t/A_t' will be calculated for the recuperator of a gas turbine with a pressure ratio of 20, flow velocities of 50 ft/sec (15.2 m/sec) on the hot and cold sides, a flow passage hydraulic diameter of 0.2 in. (0.00508 m), a mean hot side temperature of 1520°F (827°C), and a mean cold side temperature of 980°F (527°C). The hot and cold side fluids flow parallel to the axes of tubular heat transfer surfaces in the conventional heat exchanger, and normal to the heat pipe axes in the heat pipe heat exchanger.

The hot side and cold side heat transfer coefficients can be determined from this information, but only calculational results are presented here. Methods and data for calculating heat transfer coefficients and friction factors are discussed in the heat exchanger design example of Chapter 8. Data for flow normal to tube banks and inside tubes is given in Ref. 6.14. The inside-tube data can be applied to flow parallel to and outside tubes by using the hydraulic diameter for the external flow passages in calculating the Reynolds number.

The resulting values of the various heat transfer coefficient ratios are:

$$h_h/h_c = 0.154$$

$$h_h'/h_c' = 0.105$$

$$h_h/h_h' = 2.74$$

$$h_c/h_c' = 1.87$$

Thus, the heat transfer coefficients for the heat pipe heat exchanger are almost two to three times those of the conventional heat exchanger.

The calculation was carried out for counterflow heat exchange, equal unit thermal capacity rates on the hot and cold sides, ten heat pipe stages, and an overall heat exchanger effectiveness of 0.8. That is,

$$\Omega_{tc} = 1$$

$$n = 10$$

$$\varepsilon = 0.8$$

The cold side stage effectiveness ε_c was taken to be 0.8, for which the corresponding hot side stage effectiveness ε_h is 0.444. Then, using Eq. (6.105), $A_t/A_t' = 0.689$. If the same calculation is carried out for the case when $\varepsilon_h = \varepsilon_c = 0.571$, $A_t/A_t' = 0.806$.

Thus, the core size of the heat pipe heat exchanger is 70–80% of that for the conventional heat exchanger, depending on the extent to which the hot and cold side stage efficiencies are optimized.

Although the heat pipe heat exchanger core is smaller for the specified design conditions, its hot side friction factor is 8.8 times larger than that of the conventional heat exchanger, and its cold side friction factor is 6.6 times larger. Whether these higher friction factors and their resulting effect on the core pressure drop are acceptable, or whether design changes would be needed to reduce the friction factors at the expense of smaller heat transfer coefficients and a larger heat exchanger, are questions requiring additional study.

Although the case 3 example is in no way optimized, it does serve to show that a heat pipe heat exchanger core can be smaller than its conventional counterpart, and that the smaller size may well be accompanied by a substantially larger frictional resistance to fluid flow.

6.7. REFERENCES

6.1. Timoshenko, S., and Goodier, J. N., *Theory of Elasticity,* 2nd Ed., McGraw-Hill, New York, 1951, pp. 412–416.

6.2. Silverstein, C. C., "A Feasibility Study of Heat-Pipe-Cooled Leading Edges for Hypersonic Cruise Aircraft," CR-1857, NASA, Washington, D.C., November 1971, pp. 65, 66.

6.3. Penning, F. A., "Experimental Buckling Modes of Clamped Shallow Spherical Shells Under Concentrated Load," *Journal of Applied Mechanics,* June 1966, pp. 297–304.

6.4. Weinstein, I., "Heat Transfer and Pressure Distributions on Hemisphere-Cylinders in Methane-Air Combustion Products at Mach 7," TN D-7104, NASA, Washington, D.C., January 1973.

6.5. Timoshenko, S., and Gere, J. M., *Theory of Elastic Stability,* 2nd Ed., McGraw-Hill, New York, 1961, pp. 457–473.

6.6. Gorring, R. L., and Churchill, S. W., "Thermal Conductivity of Heterogeneous Materials," *Chemical Engineering Progress,* Vol. 57, No. 7, July 1961, pp. 53–59.

6.7. Chang, W. S., "Effective Thermal Conductivity of Wire Screens," Presented at the 24th ASME/AIChE National Heat Transfer Conference, Pittsburgh, Pennsylvania, August 1987.

6.8. Soliman, M. M., Graumann, D. W., and Berenson, P. J., "Effective Thermal Conductivity of Dry and Liquid-Saturated Sintered Fiber Metal Wicks," Paper 70-HT/SpT-40. Presented at the ASME Space Technology and Heat Transfer Conference, Los Angeles, California, June 1970.

6.9. Berger, M. E., and Feldman, K. T., "Analysis of Circumferentially Grooved Heat Pipe Evaporators," Paper 73-WA/HT-13. Presented at the ASME Winter Annual Meeting, Detroit, Michigan, November 1973.

6.10. Kroliczek, E. J., and Brennan, P. J., "Axial Grooved Heat Pipes—Cryogenic Through Ambient," Paper 73-ENAs-48. Presented at the SAE-ASME-AIAA-ASMA-AIChE Intersociety Conference on Environmental Systems, San Diego, California, July 1973.

6.11. Davison, H. W., "Compilation of Thermophysical Properties of Lithium," TN D-4650, NASA, Washington, D.C., July 1968.

6.12. Silverstein, C. C., "Heat Pipe Cooling for Scramjet Engines," Contractor Report 4036, NASA, Washington, D.C., December 1986, Appendix C.

6.13. Alario, J., Brown, R., and Kosson, R., "Monogroove Heat Pipe Development for the Space Constructible Radiator System," AIAA-83-1431, Presented at the AIAA 18th Thermophysics Conference, Montreal, Canada, June 1983.

6.14. Kays, W. M., and London, A. L., *Compact Heat Exchangers,* 2nd Ed., McGraw-Hill, New York, 1964.

CHAPTER
SEVEN

HEAT PIPE DESIGN: TRANSIENT BEHAVIOR

A heat pipe may operate under steady-state or transient conditions. In the steady state, heat is removed from the heat pipe at the same rate as heat is added, and the heat pipe temperature remains constant. In the transient state, there is an imbalance between the heat addition and rejection rates, and the heat pipe temperature varies with time.

The transient condition is always encountered during heat pipe startup from ambient temperature. If the heat input rate is small compared to the heat absorption capability of the heat pipe, the startup process will be slow and gradual. Overheating and/or operational difficulties are then not likely to be encountered. On the other hand, if the heat input rate is significant compared to the heat absorption capability, startup may be quite rapid, raising concerns about possible overheating and operational difficulties.

Potential startup problems are of special concern if the ambient temperature is below the freezing point of the heat pipe liquid. The vapor pressure inside the heat pipe is then extremely low, and the heat pipe may not become operational until the temperature of the evaporator section has risen substantially above the freezing point.

When the heat pipe does reach operational status, its heat transport capability could initially be insufficient to prevent freezeout of condensate if there is good thermal coupling between the heat pipe condenser and the heat sink. Dryout of the evaporator wick, overheating, and possible destruction of the heat pipe could soon follow.

If the thermal coupling is so effective that only a portion of the condenser length is needed to dissipate the incident heat load, the heat pipe might fail to become fully operational even if freezeout of condensate did not occur. The heat pipe would then not necessarily overheat, but could remain in a highly nonisothermal state throughout startup and into the steady-state operational phase.

In such circumstances, a *transient analysis* of the heat pipe during startup is advisable. The purpose of the transient analysis is threefold:

1. To determine whether overheating or excessive thermal stress develops in the heat pipe wall.
2. To determine whether heat pipe heat transport limits are exceeded.
3. To determine whether startup difficulty is encountered because of excessive thermal coupling between the heat pipe condenser and the heat sink.

These objectives are achieved by calculating the time dependence of the temperature at key heat pipe locations, the actual heat transport rate, and the position of the continuum front that separates heated vapor from the rest of the condenser. The adequacy of the heat pipe design during startup can then be established by comparing the temperatures and heat transport rate with temperature and heat transport limitations, and by observing whether the continuum front has reached the rear of the condenser.

The starting point for the transient analysis is the steady-state design. The steady-state design may be based on the heat load at the steady-state operating condition (if there is one), or on the maximum expected heat load during the transient. If the maximum heat load occurs after the entire heat pipe has become functional, peak transient heat pipe temperatures will probably not differ appreciably from steady-state temperatures based on the maximum heat load. If the peak heat load occurs before the entire heat pipe has become functional, predicted heat pipe temperatures based on the maximum heat load at steady-state conditions could be substantially overstated.

In general, the heat pipe startup process is quite complex, involving temperature variations in two or three space dimensions as well as over time, vapor heat transport in the free molecular and continuum flow regimes, expansion of the heat pipe vapor into a hard vacuum accompanied by condensation, nonequilibrium expansion effects in a nominally saturated vapor, and solid-liquid as well as liquid-vapor phase changes. One approach to this problem is described in Ref. 7.1.

A simplified treatment of the heat pipe startup process is presented here. Major features observed during heat pipe startups are modeled (Ref. 7.2), but less important details such as wall and wick temperature drops in the condenser are ignored. Considerable emphasis is placed on physical aspects of startup in developing the transient heat pipe model. Results obtained from this simplified analysis should be adequate for preliminary design purposes.

The transient model was developed specifically with liquid metals in mind. The general approach is applicable to other heat pipe fluids, but modifications would be necessary to accommodate conditions that are not encountered in the startup of liquid metal heat pipes.

Specifically, for heat pipes that do not use liquid metals, the heat pipe liquid may not be frozen at ambient temperature, and continuum flow conditions may already exist at startup; the flow of vapor out of the evaporator during startup could proceed at less than the sonic limit under some conditions; and lower thermal conductivities could require consideration of temperature drops in the condenser wall/wick in addition to that in the evaporator.

The heat pipe startup process has been described in some detail in Section 2.4. Aspects of the startup process that are relevant to development of the transient model are reviewed in the following section. Then, principal features of the transient model are described, followed by a discussion of key model parameters. Next, equations for the transient model are derived, and the calculational procedure is described. Finally, considerations involved in assessing the heat pipe startup process are presented.

An example based on the transient model developed here is given in Chapter 8 for a simple heat pipe space radiator.

7.1. HEAT PIPE STARTUP

The heat pipe startup process can be divided into three distinct phases: nonoperational, startup, and operational.

7.1.1. Phase 1: Nonoperational

In this phase, the heat pipe heat transport mechanism is not functional because the vapor density is too low to support continuum flow. Heat transport out of the evaporator by the heat pipe vapor is then negligible. Heat added to the heat pipe evaporator is absorbed solely by sensible heating, with a resultant rise in temperature. The temperature rise in the evaporator will result in some axial heat transfer between the evaporator and condenser via conduction along the heat pipe wall and wick. However, this is generally small and is neglected in the transient model.

Phase 1 extends from the beginning of the startup transient until the time when the vapor temperature and density in the evaporator are high enough to sustain continuum flow.

Phase 1 conditions are most likely to exist during startup if the heat pipe liquid was initially frozen, as is usually the case with liquid metal heat pipes at ambient temperature. The vapor space is then under a hard vacuum, with a density insufficient to support continuum flow. The transient model was developed on the assumption that phase 1 conditions are valid.

If continuum flow conditions are already present prior to startup, phase 1 does not exist. The analysis would then start with phase 2. This would generally be the case with a water heat pipe, unless the ambient temperature is close to or below the freezing point of water.

7.1.2. Phase 2: Startup

In this phase, the vapor temperature is high enough to sustain continuum flow, and the heat pipe heat transport mechanism is functional. A *continuum front* separates the heated vapor issuing from the evaporator from the remainder of the heat pipe, and gradually moves into the cooler condenser (or adiabatic section if present).

Some or all of the heat added to the evaporator is transported out of the evaporator by the heat pipe vapor. Some of this heat is sensibly absorbed in the region behind the continuum front, or dissipated to an adjacent external heat sink. The remainder is sensibly absorbed at the continuum front, raising the temperature of the condenser (or adiabatic section) just ahead of the front to that of the heated vapor. This transfer of heat to the cooler material ahead of the continuum front is the mechanism by which the front advances through the heat pipe. The velocity of the continuum front depends on the heat absorption capability of the wall and wick, and is generally much lower than the sonic velocity.

Virtually all of the vapor in the vicinity of the front condenses over a relatively short distance, on the order of the heat pipe diameter. For analytical purposes, the front is idealized as a region of zero length, with the temperature behind the front equal to that of the hot vapor, and the temperature ahead of the front equal to that in the condenser (or adiabatic section) prior to passage of the front.

In some instances, the condenser as well as the evaporator may be subjected to external heating during the transient. Consider, for example, a heat pipe that cools the leading edge of an aircraft wing by transporting heat incident on the leading edge to an aft surface of the wing (the condenser) for dissipation to the environment by radiation (Ref. 7.3). During the startup phase, the aerodynamic heat flux on the aft surface might be in excess of the radiative heat flux leaving that surface. Ahead of the continuum front, there would then be a net input of heat to the condenser section, increasing its temperature above ambient.

The condenser may also be completely insulated, with heat neither entering nor leaving the outer surface. The condenser temperature then remains at the ambient value until passage of the continuum front. Heat incident on the evaporator is dissipated both by heating of the region behind the front and by extension of the front into the unheated condenser region.

An insulated condenser acts in effect as its own heat sink. This type of heat sink can be utilized for cooling a missile during a flight of limited duration. All of the incident heat is then absorbed in the heat pipe structure and any supplemental internal heat sink that may be present (Ref. 7.4).

The region between the evaporator exit and the continuum front may or may not act as a condenser during the startup transient. If more heat is being added to the region than is being transferred away from the outer surface, the difference will be supplied by the heat released by condensing vapor (the usual case). If more heat is being removed from the region than is being lost from the outer surface, the excess will be used to vaporize liquid from the heat pipe wick. If there is a balance between heat lost from the outer surface and heat added to the region, there will be no evaporation or condensation.

If vapor is condensing behind the continuum front, this region will be acting in the normal role of a condenser. The sonic velocity and sonic limit then occur at the evaporator exit. There is also a temperature drop associated with acceleration of the vapor to sonic velocity. Vapor leaving the evaporator will continue to expand and cool, becoming supersonic before condensation is completed in the vicinity of the continuum front.

If liquid is vaporizing behind the continuum front, the entire length of heat pipe behind the front will act as an evaporator. The sonic velocity and the sonic limit are then reached at the continuum front instead of at the evaporator exit.

Phase 2 extends from the end of phase 1, when continuum flow is in effect, until the time that the continuum front reaches the rear end of the heat pipe.

7.1.3. Phase 3: Operational

As the continuum front approaches the end of the heat pipe, the high-vacuum region ahead of the front disappears. The vapor velocity once again becomes subsonic. The entire heat pipe then rapidly approaches the isothermal condition characteristic of normal operation, marking the onset of phase 3.

In phase 3, the heat pipe heat transport mechanism is operational over the entire heat pipe length. All the vapor is then at relatively uniform temperature. Although the velocity of the heat pipe vapor is now well below sonic, it is generally high enough that changes in the heat input rate will be propagated rapidly over the entire heat pipe length. Then, the heat pipe is assumed to respond to changes in the heat input rate as a single entity of uniform temperature.

Phase 3 continues from the end of phase 2, when the continuum front reaches the end of the heat pipe and disappears, until the startup process has been completed. If the heat input rate becomes constant and equal to the heat rejection rate during this time, the temperature will no longer vary, and the heat pipe will be operating in the steady-state mode.

7.2. FEATURES OF THE TRANSIENT MODEL

The principal features of a simplified transient model for characterizing the startup process will now be described. The model calculates the temperature history of selected points on a heat pipe without an adiabatic section, and the position of the continuum front as a function of time.

The basic approach to formulation of the transient model assumes that, at any time, the heat pipe consists of two or three individual segments (see Figure 7.1). Segment 1 is the evaporator, segment 2 is the condenser section behind the continuum front, and segment 3 is the condenser section ahead of the continuum front.

In phase 1, the continuum front has not yet formed. The length of segment 2 is then zero, and the heat pipe consists of segments 1 and 3. In phase 2, a continuum front forms at the evaporator exit and advances the distance x into the condenser, creating and expanding segment 2 and shrinking segment 3. In phase 3, segment 3 has shrunk to zero length, and the heat pipe consists of segments 1 and 2.

The assumed temperature distribution in the heat pipe during the various phases is indicated in Figure 7.2.

Each segment is characterized by a mean temperature, and by thermal properties evaluated at that mean temperature. Axial temperature gradients in the segments are assumed to be zero. In segments 2 and 3, temperature gradients through the heat pipe

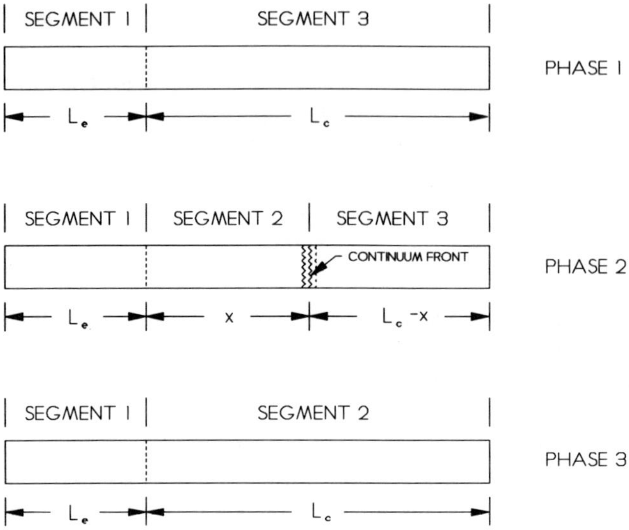

Figure 7.1. Heat pipe transient model.

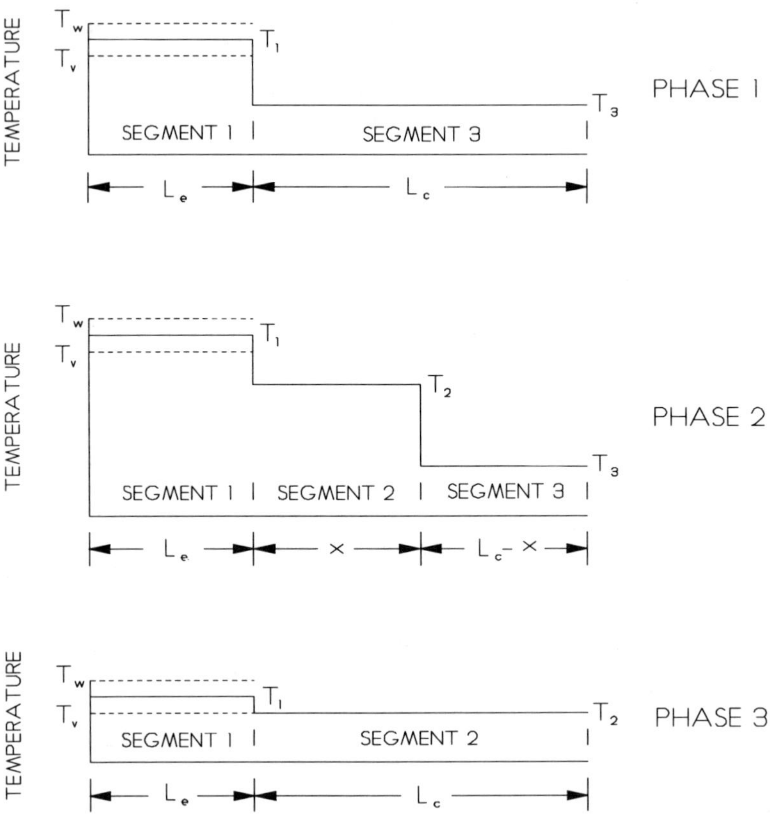

Figure 7.2. Temperature distribution in the heat pipe transient model.

wall and wick are assumed to be small. The temperature of these segments is then constant at a given time, and equal to the mean temperature.

In segment 1, where significant temperature gradients are most likely, the maximum surface temperature T_w and the vapor temperature T_v are obtained by adding and subtracting appropriate temperature increments to the mean temperature T_1. These temperature increments are derived from steady-state calculations of the temperature distribution through the evaporator wall and wick, at a point where the incident heat flux is a maximum.

The temperature drop at the liquid-vapor interface is assumed to be zero in all segments. Therefore, the vapor and segment temperatures are the same in segments 2 and 3.

In phase 1, the temperature of segments 1 and 3 are independent of each other, and vary in accordance with the incident heating rate on each segment. In phase 2, the vapor temperature in segment 2 is assumed to be below that in segment 1 by the drop in saturation temperature associated with the onset of sonic velocity in the vapor. The segment 3 temperature is equal to or greater than the ambient temperature, depending on whether segment 3 is subject to any net heating. In phase 3, the vapor temperature in segment 1 is the same as the segment 2 temperature.

The relationship between the temperatures of the heat pipe segments, the continuum front position, and the elapsed time is established from initial conditions, boundary conditions, and heat balances for each segment.

Initial conditions include the initial temperature of the heat pipe (generally ambient temperature), and the initial location of the continuum front (the evaporator exit). *Boundary conditions* specify the incident and outgoing heat fluxes as a function of surface coordinates, surface temperature, and time. *Heat balances* equate the net heat input to the change in sensible heat for each segment. The temperature dependence of relevant thermal properties is required in order to evaluate the change in sensible heat.

In phase 2, a portion of the heat input to segment 2 is used to heat sensibly a differential length of segment 3 at the continuum front to the segment 2 temperature, thus advancing the continuum front and adding to the segment 2 length.

7.3. PARAMETERS USED IN THE TRANSIENT MODEL

In this section, expressions are presented for key parameters that are used in the transient model. Topics covered include thermal capacity, temperature for continuum flow, heat transport and heat transfer rates, and evaporator surface temperatures.

7.3.1. Thermal Capacity

The change in sensible heat content of a material is described in the transient model using the concepts of volumetric thermal capacity and thermometric thermal capacity.

The *volumetric thermal capacity* C_v is equal to the heat required to increase the temperature of a volume v of material at temperature T by 1°. The *thermometric thermal capacity* C_t is the heat required to raise the temperature of a unit volume of material from T_{ref} to T. These properties are expressible by the equations

$$C_v = \rho c v \tag{7.1}$$

$$C_t(T) = \int_{T_{ref}}^{T} \rho c \, dT \tag{7.2}$$

where ρ is the density of a material at temperature T, c is its specific heat at T, and T_{ref} is a reference temperature (usually at or near the ambient temperature).

If the initial temperature of a unit volume of material is $T_a \neq T_{ref}$, the heat required to raise its temperature to T is

$$\int_{T_a}^{T} \rho c \, dT = \int_{T_{ref}}^{T} \rho c \, dT - \int_{T_{ref}}^{T_a} \rho c \, dT = C_t(T) - C_t(T_a) \tag{7.3}$$

It follows that the heat added to a given volume when its temperature T increases by dT is $C_v dT$. Similarly, the heat added to a differential volume dv when its temperature increases from T_a to T is $[C_t(T) - C_t(T_a)]\,dv$.

The *volumetric thermal capacity of a heat pipe segment* is equal to the sum of the C_v's for each component. Thus,

$$C_v = \rho_{wall} c_{wall} v_{wall} + \rho_s c_s (1 - \phi) v_{wick} + \rho_l c_l \phi v_{wick} \tag{7.4}$$

where ρ_{wall} is the wall density, c_{wall} is the wall specific heat, v_{wall} is the volume of the wall in the heat pipe segment, ρ_s is the density of the solid part of the wick, c_s is the specific heat of the solid part of the wick, ρ_l is the density of the wick liquid, c_l is the specific heat of the wick liquid, ϕ is the liquid volume fraction in the wick (mean liquid fraction for two-layer wick), and v_{wick} is the volume of the wick in the heat pipe segment.

The thermal capacity of the heat pipe vapor is generally negligible, and has not been included in the preceding equation.

A more convenient form of the volumetric thermal capacity for use in the transient model is

$$C_v = \{\rho_{wall} c_{wall} \alpha_{wallm} + [\rho_s c_s (1 - \phi) + \rho_l c_l \phi](1 - \alpha_{wallm})\} A_{wwm} L_s \tag{7.5}$$

where A_{wwm} is the average cross-sectional area of the wall and wick, L_s is the segment length, α_{wallm} is the fraction of A_{wwm} occupied by the wall, and $1 - \alpha_{wallm}$ is the fraction occupied by the wick.

If the same material is used for the wall and wick, Eq. (7.5) becomes

$$C_v = \{\rho_s c_s [\alpha_{wallm} + (1 - \phi)(1 - \alpha_{wallm})] + \rho_l c_l \phi (1 - \alpha_{wallm})\} A_{wwm} L_s \tag{7.6}$$

If additional components such as an end cap or rib are present in the heat pipe segment, the C_v of these components should be included in Eqs. (7.5) and (7.6).

The *thermometric thermal capacity at a location x in the heat pipe* is given by

$$C_t(T) = \int_{T_{ref}}^{T} \frac{\rho_{wall}\, c_{wall}\, A_{wall} + \rho_s\, c_s\, (1 - \phi) A_{wick} + \rho_l\, c_l\, \phi\, A_{wick}}{A_{ww}} \, dT$$

or

$$C_t(T) = C_{t\text{wall}}(T)\,\alpha_{\text{wall}} + [(1-\phi)\,C_{ts}(T) + \phi\,C_{tl}(T)]\,(1-\alpha_{\text{wall}}) \tag{7.7}$$

where A_{ww} is the cross-sectional area of the wall and wick at x, α_{wall} is the fraction of A_{ww} occupied by the wall and $1-\alpha_{\text{wall}}$ is the fraction occupied by the wick, $C_{t\text{wall}}$ is the thermometric thermal capacity of the wall material, C_{ts} is the thermometric thermal capacity of the wick solid, and C_{tl} is the thermometric thermal capacity of the wick liquid.

If the same material is used for the wall and wick, the above equation reduces to

$$C_t(T) = [1-\phi\,(1-\alpha_{\text{wall}})]\,C_{ts}(T) + \phi\,(1-\alpha_{\text{wall}})\,C_{tl}(T) \tag{7.8}$$

For a heat pipe of constant cross section, A_{ww} and A_{wwm} are constant and equal.

In the preceding equations, the segment volumes, areas, and area fractions are assumed to be independent of temperature and equal to their values at ambient temperature.

If the heat pipe liquid is frozen when the startup process begins, the specific heat term in the liquid component of the thermal capacity requires reinterpretation. Below the melting point, the specific heat of the frozen liquid is used.

At the melting point, the heat of fusion is substituted for the specific heat. For example, the melting point of sodium is 208°F (97.8°C), and its heat of fusion is 48.7 Btu/lb (113,300 J/kg). Therefore, the specific heat of sodium between 208 and 209°F (97.8 and 98.3°C) is taken to be 48.7 Btu/lb-°F (203,900 J/kg-°K). Similarly, the melting point of lithium is 367°F (186°C), and its heat of fusion is 186 Btu/lb (432,600 J/kg). Between 367 and 368°F (186 and 187°C), the specific heat of lithium is taken to be 186 Btu/lb-°F (778,700 J/kg-°K).

Information on ρ, c, and C_t as a function of temperature is given in Tables C.3, C.4, and C.5 of Appendix C for several materials. At the melting point of the heat pipe liquids, the heat of fusion has been substituted for the specific heat.

7.3.2. Temperature for Continuum Flow

The *continuum flow temperature* T_{cf} is the lowest vapor temperature in the evaporator at which continuum flow conditions exist. In the transient model, axial heat transport out of the evaporator is assumed to be negligible until $T_v \geq T_{cf}$. The transition from phase 1 to phase 2 occurs at that point. However, since axial heat transport predicted by phase 2 equations is quite small until $T_v \geq T_{cf}$, the startup transient can optionally begin with phase 2.

Continuum flow can be expected when the mean free path of vapor molecules is a sufficiently small fraction of the minimum vapor space dimension. Since the development of continuum flow is a gradual rather than an abrupt process, the criterion for the onset of phase 2 based on the existence of continuum flow is somewhat arbitrary.

The relationship between the mean free path λ and properties of the heat pipe vapor can be expressed as (Ref. 7.5)

$$\rho_v \lambda = \mu_v \left(\frac{\pi M}{2 R_u T_v} \right)^{1/2} \tag{7.9}$$

where R_u is the universal gas constant, M is the molecular weight, μ_v is the viscosity, and ρ_v is the density.

The product $\rho_v\lambda$ is almost independent of the vapor temperature. For sodium vapor, $\rho_v\lambda$ varies by 2.8% over the temperature range 1300 to 2100°F (704 to 1149°C). For lithium vapor, the variation is 4.5% over the temperature range 2000 to 2800°F (1093 to 1538°C). Representative values of $\rho_v\lambda$ for these fluids are:

Fluid	Vapor temperature, °F (°C)	$\rho_v\lambda$, lb/ft^2 (kg/m^2)
Sodium	1300	8.93×10^{-9}
	(704)	(4.36×10^{-8})
Lithium	2000	3.52×10^{-9}
	(1093)	(1.72×10^{-8})

A reasonable assumption is that continuum flow will be in effect when

$$\lambda = 0.01 H_v$$

where H_v is the minimum vapor space dimension.

It then follows that, for continuum flow to be reasonably well established,

$$\rho_v = \frac{\rho_v \lambda}{\lambda} = \frac{100\,(\rho_v \lambda)}{H_v} \tag{7.10}$$

Using a representative value of $\rho_v\lambda$ for the heat pipe fluid in question and an appropriate value for H_v, the vapor density at the onset of continuum flow can be found from Eq. (7.10). The corresponding temperature T_{cf} can then be obtained from a table or equation of vapor density as a function of saturation temperature.

An alternate approach is to solve directly for T_{cf} from vapor pressure–temperature relations. Using the perfect gas equation to express the vapor density in terms of the vapor pressure and temperature, along with Eq. (7.10),

$$\rho_v = \frac{P_v M}{R_u T_v} = \frac{100\,(\rho_v \lambda)}{H_v}$$

where T_v is the absolute vapor temperature. Solving for the vapor pressure P_v in the above equation,

$$P_v = \frac{100\, R_u\, (\rho_v \lambda)\, T_v}{M H_v} \tag{7.11}$$

Another relation between P_v and T_v is given by Eq. (5.100) for saturated fluids, which is repeated below:

$$\log_{10} P_v = A - \frac{B}{T_v} \tag{5.100}$$

Simultaneous solution of Eqs. (7.11) and (5.100) gives the vapor temperature $T_v \equiv$

Table 7.1. Continuum conditions for sodium and lithium

H_v, in. (m)	T_{cf}, °F (°C)	P_{cf}, psia (N/m^2)	ξ_s, Btu/in^2-sec (W/m^2)	Q_s, Btu/sec (W)
		Sodium		
0.1	911	0.0684	1.397	0.0110
(0.00254)	(488)	(472)	(2.29×10^6)	(11.61)
0.2	850	0.0328	0.706	0.0222
(0.00508)	(454)	(226)	(1.16×10^6)	(23.4)
0.5	777	0.0124	0.286	0.0562
(0.01270)	(414)	(85.5)	(4.68×10^5)	(59.3)
1.0	727	0.0059	0.1438	0.1129
(0.02540)	(386)	(40.7)	(2.35×10^5)	(119.1)
		Lithium		
0.1	1599	0.1338	5.962	0.0468
(0.00254)	(871)	(923)	(9.75×10^6)	(49.4)
0.2	1510	0.0643	3.007	0.0945
(0.00508)	(821)	(443)	(4.92×10^6)	(99.7)
0.5	1403	0.0243	1.212	0.238
(0.01270)	(762)	(167)	(1.983×10^6)	(251)
1.0	1330	0.0117	0.612	0.480
(0.02540)	(721)	(80.7)	(1.001×10^6)	(506)

T_{cf} and vapor pressure $P_v \equiv P_{cf}$ at which continuum flow is in effect. An iterative solution can be obtained in which a value of T_v is assumed, and P_v is calculated from both equations. The process is repeated with a new value of T_v until successive calculated values of P_v differ by a sufficiently small amount.

Representative values of T_{cf} and P_{cf} for sodium and lithium are given in Table 7.1 as a function of the minimum vapor space dimension H_v. The values of $\rho_v\lambda$ listed earlier were used in the calculations.

From Table 7.1 it can be seen that, in a sodium heat pipe, phase 2 starts when the vapor temperature has reached 700–900°F (371–482°C). For a lithium heat pipe, the vapor temperature rises to 1300–1600°F (704–871°C) before phase 2 is initiated.

The axial heat transport rate out of the evaporator cannot exceed the sonic limit. The sonic limit per unit of vapor flow area ξ_s and the total sonic limit Q_s corresponding to T_{cf} (assuming a circular heat pipe) are also given in Table 7.1.

It is evident that axial heat transport is quite small up to the onset of phase 2, particularly for small vapor space dimensions. If the transient were started in phase 2 instead of phase 1, the temperature response would be virtually identical until the vapor temperature reached T_{cf}, when phase 2 would become effective in any event. Therefore, another approach to calculation of the startup transient is simply to start with phase 2, ignoring the continuum temperature criterion and phase 1.

Since comparable results can be expected with either approach, whether the transient calculation is started with phase 1 or phase 2 is a matter of individual choice. However, calculational time is shorter with a phase 1 start because the phase 1 equations are fewer and simpler than those of phase 2. (See Section 7.4.)

7.3.3. Axial Heat Transport and Surface Heat Transfer

Heat is transported axially out of segment 1 (the evaporator) and into segment 2 by heat pipe vapor at the rate Q_a, given by

$$Q_a = \xi_a A_v \tag{7.12}$$

where ξ_a is the axial heat flux and A_v is the vapor space cross-sectional area at the evaporator exit.

If the heat pipe vapor is condensing in segment 2 (or if there is no phase-change heat transfer), Q_a is equal to the sonic limit $Q_s = \xi_s A_v$. If the heat pipe liquid is evaporating in segment 2, the sonic limit occurs at the continuum front. Q_a is then less than Q_s. These points are discussed in greater detail in Section 7.4.2.

The heat transfer rate Q through the outer surface A_o of a heat pipe segment at time t is given by

$$Q = \int_A q \, dA_o = q_m A_o \tag{7.13}$$

where q is the surface distribution of net heat flux (heat input minus heat output heat flux) and q_m is the average value of q.

If the heat flux is uniform over the outer segment surface,

$$Q = qA_o \tag{7.14}$$

In general, q, q_m, and Q vary with time, and may also be functions of the outer surface temperature.

7.3.4. Surface and Vapor Temperatures in the Evaporator

In the transient model, only the mean temperature T_1 of the evaporator (segment 1) is calculated directly. The maximum outer wall temperature T_{womax} at time t and the inner wick surface temperature T_v are estimated from steady-state temperature relationships.

Letting $\Delta T_{max} = T_{womax} - T_v$, T_{womax} and T_v are calculated from the relations

$$T_{womax} = T_1 + x_1 \Delta T_{max} \tag{7.15}$$

$$T_v = T_1 - x_2 \Delta T_{max} \tag{7.16}$$

where x_1 and x_2 are temperature coefficients and ΔT_{max} is the one-dimensional steady-state temperature drop across the wall and wick thicknesses (based on the average of the maximum entering heat flux q_{1max} and the maximum leaving heat flux q_{amax}). Since the liquid-vapor interface temperature drop is assumed to be negligible, the inner wick surface and the heat pipe vapor have the same temperature.

In the transient model, only the total heat rate Q_a leaving the evaporator is calculated. The maximum heat flux leaving the evaporator inner wick surface q_{amax} is estimated by multiplying the average surface heat flux over the inner surface q_a by the ratio of q_{1max} to the average heat flux q_{1m} on the outer surface. Thus,

$$q_{amax} = \left(\frac{Q_a}{2 \pi r_v L_e} \right) \left(\frac{q_{1max}}{q_{1m}} \right) \tag{7.17}$$

The average heat flux q_{1a} to be used in calculating ΔT_{max} is then

$$q_{1a} = \frac{q_{1max} + q_{amax}}{2} \tag{7.18}$$

The difference between q_{1a} and q_{1max} is greatest in phase 1, when Q_a and q_{amax} are zero. Then, $q_{1a} = q_{1max}/2$. As Q_a increases and becomes comparable to Q_1 in phases 2 and 3, the approximation that $q_{1a} = q_{1max}$ becomes increasingly good. In phase 3 of the transient model, it is assumed that $q_{1a} = q_{1max}$.

Optionally, it may be assumed that $q_{1a} = q_{1max}$ in all phases of the startup transient, at the expense of greater inaccuracy in calculating the evaporator outer and inner surface temperatures during the earlier phases.

Calculation of Temperature Coefficients. The temperature coefficients will now be obtained for a cylindrical heat pipe with a constant heat flux q_1 on the evaporator outer surface. The steady-state temperature distribution through the wall/wick is then a function of the radius r only, and can be expressed as

$$T = T_v + \frac{q_1 r_w}{k_{1m}} \ln \frac{r}{r_v} \tag{7.19}$$

where r_w is the radius of the wall outer surface, r_v is the radius of the wick inner surface, r is the radius, k_{1m} is the mean wall/wick thermal conductivity of the evaporator (segment 1), and T_v is the temperature of the heat pipe vapor and wick inner surface.

The total temperature drop across the wall and wick is obtained from Eq. (7.19) by letting $r = r_w$. The corresponding $T = T_{womax}$. Then

$$\Delta T_{max} = T_{womax} - T_v = \frac{q_1 r_w}{k_{1m}} \ln \frac{r_w}{r_v} \tag{7.20}$$

Now $r_w = r_v + l$, where l is the sum of the wall and wick thicknesses.

The mean wall/wick temperature T_1 is obtained by weighting the temperature T at radius r by the differential volume $2\pi r L_e dr$, integrating over the total thickness, and dividing by the total wall/wick volume $\pi(r_w^2 - r_v^2)L_e$, where L_e is the evaporator length.

Upon performing the integration using Equation (7.19), substituting $r_v + l$ for r_w, letting $z = l/r_v$, and simplifying,

$$T_1 = T_v + \left[\frac{(1+z)^2}{z\,(2+z)} - \frac{0.5}{\ln\,(1+z)}\right] \Delta T_{max} \tag{7.21}$$

On substituting this equation into Eq. (7.16) and solving for x_2,

$$x_2 = \frac{(1+z)^2}{z\,(2+z)} - \frac{0.5}{\ln\,(1+z)} \tag{7.22}$$

Also, letting $T_{womax} = T_v + \Delta T_{max}$ in Eq. (7.15), solving for x_1, and using Eq. (7.16),

$$x_1 = 1 - x_2 \tag{7.23}$$

Equations (7.22) and (7.23) represent the desired temperature coefficients for cylindrical heat pipe geometry.

Table 7.2. Temperature coefficients in cylindrical geometry

z	x_1	x_2
0.00	0.5000	0.5000
0.05	0.4919	0.5081
0.10	0.4841	0.5159
0.15	0.4767	0.5233
0.20	0.4697	0.5303

In the limit as $z \to 0$, it can be shown that Eq. (7.21) reduces to

$$T_1 = T_v + 0.5\,\Delta T_{\max}$$

It then follows from Eqs. (7.15) and (7.16) that

$$x_1 = 0.5 \qquad x_2 = 0.5$$

These are just the temperature coefficients for heat transfer through a slab in rectangular geometry. Thus, as expected, heat pipe cylindrical geometry may be approximated by rectangular geometry when the wall and wick thicknesses are small compared to the heat pipe radius.

Values of the temperature coefficients x_1 and x_2 for cylindrical geometry are given in Table 7.2.

If the heat flux on the evaporator surface varies with circumferential position and/or axial length, the temperature distribution through the evaporator will be two- or three-dimensional. Calculational methods for two- and three-dimensional heat transfer are described in Ref. 7.6. The temperature coefficients will then differ from those in Table 7.2, which are based on one-dimensional heat transfer.

For maximum accuracy, the temperature coefficients should be based on the actual steady-state temperature distribution. However, the one-dimensional coefficients, with $\Delta T_{\max}$ evaluated at the maximum surface heat flux, may be used to estimate evaporator surface temperatures. Such calculations should be adequate for preliminary design purposes, particularly if the wall/wick ΔT is not too large (on the order of 100°F [56°C]).

Calculation of $\Delta T_{\max}$. For rectangular geometry, the steady-state surface heat flux remains constant throughout the thickness of the wall/wick. Then, $\Delta T_{\max}$ may be calculated from the equation

$$\Delta T_{\max} = \frac{q_{1a}\, l}{k_{1m}}$$

For cylindrical geometry, the steady-state surface heat flux through the wall/wick thickness varies inversely with the radius. In Eq. (7.20), $\Delta T_{\max}$ is expressed in terms of the heat flux q_1 on the evaporator surface. It may readily be shown that, in terms of the linearly averaged heat flux q_{1a},

$$q_1 = \left[\frac{2}{1 + (r_w/r_v)}\right] q_{1a}$$

Therefore, $\Delta T_{\max}$ may be calculated from Eq. (7.20), using this expression for q_1.

Comparison of Actual and Estimated Transient Temperatures. Actual and estimated evaporator surface temperatures will now be compared for phase 1, when the inner wick surface is in effect thermally insulated. All of the incident heat is then absorbed in raising the wall/wick temperature, and the heat flux at the inner wick surface is zero. The difference between actual and estimated temperatures is greatest in phase 1.

A constant heat flux q_1 is assumed to be applied to the outer evaporator surface. The thickness of the combined wall/wick is l, and the effective thermal conductivity is k_{1m}. The heat flux q_a at the inner wick surface is zero. Rectangular geometry is assumed ($z = 0$), and the intial temperature is assumed to be 0°. The situation is then as shown in Figure 7.3.

Actual surface temperatures can be determined from an analytical solution (Ref. 7.7). After a period of time in which exponential terms initially present die away, the transient temperature distribution takes the form

$$T = \frac{q_1 l}{k_{1m}} \left(\frac{k_{1m} t}{\rho_{1m} c_{1m} l^2} + \frac{3y^2 - l^2}{6 l^2} \right) \tag{7.24}$$

where y is the distance from the inner wick surface, ρ_{1m} is the mean density of the wall/wick, c_{1m} is its mean specific heat, and t is the elapsed time.

From Eq. (7.24), it follows that

$$T_{wo\max} = T(l) = \frac{q_1 l}{k_{1m}} \left(\frac{k_{1m} t}{\rho_{1m} c_{1m} l^2} + \frac{1}{3} \right)$$

$$T_v = T(0) = \frac{q_1 l}{k_{1m}} \left(\frac{k_{1m} t}{\rho_{1m} c_{1m} l^2} - \frac{1}{6} \right)$$

$$\Delta T_{\max} = T_{wo\max} - T_v = \frac{q_1 l}{2 k_{1m}} \tag{7.25}$$

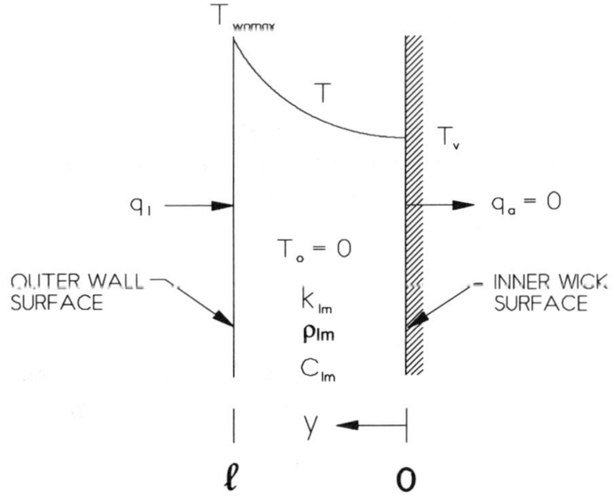

Figure 7.3. Model of evaporator in phase 1.

$$T_1 = \int_0^l \frac{T}{l}\,dx = \frac{q_1\,t}{\rho_{1m}\,c_{1m}\,l} \tag{7.26}$$

It should be noted from Eqs. (7.25) and (7.26) that ΔT_{max} is constant, and that T_1 varies linearly with time.

Using the last two equations, T_{womax}, T_v, and T may be expressed in terms of T_1 and ΔT_{max} as

$$T_{womax} = T_1 + \frac{2}{3}\,\Delta T_{max} \tag{7.27}$$

$$T_v = T_1 - \frac{1}{3}\,\Delta T_{max} \tag{7.28}$$

$$T = T_1 + \left[\left(\frac{y}{l}\right)^2 - \frac{1}{3}\right]\Delta T_{max} \tag{7.29}$$

When $T = T_1$, $y/l = 0.577$, as can be established from Eq. (7.29).

Next, solutions for the estimated evaporator surface temperatures will be considered.

For steady-state heat transfer with a heat flux equal to the mean transient heat flux, the temperature drop is

$$\Delta T_{max} = \frac{q_{1a}\,l}{k_{1m}}$$

where q_{1a} is the mean surface heat flux in the wall/wick.

From Eq. (7.18),

$$q_{1a} = \frac{q_1 + 0}{2} = \frac{q_1}{2}$$

Substituting $q_1/2$ for q_{1a} in the preceding equation for ΔT_{max},

$$\Delta T_{max} = \frac{q_1\,l}{2\,k_{1m}} \tag{7.30}$$

From Eqs. (7.25) and (7.30), it is evident that the steady-state temperature drop based on the mean transient heat flux is identical to the transient temperature drop.

From Table 7.2, $x_1 = x_2 = 0.5$ when $z = 0$. Therefore, the estimated outer and inner surface temperatures are, using Eqs. (7.15) and (7.16),

$$T_{womax} = T_1 + 0.5\,\Delta T_{max} \tag{7.31}$$

$$T_v = T_1 - 0.5\,\Delta T_{max} \tag{7.32}$$

The estimated surface temperatures as determined from Eqs. (7.31) and (7.32) may be compared to the actual surface temperatures as given by Eqs. (7.27) and (7.28). For a representative situation in which T_1 at a given time is 700°F (371°C) and ΔT_{max} is 100°F (56°C), the actual and estimated surface temperatures are:

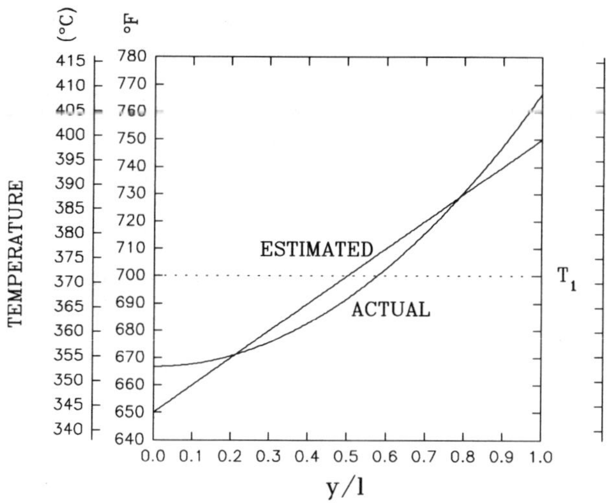

Figure 7.4. Representative actual and estimated temperature distributions through evaporator wall/wick in phase 1.

	Actual	Estimated
T_{womax}, °F (°C)	767	750
	(408)	(399)
T_v, °F (°C)	667	650
	(353)	(343)

The actual and estimated temperature distributions in the evaporator wall/wick are shown in Figure 7.4.

In this example, the steady state–based approximation underestimates the outer and inner surface temperatures by 17°F (9–10°C).

7.4. TRANSIENT EQUATIONS

The equations used to calculate the heat pipe segment temperatures and the continuum front position during the startup transient will now be derived. The equations are the result of heat balances applied to each of the heat pipe segments. Each phase of the transient is considered separately.

In all phases, the maximum outer surface temperature T_{womax} and the inner wick surface temperature T_v of segment 1 are obtained from Eqs. (7.15) and (7.16). An expression for ΔT_{max} is used that is applicable to the heat pipe geometry under consideration, and evaluated at the maximum surface heat flux.

7.4.1. Phase 1

In phase 1, only segments 1 and 3 are present, with fixed lengths L_e and L_c. All the heat incident on the surface of segment 1 (the evaporator) is absorbed by the wall and wick.

If there is net heating of Segment 3 (the condenser), all the incident heat will be similarly absorbed. Heat added to a segment in the time interval dt is equal to the increase in thermal capacity of the segment. A heat balance for segments 1 and 3 then yields

$$Q_1 dt = C_{v1} dT_1$$

$$Q_3 dt = C_{v3} dT_3$$

where Q_1 and Q_3 are the heat transfer rates to segments 1 and 3 at time t.

From these equations, the differential changes in the mean segment temperatures T_1 and T_3 are

$$dT_1 = \frac{Q_1\, dt}{C_{v1}} \tag{7.33}$$

$$dT_3 = \frac{Q_3\, dt}{C_{v3}} \tag{7.34}$$

The mean segment temperatures T_1 and T_3 in phase 1 are found from Eqs. (7.33) and (7.34), and the segment 1 surface temperatures from Eqs. (7.15) and (7.16).

7.4.2. Phase 2

In phase 2, the hot continuum front has moved a distance x from the evaporator exit. Three heat pipe segments are then present: segment 1 (the evaporator) of length L_e, segment 2 (the condenser section behind the continuum front) of length x, and segment 3 (the condenser length ahead of the continuum front) of length $L_c - x$. The segment temperatures T_1, T_2, T_3, T_{womax}, and T_v are to be calculated, along with the continuum front position x.

The heat balances on segments 1 and 2 now include heat transport between the segments by the heat pipe vapor. Within segment 2, the latent heat content of the vapor may be reduced by condensation on the wick, or augmented by vaporization of liquid from the wick. The vapor condenses completely at the continuum front. In the process, a differential length dx of condenser in segment 3 is heated to the segment 2 temperature, thereby becoming part of that segment.

Heat balances for the three segments will now be considered.

For segment 1, heat added during a time increment dt is equal to the increase in heat content of the segment plus the heat that is transported out of the segment by the heat pipe vapor. Therefore,

$$Q_1 dt = C_{v1} dT_1 + Q_a dt$$

where Q_a is the heat transport rate of the vapor out of segment 1.

For segment 2, heat is transferred through the outer surface and by the flow of vapor from segment 1. The heat is absorbed by sensible heating of segment 2 and by extending the length of the segment at the continuum front by dx. Then

$$Q_a dt + Q_2 dt = C_{v2} dT_2 + [C_t(T_2) - C_t(T_3)] A_{ww} dx$$

If heat is rejected from the outer surface of segment 2, Q_2 is negative.

If there is no heat transfer into the segment 3 outer surface, T_3 is equal to the ambient temperature, and $C_t(T_3) = C_t(T_a)$. If the ambient temperature is the same as the reference temperature, $C_t(T_a) = 0$.

For segment 3, heat added is equal to the increase in heat content of the segment. Thus,

$$Q_3 dt = C_{v3} dT_3$$

Solving these equations for dT_1, dx, and dT_3,

$$dT_1 = \frac{(Q_1 - Q_a)\, dt}{C_{v1}} \tag{7.35}$$

$$dx = \frac{(Q_a + Q_2)\, dt - C_{v2}\, dT_2}{[C_t\,(T_2) - C_t\,(T_3)]\, A_{ww}} \tag{7.36}$$

$$dT_3 = \frac{Q_3\, dt}{C_{v3}} \tag{7.37}$$

In order to solve Eqs. (7.35) and (7.36), expressions are needed for the terms Q_a and dT_2. These expressions will now be derived. First, however, a method for calculating the segment 2 temperature will be established.

Segment 2 Temperature. The temperature T_2 of segment 2 is assumed to be equal to the saturation temperature corresponding to the vapor pressure P_2 at the sonic limit. P_2 is related to the vapor pressure P_v in segment 1 (the evaporator) by

$$P_2 = \gamma P_v \tag{7.38}$$

where γ is a critical pressure ratio whose value is determined by the heat pipe vapor.

The temperature T_2 of segment 2 can be found for a given vapor temperature T_v in segment 1 from the following steps:

1. Calculate P_v from Eq. (5.100).
2. Calculate P_2 from Eq. (7.38).
3. Calculate T_2 from Eq. (5.101).

On the assumption that the heat pipe vapor acts as a monatomic perfect gas, γ can be found from Eq. (5.79) for the critical pressure ratio at the sonic limit, with FM = 0 (ignoring friction). Using a value of 5/3 for the specific heat ratio k_s (see Section 5.2.6),

$$\gamma = \frac{P_2}{P_v} = \frac{1}{1 + k_s} = 0.375$$

The following alternate method can also be used to determine γ. From Eqs. (5.68) and (5.116), the axial heat flux at the sonic limit ξ_s can be expressed as

$$\xi_s = \rho_{vs}\, V_s\, \kappa_m = A_s\, P_v^{\alpha_s} \tag{7.39}$$

where ρ_{vs} is the vapor density at the sonic limit, V_s is the sonic velocity, and κ_m is the mean heat of vaporization between T_v and T_2. By using the relationships

Table 7.3. Drop in saturation temperature at sonic limit ($\gamma = 0.375$)

Fluid	Vapor temperature, °F (°C)	Temperature drop, °F (°C)
Cesium	800 (427)	94 (52)
Potassium	1200 (649)	140 (78)
Sodium	1600 (871)	176 (98)
Lithium	2400 (1316)	221 (123)

$$\rho_{vs} = \frac{P_2}{R_g T_2} \tag{5.76}$$

$$V_s = \sqrt{k_s R_g T_2} \tag{5.77}$$

$$P_2 = \gamma P_v \tag{7.38}$$

$$T_2 = \frac{B}{A - \log_{10} P_2} \tag{5.101}$$

where R_g is the gas constant = R_u/M, R_u is the universal gas constant, and M is the molecular weight, Eq. (7.39) can be written

$$\kappa_m \gamma P_v \left[\frac{k_s}{R_g B} (A - \log_{10} \gamma P_v) \right]^{1/2} = A_s P_v^{\alpha_s} \tag{7.40}$$

For a given P_v, Eq. (7.40) can be solved for γ by trial and error. κ_m can be evaluated at T_v with little error. When this procedure is performed for sodium and lithium over a range of representative vapor pressures, γ for both fluids is found to be approximately constant and equal to 0.424. This value is believed to be more representative of γ because, in its solution, T_2 is obtained from the saturation temperature–pressure equation rather than the perfect gas equation.

Representative values of the drop in saturation temperature $T_v - T_2$ with $\gamma = 0.375$ are given in Table 7.3 for several liquid metals. Values of $T_v - T_2$ obtained using $\gamma = 0.424$ are about 12% smaller.

Differential Change in Segment 2 Temperature. Using Eq. (5.101), we can write for the segment 2 vapor temperature T_2 (same as the segment 2 temperature) and the segment 1 vapor temperature T_v,

$$T_2 = \frac{B}{A - \log_{10} P_2} \tag{7.41}$$

$$T_v = \frac{B}{A - \log_{10} P_v} \tag{7.42}$$

Also, from Eq. (7.38), $P_2 = \gamma P_v$. It follows that $dP_2 = \gamma dP_v$.

Taking the differential of each equation, substituting the above relations for P_2 and dP_2, and dividing the resulting equations, we obtain

$$\omega = \frac{dT_2}{dT_v} = \left(\frac{A - \log_{10} P_v}{A - \log_{10} \gamma P_v} \right)^2 \tag{7.43}$$

Using Eq. (7.42), the above equation can be written

$$\omega = \frac{dT_2}{dT_v} = \left(\frac{B}{B - T_v \log_{10} \gamma} \right)^2 \tag{7.44}$$

It also follows that

$$dT_2 = \omega dT_v \tag{7.45}$$

The differential change in the segment 1 vapor temperature T_v is obtained by calculating T_v at the beginning and end of a time interval dt.

With $\gamma = 0.424$, the value of ω for sodium ranges from 0.900 to 0.836 as T_v is varied from 900 to 1900°F (482 to 1038°C). For lithium, ω ranges in value from 0.898 to 0.856 as T_v is varied from 1700 to 2700°F (927 to 1482°C).

Vapor Heat Transport Rate. The heat transport rate Q_a out of segment 1 depends on whether heat transferred through the segment 2 outer surface $Q_2 dt$ is less than, equal to, or greater than the change in sensible heat $C_{v2} dT_2$ in that segment. The value of Q_a for various conditions will now be considered.

Case 1: $Q_2 dt \le C_{v2} dT_2$. This condition is applicable for the following circumstances:

a. The sensible heat increases by a greater amount than the heat transferred into the segment ($Q_2 > 0$, $dT_2 > 0$).
b. The sensible heat increases while heat is transferred out of the segment ($Q_2 < 0$, $dT_2 > 0$).
c. The heat transferred out of the segment is greater than the decrease in sensible heat ($Q_2 < 0$, $dT_2 < 0$).
d. The increase (or decrease) in sensible heat is equal to the heat transferred into (or out of) the segment ($Q_2 > 0$, $dT_2 > 0$ or $Q_2 < 0$, $dT_2 < 0$).

The heat deficiency in situations (a), (b), and (c) must be supplied by the heat release associated with condensing vapor. In situation (d), there is no phase change heat transfer.

Therefore, for case 1, either condensation or no phase change is occurring in segment 2. The vapor velocity then becomes sonic at the segment 1 (evaporator) exit. It follows that

$$Q_a = Q_s \tag{7.46}$$

Case 2: $\boldsymbol{Q_2 dt > C_{v2} dT_2}$. This condition is applicable for the following circumstances:

a. The heat transferred into the segment is greater than the increase in sensible heat ($Q_2 > 0$, $dT_2 > 0$).
b. The sensible heat decreases while heat is being transferred into the segment ($Q_2 > 0$, $dT_2 < 0$).
c. The heat transferred out of the segment is less than the decrease in sensible heat ($Q_2 < 0$, $dT_2 < 0$).

The heat excess in each of these situations must be dissipated by vaporization of the heat pipe liquid. Thus, for case 2, vaporization is occurring in segment 2.

For this circumstance the vapor velocity does not become sonic until the continuum front is reached. Therefore, Q_a is less than Q_s. The increase in the heat content of the vapor resulting from vaporization in segment 2 is

$$Q_2 dt - C_{v2} dT_2$$

Performing a heat balance on the vapor in segment 2,

$$Q_a dt + (Q_2 dt - C_{v2} dT_2) = Q_s dt$$

Upon solving for Q_a,

$$Q_a = Q_s - Q_2 + C_{v2} \frac{dT_2}{dt} \tag{7.47}$$

Equation (7.47) is the desired expression for Q_a when vaporization is occurring in segment 2.

Final Form of Equations for $\boldsymbol{dT_1}$ and $\boldsymbol{dx}$. If dT_2 from Eq. (7.45) and Q_a from Eq. (7.46) or (7.47) are substituted into Eqs. (7.34) and (7.35), the final forms for dT_1 and dx in phase 2 are obtained. The equations can be expressed in the general form

$$dT_1 = \frac{(Q_1 - Q_s + a_c\, Q_2)\, dt - a_c\, C_{v2}\, \omega\, dT_v}{C_{v1}} \tag{7.48}$$

$$dx = \frac{(Q_s + b_c\, Q_2)\, dt - b_c\, C_{v2}\, \omega\, dT_v}{[C_t\,(T_2) - C_t\,(T_3)]\, A_{ww}} \tag{7.49}$$

In Eqs. (7.48) and (7.49), if $Q_2 dt \leq C_{v2} \omega dT_v$, then condensation or no phase change is occurring in segment 2, and $a_c = 0$ and $b_c = 1$. Also, if $Q_2 dt > C_{v2} \omega dT_v$, then vaporization is occurring in segment 2, and $a_c = 1$ and $b_c = 0$.

In the above conditional statements, dT_2 has been replaced by ωdT_v [see Eq. (7.45)].

Equations (7.48), (7.49), (7.37), (7.15), and (7.16), along with the calculational procedure described under "Segment 2 Temperature," constitute the equations and methods needed to evaluate the segment temperatures and the continuum front position in phase 2.

7.4.3. Phase 3

Phase 3 starts when the continuum front reaches the end of the heat pipe. At this time, segment 3 shrinks to zero and disappears. The heat pipe then consists of segment 1 (the evaporator) of length L_e and segment 2 (the condenser) of length L_c. The vapor velocity becomes subsonic, the vapor temperature and pressure differences between segments 1 and 2 disappear, and the vapor becomes essentially isothermal over the entire length of the heat pipe. Then, $T_2 = T_v$. The entire heat pipe then acts as a single thermal entity.

It is assumed implicitly that in phase 3 the actual heat transport rate is a small fraction of the sonic limit.

A heat balance on the entire heat pipe yields

$$(Q_1 + Q_2)dt = C_{v1}dT_1 + C_{v2}dT_2 = (C_{v1} + C_{v2})\,dT_1$$

Since the heat pipe is now a single thermal entity,

$$dT_2 = dT_1$$

Solving for dT_1,

$$dT_1 = \frac{(Q_1 + Q_2)\,dt}{C_{v1} + C_{v2}} \tag{7.50}$$

Again, if heat is being rejected from the condenser, Q_2 is negative.

Equation (7.50), and Eqs. (7.15) and (7.16) for the segment 1 surface temperatures, are the defining equations for phase 3.

7.5. CALCULATIONAL PROCEDURE

7.5.1. Overview

The time dependence of the heat pipe temperatures and the continuum front position is calculated by converting the equations developed in Section 7.4 to finite-difference form (Refs. 7.6, 7.8). The transformation is accomplished by replacing differential increments of temperature, distance, and time in the equations with finite increments. Temperatures and the continuum front position after a given time increment are estimated from values existing at the start of that increment.

In this section, equations from Section 7.4 that have been converted to finite difference form are identified by the addition of the suffix c to the equation numbers.

It is necessary to specify the initial conditions (initial temperature distribution and continuum front position) and boundary conditions (the heat flux distribution over the heat pipe surface as a function of time and, in some cases, the temperature of the outer heat pipe surface). The time increment must also be specified.

The following information is also required: the volume of the heat pipe segments; values of the density, specific heat, and thermal conductivity of the various heat pipe materials as a function of temperature; and appropriate expressions for calculating the mean thermal conductivity of the heat pipe wall/wick combination.

The solution is "marched out," time increment by time increment, until the heat pipe temperatures and continuum front position have been determined for the total time interval of interest.

The selected time increment is an important calculational parameter. It must be small enough to assure that the solution is stable (i.e., that calculated temperatures do not start to fluctuate and grow exponentially in magnitude). Also, the accuracy of the calculation generally improves as the time increment is reduced in size. On the other hand, the selection of too small a time increment can result in excessively long solution time.

Generally, small time increments are required when the incident heat flux and/or heat pipe temperatures are changing rapidly. When the heat flux and/or temperatures are changing at a slower rate, the time increment may be increased without destabilizing the solution. The best way to establish the appropriate time increment is by trial and error, varying the time increment until a stable solution is obtained that does not change significantly with subsequent decreases in the time increment.

As a rough guide, a time increment on the order of 1/10,000th of the total transient time is suggested for initial evaluation. If the incident heat flux changes relatively slowly, a considerably larger time increment may be tried. The initially selected time increment may be increased or decreased in subsequent calculations, as may be required to maintain stability of the solution and minimal calculation time.

Because of the large number of calculations, the use of a computer is essential. In general, a personal computer will be adequate to run the transient analysis that has been described in this chapter.

In the transient model, Δt, ΔT, and Δx are the time increment, temperature increment, and increment in the continuum front position, respectively. The time increment is usually held constant, but may be changed after various time intervals, if desired. The temperature and continuum front position increments are functions of time.

Conditions at the end of the ith time increment are denoted by i, and those at the end of the next time increment by $i + 1$. Then,

$$t^{i+1} = t^i + \Delta t = (i + 1)\Delta t$$

$$T^{i+1} = T^i + \Delta T^i$$

$$x^{i+1} = x^i + \Delta x^i$$

Superscripts are used to identify the time increments to avoid confusion with the subscripts of subscripted variables.

Generally, $t^0 = 0$, $T^0 = T_a$, and $x^0 = 0$.

7.5.2. Input Data

The following input data are required for the transient analysis:

1. Heat pipe materials (wall, wick, heat pipe fluid).
2. Density, specific heat, thermal conductivity, and thermometric thermal capacity for each material as a function of temperature.
3. Heat pipe configuration and dimensions.

4. Incident heat flux on segment outer surface as a function of location, surface temperature (if applicable), and time.
5. Initial temperature (generally ambient temperature).
6. Initial location of continuum front (at evaporator exit, where $x = 0$).
7. Mean liquid fraction in heat pipe wick.
8. Ratio γ of vapor pressure at sonic limit to initial vapor pressure.
9. Total elapsed time t_{max}.
10. Time increment Δt.

7.5.3. Parameters for Transient Equations

From the input data, the following parameters used in the transient equations are calculated:

1. Volumes, surface areas, cross-sectional areas, and mean cross-sectional areas. These parameters are constant for segment 1 (the evaporator), and may be functions of the continuum front position x for the other two segments.
2. Volumetric thermal capacity. For segment 1, C_v is a function of temperature only. For the other segments, C_v may vary with both temperature and the continuum front position x.
3. Thermometric thermal capacity. C_t is a function of temperature, and may also vary with x if the wall and/or wick cross sections vary with x.
4. Net heating rate on the surface of the heat pipe segments. Q is generally a function of time, and may also depend on the surface temperature.
5. Ratio ω of the differential change in segment 2 temperature T_2 to the vapor temperature T_v in segment 1.
6. Sonic limit Q_s. If the sonic limit occurs at the evaporator exit, it is a function of the segment 1 vapor temperature T_v only. If Q_s occurs at the continuum front, it may depend on the continuum front position x as well as T_v (if the vapor space cross-sectional area A_v is a function of x).

The time- and temperature-dependent parameters are recalculated for each time interval.

7.5.4. Continuum Flow Criterion

The temperature T_{cf} at which continuum flow has been established is calculated at the beginning of the analysis. When $T_v = T_{cf}$, phase 1 has been completed and phase 2 begins. T_{cf} can be calculated from Eqs. (7.11) and (5.100). Alternatively, if the decision has been made to start the transient with phase 2, there is no need to calculate T_{cf}. However, calculational time will be longer if the latter option is selected.

7.5.5. Outer Wall and Inner Wick Temperatures

As has been discussed previously, the segment 1 surface temperatures are obtained from the mean segment 1 temperature using the equations

$$T_{womax}^{i+1} = T_1^{i+1} + x_1 \, \Delta T_{max}^{i} \tag{7.15c}$$

$$T_v^{i+1} = T_1^{i+1} - x_2 \, \Delta T_{max}^{i} \tag{7.16c}$$

where x_1 and x_2 are temperature coefficients derived from steady-state heat calculations and ΔT_{max}^{i} is the maximum steady-state one-dimensional temperature drop across the wall/wick thickness at t^i (evaluated at the mean evaporator temperature $T_1{}^i$ and the mean heat flux q_{1a}^i in segment 1).

Values of x_1 and x_2 for a cylindrical heat pipe with a uniform heat flux on the evaporator surface are given in Table 7.2. If the wall plus wick thickness is a small fraction of the vapor space radius, x_1 and x_2 are approximately equal to 0.5.

When the incident heat flux on the evaporator surface is nonuniform, the values of x_1 and x_2 based on uniform heating may be used to approximate surface temperatures. More accurate values of the temperature coefficients require a two- or three-dimensional steady-state analysis of the temperature distribution in the segment 1 wall/wick, at the expense of increased calculational time.

The average heat flux q_{1a}^i is given by

$$q_{1a}^i = \frac{q_{1max}^i + \varepsilon_{pa}^i \, Q_a{}^i \, / \, (2 \, \pi \, r_v \, L_e)}{2} \tag{7.18c}$$

where q_{1max}^i is the maximum net incident heat flux on segment 1, $Q_a{}^i$ is the exit heat transport rate, r_v is the vapor space radius, and ε_{pa}^i is the ratio of the peak heat flux to the average heat flux on the outer surface of segment 1. $Q_a{}^i$ is zero during phase 1. In phase 2, $Q_a{}^i$ is calculated using the procedure given in Section 7.5.7. [See Eqs. (7.46–7.47c).] In phase 3, it is assumed that $q_{1a}^i = q_{1max}^i$. The heat flux ratio ε_{pa}^i may or may not be constant.

As was mentioned in Section 7.3.4, q_{1a}^i may also optionally be assumed equal to q_{1max}^i in all phases of the startup transient, at the expense of less accurate calculations of segment 1 surface temperatures during the early phases.

The one-dimensional steady-state temperature drop across the wall and wick of a cylindrical evaporator is given by

$$\Delta T_{max}^i = \frac{q_{1a}^i \, r_w}{k_{1m}^i} \left[\frac{2}{1 + (r_w / r_v)} \right] \ln \frac{r_w}{r_v} \tag{7.20c}$$

where k_{1m}^i is the mean wall/wick thermal conductivity evaluated at $T_1{}^i$ and r_w is the outer radius. k_{1m}^i is evaluated using the methods of Section 6.5.1. Note that q_1 in Eq. (7.20) has been replaced by the expression developed in Section 7.3.4, "Calculation of ΔT_{max}."

The outer wall and vapor (inner wick) temperatures of segment 1 are calculated in this manner for all phases of the startup transient.

7.5.6. Phase 1

In phase 1, the heat pipe heat transport mechanism is inactive. Only segments 1 (the evaporator) and 3 (the condenser) exist, and there is no interchange of heat between these phases. The mean temperature in each segment is given by the equations

$$T_1^{i+1} = T_1^i + \frac{Q_1^i \Delta t}{C_{v1}^i} \tag{7.33c}$$

$$T_3^{i+1} = T_3^i + \frac{Q_3^i \Delta t}{C_{v3}^i} \tag{7.34c}$$

where Q^i is the net heat transfer rate into a segment at time t^i, and $C_v{}^i$ is the volumetric thermal capacity of the segment at temperature T^i.
The maximum wall temperature T_{womax}^{i+1} and the vapor temperature $T_v{}^{i+1}$ in segment 1 are calculated using Eqs. (7.15c), (7.16c), (7.18c), and (7.20c).

The calculation starts at $i = 0$, when $t = t^0 = 0$, $x = x^0 = 0$, $T_1 = T_1{}^0$, and $T_3 = T_3{}^0$. $T_1{}^0$ and $T_3{}^0$ are generally equal to the ambient temperature. After each calculation, i is incremented by 1. The calculations are repeated until $T_v{}^{i+1} \geq T_{cf}$, marking the end of phase 1. The calculations then continue using the phase 2 equations.

7.5.7. Phase 2

In phase 2, the continuum front starts to move away from the evaporator exit into the condenser. There are now three segments: segment 1 (the evaporator), segment 2 (the portion of the condenser behind the continuum front of length x), and segment 3 (the portion of the condenser ahead of the front of length $L_c - x$). Three mean segment temperatures, the segment 1 outer surface temperature, the segment 1 vapor temperature, and the position of the continuum front must now be calculated.

The segment temperatures are calculated using the following equations.

$$\omega^i = \left(\frac{B}{B - T_v{}^i \log_{10} \gamma} \right)^2 \tag{7.43c}$$

$$T_1^{i+1} = T_1^i + \frac{(Q_1^i - Q_s^i + a_c Q_2^i) \Delta t - a_c C_{v2}^i \omega^i (T_v^{i+1} - T_v^i)}{C_{v1}^i} \tag{7.48c}$$

$$T_3^{i+1} = T_3^i + \frac{Q_3^i \Delta t}{C_{v3}^i} \tag{7.37c}$$

$$\log_{10} P_v^{i+1} = A - \frac{B}{T_v^{i+1}} = \varsigma^{i+1} \tag{5.100c}$$

$$P_v^{i+1} = 10^{\varsigma^{i+1}}$$

$$P_2^{i+1} = \gamma P_v^{i+1} \tag{7.38c}$$

$$T_2^{i+1} = \frac{B}{A - \log_{10} P_2^{i+1}} \tag{5.101c}$$

The maximum wall temperature T_{womax}^{i+1} and the vapor temperature $T_v{}^{i+1}$ in segment 1 are again calculated using Eqs. (7.15c), (7.16c), (7.18c), and (7.20c).

The continuum front position is calculated from the equation

$$x^{i+1} = x^i + \frac{(Q_s^{\,i} + b_c\, Q_2^{\,i})\, \Delta t - b_c\, C_{v2}^{\,i}\, \omega^i\, (T_v^{\,i+1} - T_v^{\,i})}{[C_t\,(T_2^{\,i}) - C_t\,(T_3^{\,i})]\, A_{ww}^{\,i}} \tag{7.49c}$$

The heat transport rate $Q_a^{\,i}$ out of segment 1 is needed to calculate $q_{1a}^{\,i}$ in Eq. (7.18c). It is given by

$$Q_a^{\,i} = Q_s^{\,i} - a_c\, Q_2^{\,i} + a_c\, C_{v2}^{\,i}\, \frac{(T_2^{\,i} - T_2^{\,i-1})}{\Delta t} \tag{7.46–47c}$$

The values of a_c and b_c in Eqs. (748c), (7.49c), and (7.46–47c) are established by the following criteria: If $Q_2^{\,i}\, \Delta t \le C_{v2}^{\,i} \omega^i (T_v^{\,i+1} - T_v^{\,i})$, then $a_c = 0$, $b_c = 1$. If $Q_2^{\,i}\, \Delta t > C_{v2}^{\,i} \omega^i (T_v^{\,i+1} - T_v^{\,i})$, then $a_c = 1$, $b_c = 0$.

The calculations start at the time t^{i+1} when $T_v^{\,i+1} \ge T_{cf}$, or at time zero if the decision has been made to start with the phase 2 equations. After each calculation, i is incremented by 1. The calculations are repeated until $x \ge L_c$, marking the end of phase 2. The calculations then continue using the phase 3 equations.

7.5.8. Phase 3

Phase 3 starts at the time t^{i+1} when $x \ge L_c$. At that time, segment 3 disappears, and only segments 1 (the evaporator) and 2 (the complete condenser) remain. $Q_2^{\,i}$ and $C_{v2}^{\,i}$ are then evaluated at $x = L_c$. The heat pipe heat transport mechanism is now fully effective, and the transported heat is rapidly distributed over the entire heat pipe. The heat pipe can then be approximated as a single thermal entity, with the segment 1 and 2 temperatures changing at the same rate.

The following equations are used to calculate the segment temperatures in phase 3:

$$T_1^{\,i+1} = T_1^{\,i} + \left(\frac{Q_1^{\,i} + Q_2^{\,i}}{C_{v1}^{\,i} + C_{v2}^{\,i}} \right) \Delta t \tag{7.50c}$$

$$T_2^{\,i+1} = T_v^{\,i+1}$$

Again, $T_{womax}^{\,i+1}$ and $T_v^{\,i+1}$ are evaluated using Eqs. (7.15c), (7.16c), (7.18c), and (7.20c). The calculations are terminated when $t^{i+1} \ge t_{max}$, where t_{max} is the total startup time.

7.6. ASSESSMENT OF HEAT PIPE STARTUP

7.6.1. Results of Startup Transient

From the results of the heat pipe startup transient, it can be established whether the peak temperature is within acceptable limits, and whether the thermal coupling between the condenser and the heat sink is causing startup difficulty.

If excessively high temperatures are encountered, design modifications to correct the situation will be necessary. These could include lowering the incident heat flux, increasing the thermal capacity of the heat pipe by adding more material, or using a material of higher thermal capacity.

If an excessively long time is required to complete phase 2, or if the continuum front "stalls" so that phase 2 is never completed, the thermal coupling between the condenser and the heat sink may be excessively strong. Then, most or all of the heat entering segment 2 would be dissipated from the outer surface to the heat sink, with little or no heat available to extend the length of segment 2 at the continuum front.

Difficulty in completing phase 2 is a signal that startup to full operational conditions (characterized by relatively isothermal operation) may not be possible with the current design. Instead, the heat pipe could be operating in a highly nonisothermal mode. This situation could prevail even though the mean segment 2 temperature remained above the freezing point of the heat pipe liquid.

Under some circumstances the continuum front could reverse direction, moving back toward the evaporator and shrinking segment 2. This would be an indication that the segment 2 temperature calculated by the transient model is too high, leading to an excessive heat transfer rate to the heat sink. The actual segment 2 temperature could be much lower than the calculated value, possibly dropping below the freezing point of the heat pipe liquid. A freezeout situation could then develop, with subsequent dryout and structural damage in the evaporator.

Although the transient model cannot predict the onset of freezeout, shrinkage of segment 2 during phase 2 is an indication that freezeout could be (but not necessarily is) a problem during startup. Modification of the heat pipe design to eliminate segment 2 shrinkage is then advisable. Such a modification could consist of adding thermal insulation to reduce the thermal coupling between the condenser and the heat sink.

In addition to the preceding information, which is obtainable directly from the transient heat transfer analysis, the following additional heat pipe characteristics should be evaluated during the startup transient: heat transport limits, minimum vapor and liquid pressure, and stress in the heat pipe wall.

7.6.2. Heat Transport Limits

The equations for steady-state heat transport limits that were presented in Chapter 5 may be used to calculate heat transport limits at selected times during the startup transient. Since the heat pipe is not functional and there is no axial heat transport during phase 1, only the heat transport limits during phases 2 and 3 are of interest. Application and interpretation of the heat transport limit equations during phase 2 requires some modification, because the departure from steady-state conditions is most extreme during this phase. The necessary modifications are discussed subsequently. Once the transient has moved into phase 3, the unmodified steady-state equations may be used.

The transition from phase 2 to phase 3 is not considered when calculating the heat transport limits. Therefore, abrupt rather than smooth changes in heat transport limits and the actual heat transport rate can be associated with the transition.

For reasons that were explained in the introduction to Chapter 5, either the heat transport limits may be calculated for a specified heat pipe design, or a key design parameter may be evaluated for the situation when a specified heat transport limit is set equal to the actual heat transport rate. In the case of the capillary pumping limit, it is particularly convenient to determine the diameter of curvature in the heat pipe wick corresponding to the actual heat transport rate. Adequate capillary pumping capability is then assured if the calculated diameter of curvature is greater than the effective diameter of the pores in the capillary pumping layer of the heat pipe wick.

The heat transport limits (or the equivalent design parameters) are calculated at various times during the transient when limiting heat transport conditions are most likely to be encountered. Times at which the incident heat flux, the vapor temperature, and any acceleration present have peaked are of particular interest. If the actual heat transport rate should exceed the heat pipe heat transport capability at any time, heat transport could be disrupted and overheating leading to heat pipe damage or destruction could occur.

The design would then be modified to insure that the heat transport limits are always equal to or greater than the actual heat transport rate. Such modifications could include use of a heat pipe fluid with greater heat transport capability, a reduction in the incident heat load during startup, an increase in vapor and/or wick cross-sectional dimensions, or a smaller pore size in the capillary pumping layer.

Actual Heat Transport Rate. The actual heat transport rate Q_{ht} is the maximum rate at which latent heat is being transported by the heat pipe vapor. In phase 2, it is equal to the sonic limit. If the heat pipe vapor is condensing in segment 2, or if there is no phase-change heat transfer, the sonic limit occurs at the exit of segment 1. If there is evaporation in segment 2, the sonic limit occurs at the continuum front.

In phase 3, the actual heat transport rate is equal to or less than the sonic limit, and occurs at the exit of segment 1. It is equal to the difference between the net rate at which heat is being transferred to the surface of segment 1 and the rate of sensible heating in the segment. That is,

$$Q_{ht}^{i} = Q_1^{\,i} - \frac{C_{v1}^{i}\,(T_1^{i+1} - T_1^{\,i})}{\Delta t} \tag{7.51}$$

with the restriction that $Q_{ht}^{i} \le Q_s^{\,i}$. If the value of Q_{ht}^{i} calculated from Eq. (7.54) should exceed $Q_s^{\,i}$, it should be replaced by $Q_s^{\,i}$ in the heat transport calculations.

Sonic Limit. Since the sonic limit is evaluated at the maximum vapor temperature in the heat pipe, the sonic limit can be calculated from Eq. (5.116) without modification in all phases of the startup transient. The segment 1 vapor pressure P_v is used in the calculation.

Capillary Pumping Limit. The capillary pumping limit equation was derived on the assumption that the vapor density is constant along the heat pipe length. This implies that the vapor temperature and pressure drops are small, and that the vapor velocity remains well below the sonic value.

In phase 2, the vapor reaches sonic velocity and, if condensation is occurring in segment 2, supersonic velocity. Since the flow of the heat pipe vapor in phase 2 is accompanied by large changes in density, temperature, and pressure, considerable uncertainty is associated with the vapor pressure drop component of the capillary pumping limit equation.

The vapor pressure and temperature in phase 2 are substantially lower than when the heat pipe is operating under the steady-state design condition. This is because the cool portion of the condenser into which the continuum front advances serves as an excellent internal heat sink, absorbing heat that would otherwise have to be dissipated to the external heat sink.

Since the vapor pressure drop cannot exceed the maximum vapor pressure, the vapor pressure drop during phase 2 is usually much smaller than that during steady-state operation. Uncertainties in the magnitude of the vapor pressure drop then have relatively less impact on the capillary pumping limit or the diameter of curvature required for capillary pumping.

If the vapor pressure drop is small compared to the liquid pressure drop, the impact of uncertainties in the vapor pressure drop calculation on capillary pumping capability will be minimal. However, if the vapor pressure drop is comparable to or greater than the liquid pressure drop, the impact on capillary pumping capability will be greater. Conservative design practice for this situation suggests that the vapor pressure drop be set equal to the maximum vapor pressure in establishing capillary pumping capability.

The following points should be kept in mind when using the capillary pumping limit equation in phase 2:

1. The nonisothermal character of the heat transport process can be roughly accounted for by evaluating fluid properties at T_2 rather than T_v.
2. The heat transport rate is equal to the sonic limit.
3. The effective heat pipe length is $L_e + x$ (the distance to the continuum front position).
4. Much if not most of the vapor does not condense until the continuum front is reached. Fluid velocities are then significant at the effective end of the heat pipe, instead of dropping off to zero as in the steady-state situation. Therefore, fluid velocities conservatively should be assumed constant in segment 2 (i.e., most of segment 2 should be considered adiabatic). An adiabatic length of $0.999x$ and a condenser length of $0.001x$ are suggested.

Introduction of both points 1 and 4 could lead to an overly conservative approach to prediction of the capillary pumping limit during phase 2. Therefore, interpretation of calculated results may require some measure of engineering judgment.

Entrainment Limit. From Eq. (5.121), the entrainment limit is proportional to the square root of the vapor density. It should therefore be evaluated where the vapor density is lowest. Since the transient model does not calculate the actual distribution of vapor properties along the heat pipe in phase 2, the minimum vapor density and the entrainment limit are not known precisely.

For design purposes, the entrainment limit in phase 2 should be evaluated at the temperature T_2 of segment 2, at which the vapor density is a minimum. (Although the segment 3 temperature is lower, it contains no vapor.) Because of the uncertainty associated with calculation of the entrainment limit in phase 2, it conservatively should be a factor of 2 or more larger than the actual heat transport rate (which is the sonic limit).

If the margin of safety is less than this, the design should be modified to raise the calculated entrainment limit. From Eq. (5.121), these modifications could take the form of a reduction in the size of the capillary pumping layer pores or use of a heat pipe fluid with a higher heat of vaporization, surface tension, and/or vapor density. Although the entrainment limit can also be increased by enlarging the vapor space flow area, the actual heat transport rate (the sonic limit) would rise by a similar factor.

Boiling Limit. The boiling limit is the heat flux at the outer wick surface required to initiate boiling in the segment 1 wick. It can be evaluated during phase 2 using the methods of Section 5.4. Basically, the superheat for boiling is calculated, and then the heat flux on the wick outer surface necessary to produce a wick temperature drop equal to the superheat for boiling is determined.

The equivalent boiling heat flux on the outer surface of segment 1 should then be calculated and compared with the maximum net heat flux on the segment 1 surface at the time in question. If the equivalent boiling heat flux is larger, boiling will not occur.

The boiling limit calculation is quite approximate because it is based on steady-state heat transfer, and is being applied to a transient situation. However, the relatively low vapor temperatures that are characteristic of phase 2 yield boiling limits that can be quite large compared to the incident heat flux. There is thus room for a large margin of error in calculation of the boiling limit before the usual conclusion that boiling is unlikely in phase 2 would be questionable.

If calculations indicate that the boiling limit is exceeded at some time during phase 2, steps can be taken to correct the situation. These could include reduction of the wick thickness to reduce the wick superheat, reduction of the incident heat flux, reduction of the effective nucleation site radius, and selection of an alternate heat pipe fluid with greater resistance to boiling.

7.6.3. Minimum Vapor and Liquid Pressures

The minimum vapor pressure in the heat pipe during the startup transient can be calculated from the equation

$$P_{v\min} = P_v - \Delta P_v \tag{7.52}$$

where, at a given time, P_v is the maximum (segment 1) vapor pressure and ΔP_v is the vapor pressure drop. ΔP_v is obtainable from the capillary pumping limit calculation.

Calculation of the minimum liquid pressure $P_{l\min}$ is more involved. It depends on the location of the wet point, whether there is a liquid pressure drop or pressure rise, and, for a liquid pressure rise, the absolute magnitude of that rise compared to the vapor pressure drop. The minimum liquid pressure can be calculated from Eqs. (5.58) in the "Vapor Pressure Drop and Minimum Liquid Pressure" section of Section 5.1.2.

In the early part of the transient, vapor temperatures and hence vapor pressures are substantially below design values. Hence, the possibility that $P_{v\min}$ or $P_{l\min}$ could become negative should be examined.

Since the condition $P_{v\min} < 0$ is physically impossible, its occurrence is an indication that the startup heat transport cannot be handled. Heat would then be transported out of the evaporator at a lower rate corresponding to $P_{v\min} \geq 0$, with a resultant more rapid heating of the evaporator. This situation would continue until heat transport at the rate predicted by the transient model no longer produces a negative vapor pressure.

The transient model is not designed to function properly if the vapor pressure is negative. The occurrence of negative vapor pressure may be avoided by a reduction in the vapor pressure drop. The vapor pressure drop can be reduced by enlarging the vapor flow area or by slowing the rate of advance of the continuum front into the condenser, thereby reducing the effective heat pipe length at a given time in phase 2. The latter can be accomplished by thickening the heat pipe wall or substituting a wall or wick material with a higher specific heat to add thermal capacity. An alternative approach to the avoidance of negative vapor pressure would be to substitute a heat pipe fluid with a higher vapor pressure.

The condition $P_{l\min} < 0$ indicates that the liquid is under tension. Although heat pipes can function when the liquid is under tension (see Section 2.1.4), design modifications to minimize or eliminate the occurrence of tension are advisable. Such modifications could include a reduction in acceleration, use of a lower-density heat pipe liquid, or an increase in the thermal capacity of the evaporator (to retard the rate at which the continuum front advances, and hence the length of the liquid column that is subject to acceleration).

If the heat pipe is subjected to high acceleration during the initial part of phase 2, as may be the case for a missile cooling application, the resulting liquid static pressure drop may make it impossible to avoid placing the heat pipe liquid under tension. In that event, startup capability under the expected acceleration would require experimental verification.

7.6.4. Stress in the Heat Pipe Wall

The heat pipe wall is subject to thermal stress from the temperature drop across the wall, and to hoop or bending stress from the pressure differential that usually exists between the heat pipe interior and the external environment. These stresses should be examined to determine whether excessive wall stress develops during the startup transient.

Calculation of the thermal stress at a given time requires that the temperature drop through the segment 1 wall be known. This information is not provided by the transient model. The wall ΔT can be estimated from the steady-state heat transfer equation (5.140) and Table 5.7. The thermal stress is then calculated from Eq. (6.2) as described in Section 6.4.1, evaluating material properties at the mean wall temperature.

Hoop or bending stresses can be calculated as described in Section 6.4.2, with the internal heat pipe pressure taken equal to the segment 1 vapor pressure P_v.

The thermal and mechanical stresses are then combined algebraically (tensile stress positive, compressive stress negative) to obtain the net stress at the inner and outer wall surfaces. The net stress is then compared with the allowable stress (generally some fraction of the yield stress) of the wall material.

If wall stress is found to be excessive, it can be reduced by appropriate changes in wall thickness. A thinner wall will reduce the wall ΔT and thermal stress, but the hoop or bending stress will be increased. Bending stress in the wall can be reduced without affecting the thermal stress by reducing the distance between supporting ribs in the heat pipe interior.

Alternatively, if the heating rate on segment 1 can be reduced at the time when peak wall stress occurs, the segment 1 wall will run cooler. The allowable stress in the wall will then increase, hopefully to a level above the imposed wall stress.

If the peak transient stresses persist over a long period of time, or the application requires repeated startup of the heat pipe, the transient stresses should be compared with the creep stress as well as the yield stress of the wall material.

7.7. REFERENCES

7.1. Colwell, G. T., Jang, J. H., and Camarda, C. J., "Modeling of Startup from the Frozen State." Presented at the Sixth International Heat Pipe Conference, Grenoble, France, May 1987.

7.2. Cotter, T. P., "Heat Pipe Startup Dynamics." Presented at the 1967 Thermionic Conversion Specialist Conference, Palo Alto, California, October 1967.

7.3. Silverstein, C. C., "A Feasibility Study of Heat-Pipe-Cooled Leading Edges for Hypersonic Cruise Aircraft," CR-1857, NASA, Washington, D.C., November 1971.

7.4. Silverstein, C. C., "Heat Pipe Cooling System with Sensible Heat Sink." Presented at the 15th Space Simulation Conference, Williamsburg, Virginia, November 1988.

7.5. Eckert, E. R. G., and Drake, R. M., Jr., *Heat and Mass Transfer,* 2nd Ed., McGraw-Hill, New York, 1959, p. 274.

7.6. Kreith, F., *Principles of Heat Transfer,* 2nd Ed., International Textbook, Scranton, Pennsylvania, 1967, pp. 76–116, 175–188.

7.7. Carlslaw, H. S., and Jaeger, J. C., *Conduction of Heat in Solids,* Oxford University Press, London, 1948, p. 104.

7.8. Myers, G. E., *Analytical Methods in Conduction Heat Transfer,* Genium Publishing, Schenectady, New York, 1987, Chapter 8.

CHAPTER

EIGHT

DESIGN EXAMPLES

In this chapter, the following three representative examples of heat pipe designs are reviewed:

- Steady state design of a heat pipe space radiator.
- Transient analysis of a heat pipe space radiator.
- Steady-state design of a heat pipe recuperator.

These examples draw on the design procedures presented in Chapters 5, 6, and 7. Representative heat pipe design characteristics are calculated, and modified as necessary to conform to stated design criteria. For cases in which calculations are quite involved and best performed with the aid of a computer, detailed calculational steps are not presented. This is the case with the heat transport limits and the transient analysis. In other situations in which the calculations are more straightforward, as with stresses, the relevant equations are identified.

Methods for handling changes in density of the heat pipe liquid with temperature, as described in Section 6.5.2, and the calculation of heat pipe weight and volume, have not been considered in the design examples.

8.1. HEAT PIPE SPACE RADIATOR

A cylindrical heat pipe is to be designed to remove heat from the cold junction of a high-temperature radioisotope-powered thermionic generator, and radiate that heat to

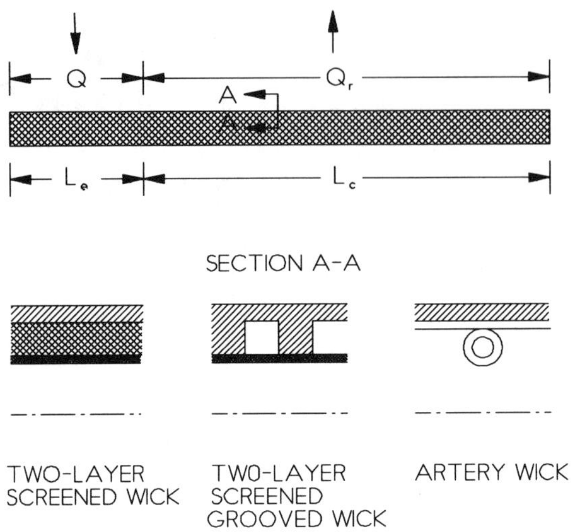

Figure 8.1. Schematic of heat pipe space radiator.

space. In this case, a single heat pipe without a radiating fin is adequate to meet design requirements. A schematic of the arrangement is shown in Figure 8.1, along with cross sections of wick options.

An annular heat source (the cold junction of the thermionic generator) surrounds and contacts the heat pipe evaporator and adds heat uniformly over the surface at the rate Q. Heat is removed at the same rate by radiation from the surface of the heat pipe condenser. The heat pipe vapor temperature is T_V. Contact resistance between the heat source and the evaporator surface is assumed to be negligible. The temperature drop between the heat pipe vapor and the outer surface of the condenser is assumed to be negligible, so that the radiation temperature is also T_V.

Applicable design criteria are listed in Table 8.1.

The vapor temperature of 1400°F (760°C) immediately signals that sodium should be considered as the heat pipe fluid, and that stainless steel or a superalloy should be used as the wall and wick material. The properties of the superalloy Haynes 188 are used in this example. (See Tables C.5 and C.6 of Appendix C.)

Table 8.1. Design criteria for heat pipe space radiator

Nominal heat transport rate	2 Btu/sec (2110 W)
Nominal cold junction temperature	1420°F (771°C)
Heat pipe vapor temperature	1400°F (760°C)
Evaporator length	8 in. (0.203 m)
Condenser length	32 in. (0.813 m)
Overall heat pipe length	40 in. (1.016 m)
Nominal heat pipe diameter	0.6 in. (0.0152 m)
Thickness of heat pipe wall	0.020 in. (0.000508 m)

8.1.1. Heat Transfer

The first step is to establish heat pipe dimensions that will allow heat to be radiated away from the condenser surface at the indicated nominal rate of 2 Btu/sec (2110 W), without exceeding the nominal maximum heat pipe diameter of 0.6 in. (0.0152 m). The approach taken here is to assume certain heat pipe dimensions, and vary them as necessary to satisfy the heat transfer requirement. Further changes in the dimensions may be necessary as other aspects of the heat pipe design are considered.

The condenser surface area A_c is given by

$$A_c = 2\pi(r_v + t_{\text{wick}} + t_{\text{wall}})\, L_c \tag{8.1}$$

where r_v is the vapor space radius, t_{wick} is the wick thickness, t_{wall} is the wall thickness, and L_c is the condenser length.

The rate Q_r at which heat is radiated away from the condenser (assuming that heat radiated or reflected to the condenser from other bodies in space is small compared to the condenser heat rejection rate) is given by

$$Q_r = \varepsilon_s \sigma_r T_v^4 A_c \tag{8.2}$$

where ε_s is the emissivity of the condenser surface and σ_r is the Stefan-Boltzmann radiation constant. Since the temperature drop through the condenser wall and wick has been neglected, the condenser surface temperature is the same as the heat pipe vapor temperature T_v.

In Eq. (8.2), ε_s is given the value of 0.8, and $\sigma_r = 0.476 \times 10^{-12}$ Btu/sec-ft^2-°R^4 (5.67 $\times 10^{-8}$ W/m^2-°K^4).

With the following preliminary values selected for the as yet unspecified dimensions,

$$r_v = 0.250 \text{ in. } (0.00635 \text{ m})$$

$$t_{\text{wick}} = 0.040 \text{ in. } (0.00102 \text{ m})$$

and using the dimensions given in Table 8.1, the calculated value of Q_r from Eqs. (8.1) and (8.2) is 1.974 Btu/sec (2083 W). The selected dimensions satisfy the nominal criteria for the heat transport rate and the maximum heat pipe diameter. The overall heat pipe length remains 40 in. (1.016 m), and the heat pipe outer radius is 0.310 in. (0.00787 m) (slightly above the nominal value).

8.1.2. Heat Transport Limits

The next step in the design process is to calculate the various heat transport limits and compare them to the actual heat transport rate. The wick characteristics must be specified first. A two-layer screen wick was selected for initial evaluation, with the following properties.

Liquid flow channel	4 layers 100 mesh screen
Thickness:	0.032 in. (0.000813 m)
Porosity:	0.5

Capillary pumping layer	4 layers 400 mesh screen
Thickness:	0.008 in. (0.000203 m)
Porosity:	0.5

The indicated wick thicknesses represent a slight compression of the cumulative thickness of individual screen layers. Using thermal and frictional property data from Appendix C, the heat transport limits were obtained using the methods of Chapter 5.

In the calculations, the interface heat transfer coefficient was assumed to be infinite, and the interface temperature drop zero. A value of 3 microns was used for the nucleation site radius. The effective pore radius of the capillary pumping layer was taken to be 30 microns. When calculating the boiling limit, the radius of curvature was assumed to be equal to the effective pore radius. Results are summarized in Table 8.2.

From Table 8.2, the ratio of the actual heat transport rate to the sonic limit is 0.153, which is well below the maximum recommended value of 0.3 for reasonably isothermal heat pipe operation.

The actual heat transport rate is also well below the entrainment and boiling limits. The high boiling limit is characteristic of liquid metals at moderate vapor pressure. At 1400°F (760°C), the vapor pressure of sodium is 4.3 psia (29,650 N/m^2). However, the actual heat transport rate exceeds the capillary pumping limit.

8.1.3 Design Variations

Because the actual heat transport rate is larger than the capillary pumping limit, modification of the initially specified design is necessary. The design parameters listed in Table 8.3 were individually varied so that the effect of each parameter on heat pipe performance could be clearly identified.

Those parameters marked with an asterisk are not directly involved in the heat pipe space radiator design. That is, a particular vapor temperature has been specified, there is no gravity if the space application is located on an orbiting satellite, an interface heat transfer coefficient of infinity has been assumed, and the heat pipe does not have an adiabatic section. Nevertheless, these parameters were varied to provide additional understanding of their effects in situations in which they do play a significant role in heat pipe design. Also, since the magnitude of the interface heat transfer coefficient is uncertain, it is useful to assess the extent to which heat pipe performance is affected by this uncertainty.

Table 8.2. Heat transport limits for heat pipe space radiator, Btu/sec (W)

Actual heat transport rate	1.97 (2078)
Sonic limit	12.94 (13,650)
Entrainment limit	7.46 (7870)
Boiling limit	41.19 (43,460)
Capillary pumping limit	1.57 (1656)

Table 8.3. Design parameters varied in example 1

Design parameter	Range
*Vapor temperature	1200–1600°F (649–871°C)
*Gravity	0, +1*g*, −1*g*
*Interface heat transfer coefficient	$0.1h_{imax} - h_{imax}$, infinite
*Adiabatic length	0–30 in. (0–0.762 m)
Vapor space radius	0.10–0.30 in. (0.00254–0.00762 m)
Wick type	Two-layer screen Two-layer screened groove Artery
Liquid channel thickness (two-layer wicks)	0.008–0.040 in. (0.000203–0.001016 m)
Liquid channel radius (single-artery wick)	0.008–0.032 in. (0.000203–0.00813 m)
Number of arteries (artery wicks)	1–4

Vapor Temperature. The effect of vapor temperature on heat transport is shown in Figure 8.2. The actual heat transport rate increases with vapor temperature, since radiation heat transfer from the condenser varies as the fourth power of the surface temperature (assumed the same as the vapor temperature). Even though the actual heat transport rate more than doubles, it remains well below the sonic, entrainment, and boiling limits.

The capillary pumping limit remains almost constant over the indicated vapor temperature range. This is because the vapor pressure drop, which does vary signifi-

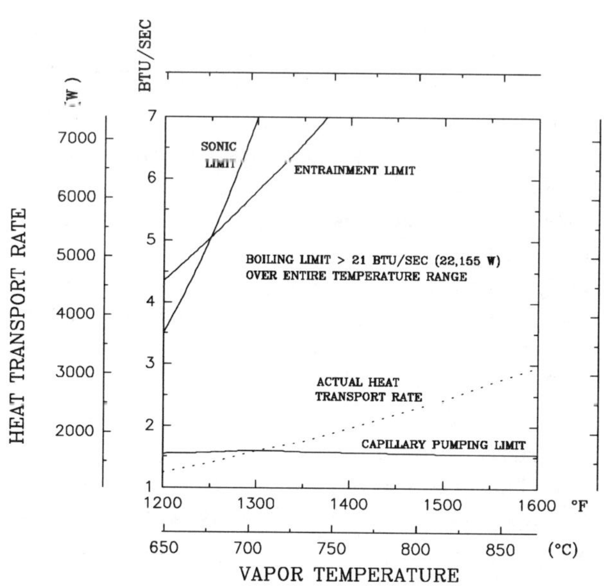

Figure 8.2. Effect of vapor temperature on heat pipe heat transport.

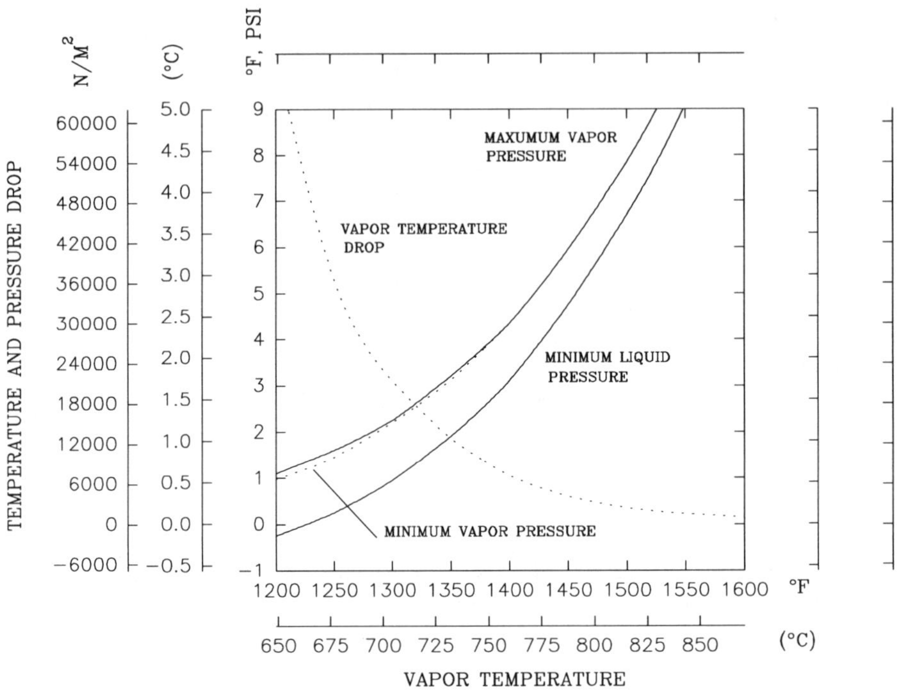

Figure 8.3. Effect of vapor temperature on pressure and temperature changes in a heat pipe.

cantly with vapor temperature, remains almost negligible compared with the liquid pressure drop. The actual heat transport rate is less than the capillary pumping limit only at temperatures below 1300°F (704°C). At higher vapor temperatures, the heat pipe is still not functional.

The associated changes in vapor and liquid pressures and vapor temperature drop at the capillary pumping limit are shown in Figure 8.3. Note that the vapor pressure and minimum liquid pressure drop off sharply with a decrease in vapor temperature, whereas the vapor temperature drop becomes larger.

If the heat pipe heat transport rate were equal to the capillary pumping limit at 1200°F (649°C), the vapor temperature drop along the heat pipe length would be 10°F (6°C), and the minimum liquid pressure would be –0.27 psia (–1862 N/m^2). That is, the heat pipe liquid would be under tension.

Gravity. Since the heat pipe is presumably for use aboard a satellite, it will not be subject to the acceleration of gravity. However, it is of general interest to evaluate the effect of gravity on a sodium heat pipe. The influence of gravity, as well as other acceleration forces, is readily found from the capillary pumping limit calculation. The other heat transport limits are not affected. The effect of placing a heat pipe

Table 8.4. Effect of gravity on capillary pumping limit

Heat pipe orientation	Gravitational acceleration	Capillary pumping limit, Btu/sec (W)
Horizontal	$0g$	1.566 (1652)
Vertical, evaporator up	$+1g$	0.149 (157)
Vertical, evaporator down	$-1g$	2.940 (3102)

based on the initial design in horizontal and vertical orientations in a $1g$ field is shown in Table 8.4.

The heat pipe remains functional regardless of orientation. However, capillary pumping capability is reduced by an order of magnitude when gravity acts to retard the return of liquid condensate to the evaporator. When gravity acts to enhance the return of condensate to the evaporator, capillary pumping capability doubles.

Interface Heat Transfer Coefficient. In the initial design, it was assumed that the interface heat transfer coefficient is infinite, with the result that the temperature drop across the liquid-vapor interface is zero. The interface heat transfer coefficient is actually finite, although of large but uncertain magnitude. It is therefore of interest to

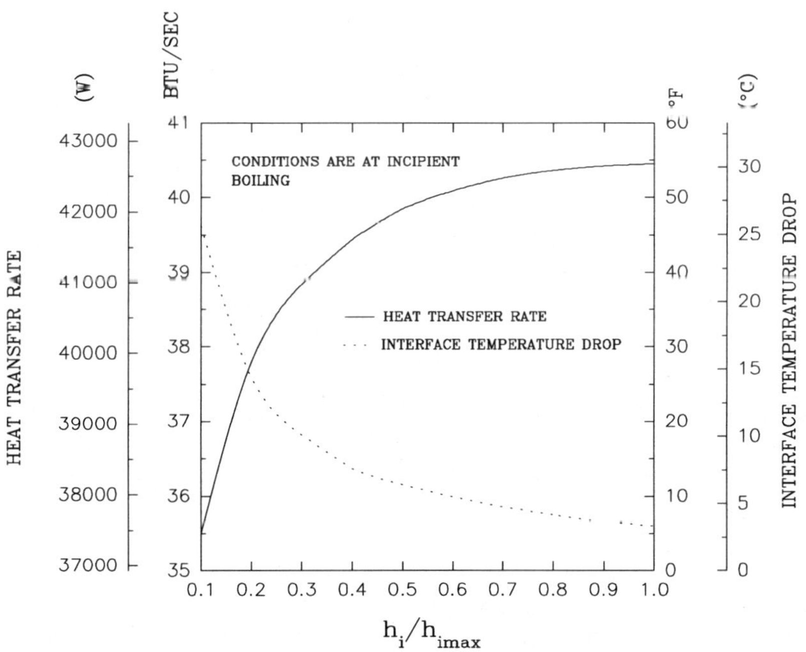

Figure 8.4. Influence of interface heat transfer coefficient on boiling limit.

assess the extent to which heat pipe performance is influenced by this parameter. Only the boiling limit is affected by the interface heat transfer coefficient.

In Figure 8.4, the variation of the boiling limit and the interface temperature drop with h_i/h_{imax} is indicated, where h_i is the interface heat transfer coefficient and h_{imax} is the magnitude of h_i at negligible vapor pressure.

If $h_i = h_{imax}$, the interface temperature drop is 6°F (3°C), and the boiling limit decreases from 41.2 to 40.4 Btu/sec (43,470 to 42,620 W). If $h_i = 0.1h_{imax}$, the interface ΔT rises to 46°F (26°C), and the boiling limit drops to 35.4 Btu/sec (37,350 W). In this example the boiling limit remains well above the actual heat transport rate, even if the thermal resistance at the liquid-vapor interface is substantial.

Adiabatic Length. The effect of replacing a portion of the condenser length by an adiabatic (thermally insulated) section is shown in Figure 8.5. Only the capillary pumping limit and the actual heat transport rate are affected. The capillary pumping limit falls off with an increase in the adiabatic length because the liquid and vapor velocities remain at their maximum values in this thermally insulated region. The actual heat transport rate falls off linearly with an increase in the adiabatic length, since the available condenser length for heat rejection decreases proportionately.

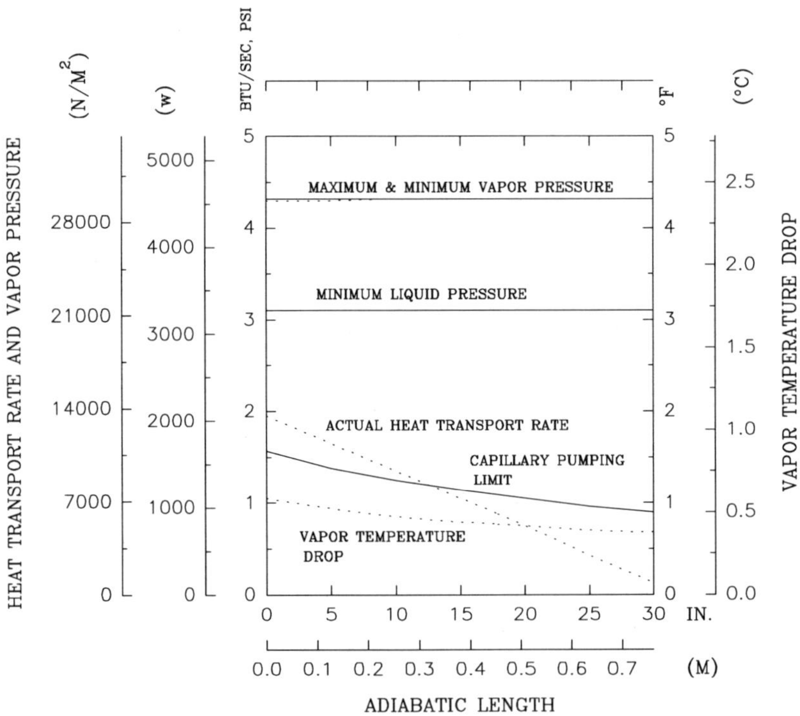

Figure 8.5. Effect of adiabatic length on heat transport.

For adiabatic lengths greater than 13 in. (0.330 m), the actual heat transport rate is below the capillary pumping limit. The heat pipe design is then operable, but at a lower than desired heat transport rate.

Vapor Space Radius. The sonic and entrainment limits rise sharply with vapor space radius, primarily because of the resulting increase in vapor flow area. The boiling limit also undergoes a significant increase and, as indicated in Figure 8.6, remains far above the actual heat transport rate.

The capillary pumping limit rises with an increase in the vapor space radius, primarily because the larger radius increases the liquid channel flow area, reducing the frictional resistance to liquid flow. However, the actual heat transport rate rises at a comparable rate, because of the larger radiating area of the condenser surface, remaining above the capillary pumping limit. Thus, as is evident from Figure 8.6, the initial design cannot be made functional by varying the vapor space radius.

Wick Type. The impact of three wick types on heat transport capability is considered next: the two-layer screen wick, the two-layer grooved wick, and the artery wick. For the two-layered wicks, the basic design variable is the thickness of the liquid flow

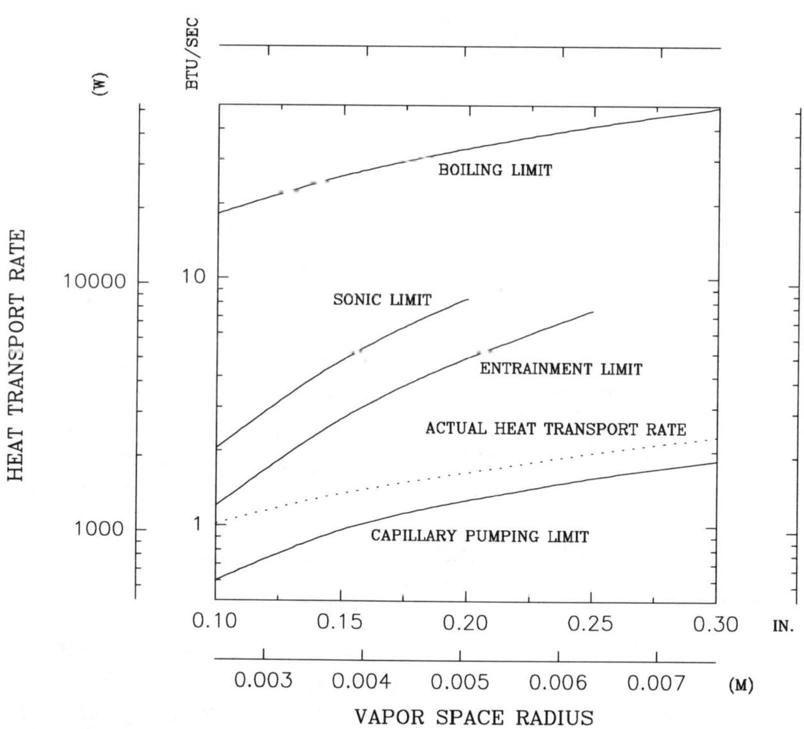

Figure 8.6. Effect of vapor space radius on heat transport.

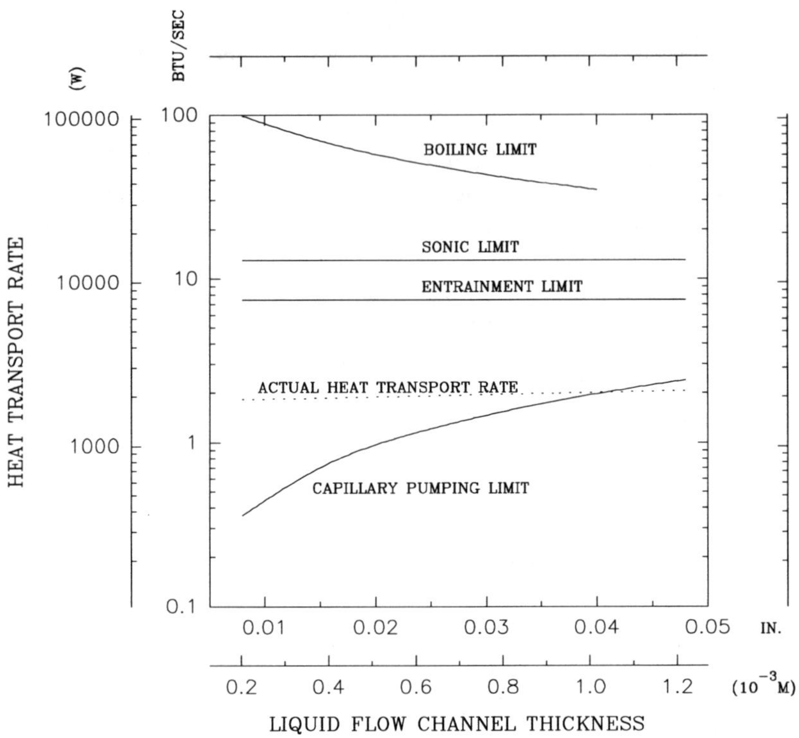

Figure 8.7. Effect of liquid channel thickness on heat transport with a two-layer screen wick.

channel. For the artery wicks, the artery radius and the number of arteries are separately varied.

Two-layer screen wick. The effect of the liquid channel thickness on heat transport for the two-layer screen wick is indicated in Figure 8.7. The design becomes operable if the liquid channel thickness is greater than 0.041 in. (0.00104 m), since the capillary pumping limit is then above the actual heat transport rate. This occurs despite the slight increase in the actual heat transport rate that accompanies the increase in the liquid channel thickness.

The boiling limit is reduced when a thicker wick is used, but it still remains an order of magnitude above the actual heat transport rate.

Two-layer grooved wick. With the two-layer grooved wick, a grooved liquid flow channel is covered by a screened capillary pumping layer (see Figure 8.1). In this example, the groove depth, groove width, and land width (space between grooves) are assumed to be equal. The capillary pumping layer is identical to that of the two-layer screened wick.

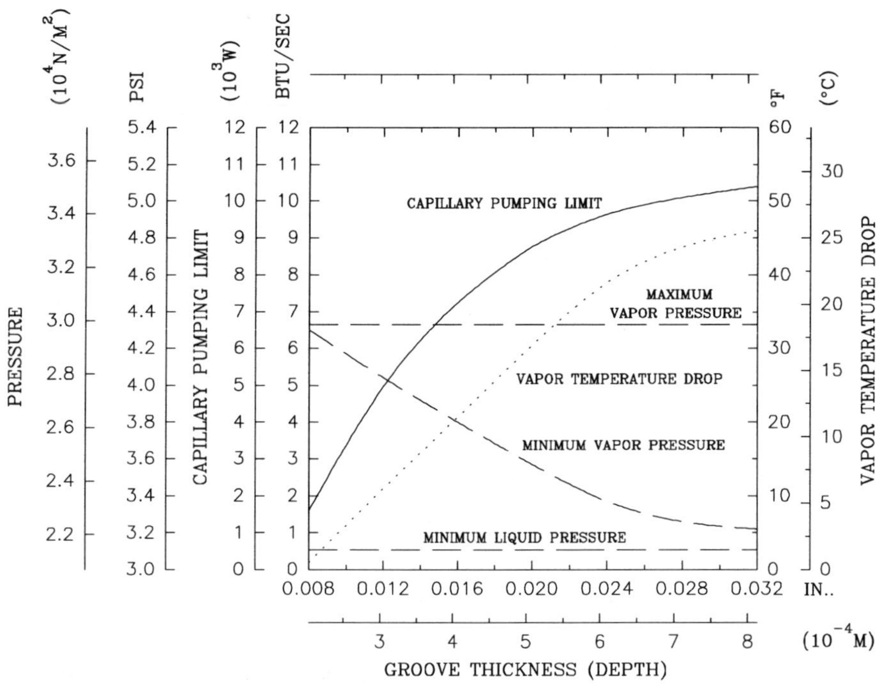

Figure 8.8. Effect of liquid channel thickness on heat transport for a two-layer grooved wick.

Figure 8.8 shows that the capillary pumping limit with a groove depth equal to 0.032 in. (0.000813 m) (the same as the liquid channel thickness in the initially selected two-layer screened wick) is greater than 10 Btu/sec (10,550 W). This value reflects the much lower liquid frictional resistance of the grooved wick.

The grooved liquid channel can be appreciably thinner than the screened channel, with the capillary pumping limit remaining well above the actual heat transport rate. For example, at a groove depth of 0.010 in. (0.000254 m), the capillary pumping limit of 3.4 Btu/sec (3590 W) still exceeds the actual heat transport rate of about 2 Btu/sec (2110 W) by an adequate margin.

Artery wick. Capillary pumping capability is also increased substantially with an artery wick (see Figure 8.1). In this type of wick, one or more arterial liquid flow channels are in contact with a thin distribution wick on the inner surface of the vapor space. The artery boundaries and the distribution wick act as the capillary pumping layer. In this example, the capillary pumping layer of the artery wick is identical to that of the two-layer screen wick.

The capillary pumping limit for a single-artery wick is shown as a function of the artery liquid channel radius in Figure 8.9. The range of sonic, entrainment, and boiling limits over the specified range in artery radius is also indicated.

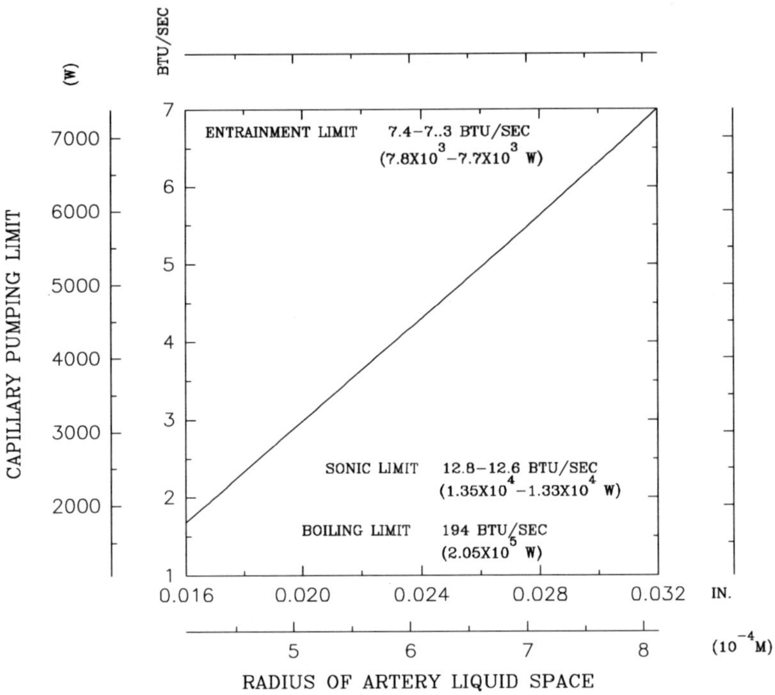

Figure 8.9. Effect of artery radius on heat transport with a single-artery wick.

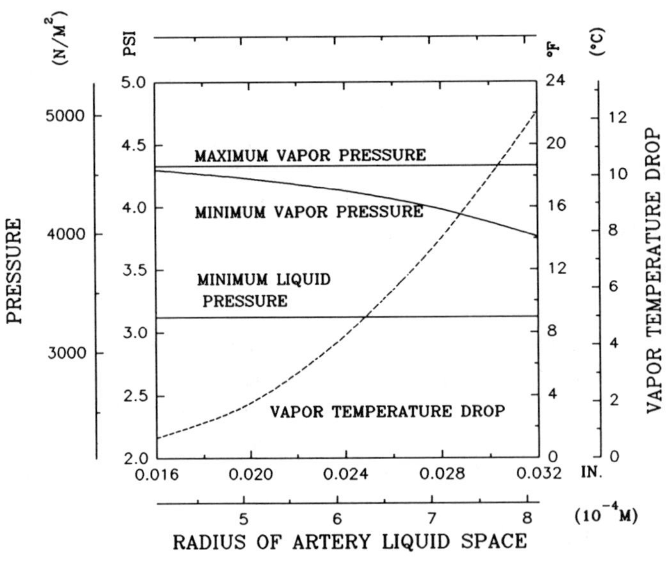

Figure 8.10. Effect of artery radius on pressures and vapor temperature drop with a single-artery wick.

The capillary pumping limit increases almost linearly with the artery radius. At a radius of 0.020 in. (0.000508 m), it is equal to 3 Btu/sec (3170 W), about 50% larger than the actual heat transport rate.

As shown in Figure 8.10, the vapor pressure drop at the capillary pumping limit becomes a significant fraction of the capillary pressure (the difference between the maximum vapor pressure and the minimum liquid pressure) as the artery radius increases. This results in a considerable departure from isothermality, as indicated by the sharp rise in the vapor temperature drop.

The vapor temperature drop at the actual heat transfer rate is much smaller. It is about 1°F (0.6°C) at a radius of 0.017 in., which corresponds to a capillary pumping limit of about 2 Btu/sec (2110 W). (See Figure 8.9.)

In Figure 8.11, the capillary pumping limit is plotted as a function of the number of arteries, each with a liquid channel radius of 0.016 in. (0.000406 m). With two arteries, the capillary pumping limit is 3.25 Btu/sec (3430 W), again giving a comfortable margin above the actual heat transport rate of about 2 Btu/sec (2110 W).

The vapor pressure drop at the capillary pumping limit increases significantly with the number of arteries, resulting in a substantial increase in the vapor temperature drop.

The range of the sonic, entrainment, and boiling limits with the range in number of

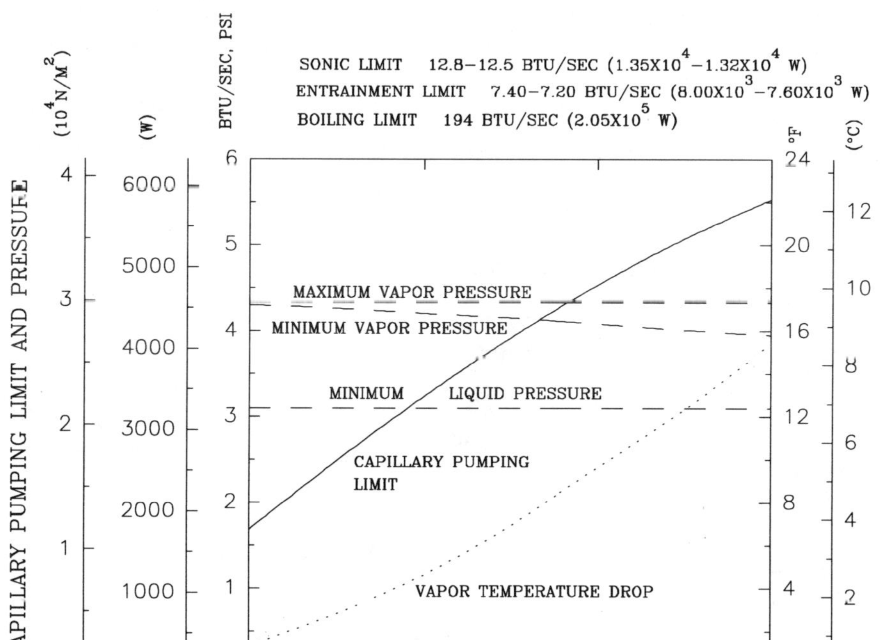

Figure 8.11. Effect of number of arteries on heat transport.

arteries is also indicated in Figure 8.11. The somewhat smaller vapor space area with the arteries present reduces the sonic and entrainment limits slightly. On the other hand, the boiling limit increases by a large factor because of the lower thermal resistance of the distribution wick.

The artery wick occupies considerably less volume and weight than the two-layer wicks.

Modified Design. The heat transport capability of the initial design was upgraded by replacing the two-layer screen wick with a two-layer grooved wick. An artery wick could also have been used effectively for this purpose. The groove depth, groove width, and land width are equal to 0.010 in. (0.000254 m).

The capillary pumping limit then increases to 3.25 Btu/sec (3430 W), providing an adequate margin over the actual heat transport rate of about 2 Btu/sec (2110 W). There is no change in the sonic and entrainment limits, but the boiling limit with the thinner wick rises to 974 Btu/sec (1.03×10^6 W).

The outer radius of the heat pipe is also reduced somewhat, resulting in a comparable reduction in the actual heat transport rate radiating away from the condenser surface. The outer radius is now 0.288 in. (0.00732 m) (below the nominal outer radius), and the actual heat transport rate is 1.834 Btu/sec (1935 W). The somewhat lower value of the actual heat transport rate is assumed to be acceptable.

The evaporator outer surface temperature (the cold junction temperature) is the sum of the vapor temperature, the wick temperature drop, and the wall temperature drop.

The temperature drop across the heat pipe wick can be found from the boiling limit calculation by setting the boiling limit equal to the actual heat transport rate of 1.834 Btu/sec (1935 W). The corresponding heat flux on the wick outer surface (radius =

Table 8.5. Modified design characteristics of heat pipe space radiator

Parameter	Value
Heat pipe fluid	Sodium
Wall and wick material	Haynes 188
Vapor temperature	1400°F (760°C)
Evaporator length	8 in. (0.203 m)
Condenser length	32 in. (0.813 m)
Total heat pipe length	40 in. (1.016 m)
Vapor space radius	0.250 in. (0.00635 m)
Wall thickness	0.020 in. (0.000508 m)
End plate thickness	0.020 in. (0.000508 m)
Wick type	Axial grooves covered by screen
Flow channel layer	Grooves
Groove thickness (depth)	0.010 in. (0.000254 m)
Groove and land width	0.010 in. (0.000254 m)
Liquid fraction (porosity)	0.5
Capillary pumping layer	400 mesh screen layers
Thickness	0.008 in. (0.000203 m)
Liquid fraction (porosity)	0.5

0.268 in. [0.00681 m]) is equal to 19.60 Btu/ft^2-sec (222,660 W/m^2). The wick temperature drop is then 4.7°F (2.6°C).

The wall ΔT can be obtained from Eq. (7.20), modified to apply to the wall rather than the wall and wick. The radii of the outer and inner wall surfaces are 0.288 and 0.268 in. (0.00732 and 0.00681 m). The heat flux on the wall outer surface is equal to 18.24 Btu/ft^2-sec (207,210 W/m^2). From Table C.5 of Appendix C, the thermal conductivity for Haynes 188 at 1400°F (760°C) is equal to 13.7 Btu/hr-ft-°F (23.7 W/m-°K). The calculated wall temperature drop is then 8.3°F (4.6°C).

The surface temperature is then 1400°F + 4.7°F + 8.3°F = 1413°F (760°C + 2.6°C + 4.6°C = 767°C), below the nominal design criterion of 1420°F (771°C).

Characteristics of the modified heat pipe space radiator design are summarized in Table 8.5.

8.1.4. Stress and Buckling in the Heat Pipe Wall

During operation, the heat pipe will be subjected to thermal and mechanical stress. Thermal stress arises from the temperature difference between the inner and outer surface of the evaporator wall. Mechanical stress arises because of the difference between the vapor pressure and that of the external environment.

The thermal stress is given by Eq. (6.2). The value of Poisson's ratio υ is taken as 0.3. From Table C.6 of Appendix C, the following values apply for the modulus of elasticity E and coefficient of thermal expansion α of Haynes 188 at 1400°F (760°C).

$$E = 24.9 \times 10^6 \text{ psi } (1.72 \times 10^{11} \text{ N/m}^2)$$

$$\alpha = 9.0 \times 10^{-6} \text{ in./in. per °F } (1.62 \times 10^{-5} \text{ m/m per °C})$$

The calculated thermal stress is then

Surface	Stress	Type
Outer	1422 psi (9.80 × 10^6 N/m^2)	Compressive
Inner	1319 psi (9.09 × 10^6 N/m^2)	Tensile

The mechanical hoop stress is obtainable from Eq. (6.3). At the operating temperature of 1400°F (760°C), the vapor pressure is 4.33 psia (29,860 N/m^2), and the external pressure (in space) is zero. The ΔP is then 4.33 psi (29,860 N/m^2), resulting in a tensile hoop stress of 60.2 psi (4.15 × 10^5 N/m^2).

Adding this stress to the thermal stress gives a net compressive stress of 1362 psi (9.39 × 10^6 N/m^2) at the outer surface and a net tensile stress of 1379 psi (9.51 × 10^6 N/m^2) at the inner surface. A reasonable allowable stress for extended elevated-temperature operation is the stress for 0.5% creep in 1000 hr. From Table C.6 for Haynes 188, this stress is 11,700 psi (8.10 × 10^7 N/m^2). The stresses in the heat pipe wall are well below this level.

The heat pipe is also subject to a compressive hoop stress and radial buckling when nonoperational at ambient temperature and pressure. Since there is no heat addition in this situation, there is no thermal stress. The pressure differential between atmospheric pressure on the outside of the heat pipe and the hard vacuum inside the sodium heat pipe is 14.7 psi (1.01×10^5 N/m^2). The compressive hoop stress is then 204 psi (1.41×10^6 N/m^2), far below the yield strength of 63,700 psi (4.64×10^8 N/m^2) for Haynes 188 at ambient temperature.

The external pressure required to buckle a cylinder is given by Eq. (6.18). The modulus of elasticity E of Haynes 188 at ambient temperature is 33.6×10^6 psi (2.32×10^{11} N/m^2). The external pressure required to buckle the heat pipe wall is then 3437 psi (2.37×10^7 N/m^2), far greater than the actual external pressure of 14.7 psia (1.01×10^5 N/m^2). Therefore, buckling of the heat pipe wall is not a problem.

8.2. TRANSIENT ANALYSIS OF THE HEAT PIPE SPACE RADIATOR

In this example, the modified heat pipe design of example 1 (Section 8.1) is subjected to a transient startup analysis, using the methods of Chapter 7.

8.2.1. Boundary and Initial Conditions

The transient response to four time-dependent heat input rates to the evaporator was determined. These input heating conditions are illustrated in Figure 8.12 and summarized below. For each case except case 1, a heat pulse of peak amplitude equal to twice the steady-state heating rate Q_{SS} of 1.834 Btu/sec (1935 W) is applied.

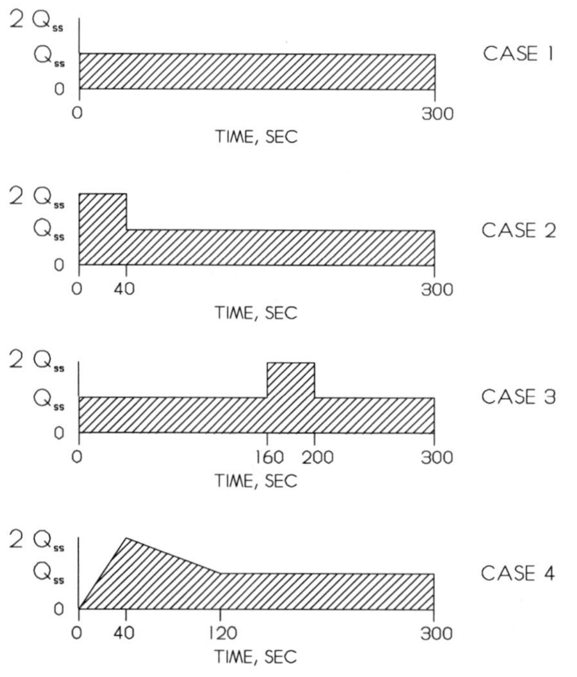

Figure 8.12. Transient heat loadings for the heat pipe space radiator.

Case 1 represents the constant heat loading Q_{ss} that develops in example 1 when the heat pipe is inserted into the cold junction of the radioisotope-powered thermionic converter.

The other cases are not relevant to this application, but have been included to illustrate how heat pipe startup characteristics can be influenced by the time-dependent heat input rate.

Case 2 represents a rectangular heat pulse applied at the beginning of startup that drops back to the steady-state heat input rate before phase 2 is completed. The heat pulse in case 3 is identical to that in case 2, but occurs within phase 3. In case 4, a ramp-type heat pulse applied at the beginning of the startup transient peaks within phase 2, but does not revert back to the steady-state heat input rate until after phase 3 has begun.

These relations between the various heat loadings and the transient phases are not intuitively obvious. They were determined following several preliminary transient calculations.

The time-dependent heat loading is generally an imposed boundary condition, and not subject to control by the designer. However, as will be shown in Section 8.2.3, the effects of heat pulses in excess of the steady-state heat input rate can be mitigated by extending phase 2 so that the heat pulse terminates in phase 2. Phase 2 can be extended by increasing the heat pipe thermal capacity. Methods for accomplishing this include use of a thicker wall and/or wick, use of materials with higher specific heats, and addition of a supplementary heat sink between the heat pipe wall and wick.

The condenser boundary conditions are the same as in the example of Section 8.1. Heat is rejected by radiation from the heated portion of the condenser, with negligible heat input from the surroundings. The temperature of the unheated portion of the condenser in phase 2 remains at the ambient value. The temperature of the heated portion of the condenser in phase 2 is T_2. In phase 3, it is T_v.

Initially, the heat pipe is assumed to be at an ambient temperature of 90°F (32°C). The initial position of the continuum front is at the evaporator-condenser interface, at a distance of 8 in. (0.203 m) from the evaporator end.

8.2.2. Objective of Transient Analysis

The objective of the transient analysis was to obtain information on the following heat pipe characteristics during startup:

- Temperature and continuum front histories (cases 1–4).
- Heat transport limits (case 1).
- Actual heat transport rates (case 1).
- Minimum liquid pressure (case 1).

The last three items were evaluated only for case 1. This the only case whose heat input rate is applicable to the example of Section 8.1.

Since thermal stress, mechanical stress, and buckling conditions are not significantly different during startup than during steady-state operation, these aspects of heat pipe design are not considered in this example. Also, it was assumed that the heat pipe is free to expand axially as the temperature behind the continuum front increases. If this should not be the case, significant axial thermal stress could develop.

8.2.3. Temperatures and Continuum Front Position

Calculational Approach. The transient analysis should be carried out over a sufficiently long time interval t_{max} to allow heat pipe temperatures to approach steady-state values closely. For purposes of this example, the steady-state condition is assumed to be reached when the heat pipe vapor temperature approaches within 1° of 1400°F (760°C), the steady-state design value. Since the startup time is not known beforehand, an initial value of t_{max} was assumed, and then modified in subsequent transient calculations.

For the problem under consideration, it was established that startup would be completed, or almost completed, within 300 sec for the various heat input cases. Therefore, 300 sec was used for t_{max}.

A time increment Δt of 0.1 sec was selected (1/3000th of t_{max}). With this value of Δt, the transient solutions were stable and could be completed in a reasonable time. Calculational results were recorded after every 20 time increments, or every 2 sec.

Calculation of the evaporator outer wall and vapor temperatures was simplified by assuming that the average heat flux in the wall/wick is equal to the incident heat flux on the evaporator surface. The calculation then proceeded as described in Section 7.5.5. The error resulting from this assumption is relatively small in this case, since the steady-state temperature drop across the wall and wick is only 13°F (7°C).

Results. Results of the heat pipe transient analysis of the modified heat pipe space radiator design are now presented. Key transient events are summarized in Table 8.6.

For the sodium heat pipe under consideration here with a vapor space radius of 0.25 in. (0.00635 m), the minimum vapor space dimension is 0.5 in (0.01270 m). The calculated value of the continuum flow temperature T_{cf} is then 777°F (414°C).

Case 1 transient characteristics. In Figure 8.13, the evaporator outer surface temperature T_{womax}, the vapor temperature T_v, the condenser temperature behind the contin-

Table 8.6. Key events during startup transient for cases 1–4

Case	Startup time, sec	End of phase 1, sec	End of phase 2, sec	Peak wall temperature, °F (°C)	Time at peak, sec	T_v when startup complete, °F (°C)
1	246	4–6	132	1412 (767)	246	1399 (759)
2	172	2–4	58	1412 (767)	172	1399 (759)
3	300+	4–6	132	1731 (944)	200	1401 (761)
4	196	14–16	70	1517 (865)	100	1401 (761)

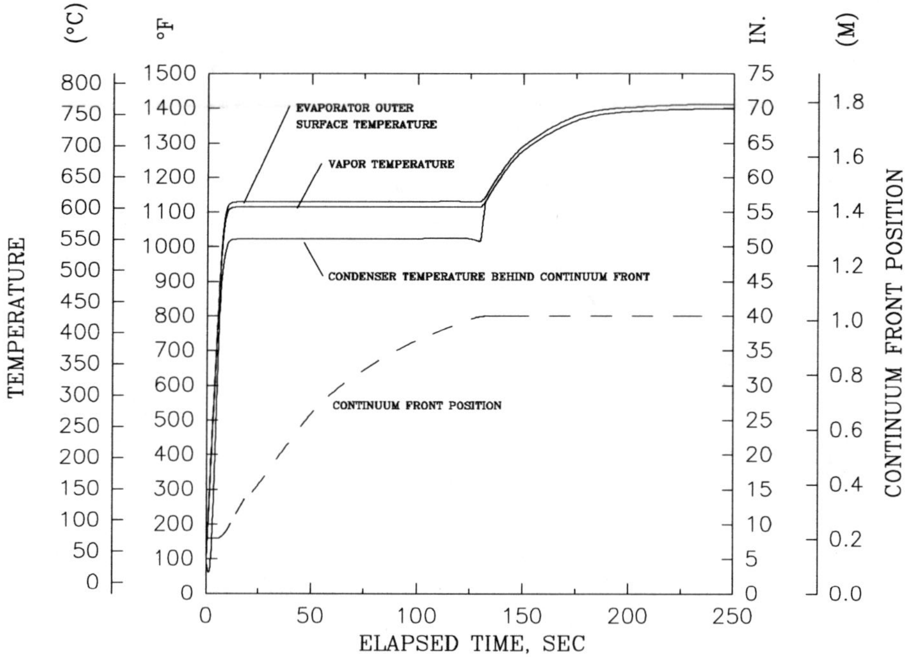

Figure 8.13. Temperature and continuum front position histories for case 1 transient.

uum front T_2, and the continuum front position x are plotted against elapsed time for the constant heat input rate of case 1.

The wall and vapor temperatures rise rapidly during phase 1, reaching the onset of phase 2 in 2 to 4 sec, when T_v becomes equal to T_{cf}. At this time, a continuum front starts advancing into the condenser from the evaporator-condenser interface.

The sonic limit increases rapidly with the vapor temperature, and becomes comparable to the heat input rate between 10 and 12 sec. The heat pipe temperature then stabilizes while the continuum front advances into the condenser at a decreasing rate.

During this period, $T_{wo\max}$ = 1129°F (609°C), T_v = 1115°F (602°C), and T_2 = 1023°F (551°C). The temperature drop $T_{wo\max} - T_v$ across the wall and wick is 14°F (8°C), and the temperature drop $T_v - T_2$ corresponding to operation at the sonic limit is 92°F (51°C).

The continuum front reaches the end of the condenser after 132 sec, ending phase 2 and marking the start of phase 3. The condenser temperature behind the continuum front T_2 then immediately becomes equal to T_v. In phase 3, the heat pipe temperature rises initially at a rapid rate, then levels off and approaches the steady-state value. The vapor reaches 1399°F (759°C), marking the effective completion of startup, after 246 sec.

With the case 1 startup, there is no "overshoot" of the heat pipe temperature, and the peak wall temperature coincides with the steady-state value of 1413°F (767°C).

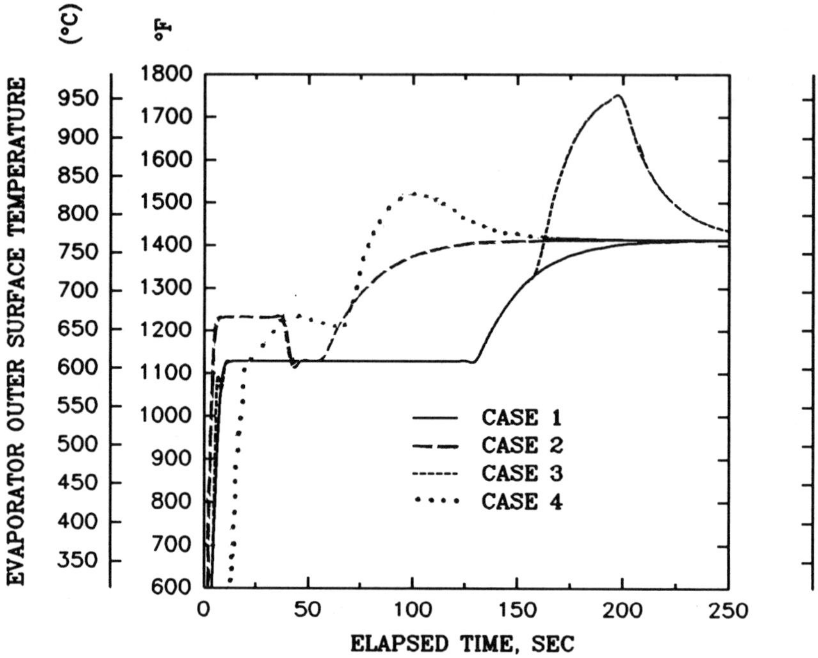

Figure 8.14. Comparison of temperature histories for cases 1–4.

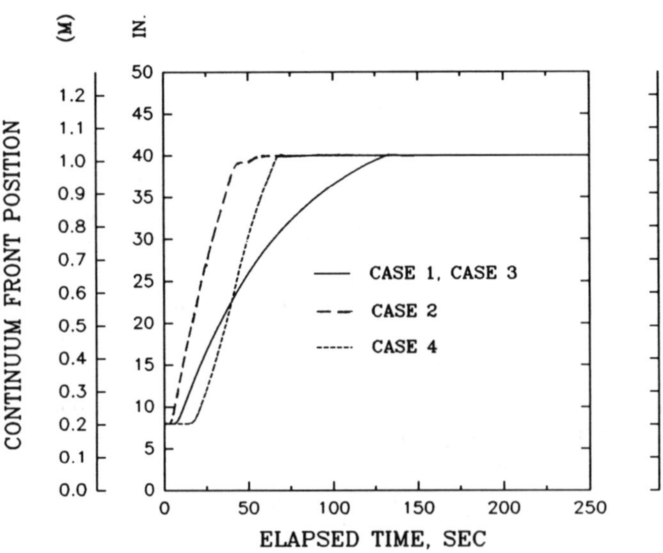

Figure 8.15. Comparison of continuum front position histories for cases 1–4.

Transient response comparison for cases 1–4. The effect of different heat loadings on heat pipe transient behavior is compared in Figures 8.14 and 8.15, where the evaporator outer wall temperature T_{womax} and the continuum front position x are plotted against elapsed time for cases 1–4.

The case 1 curves are identical to those in Figure 8.13.

With case 2, a square heat pulse of amplitude equal to twice the steady-state heating rate Q_{SS} and a length of 40 sec is introduced at the beginning of startup. Initial heatup of the heat pipe is faster, and the temperature in phase 2 stabilizes at 1233°F (667°C). Thus, only about a 100°F (56°C) rise in the phase 2 temperature is required to double the sonic limit, matching the higher initial heating rate.

At 40 sec, the heating rate reverts back to Q_{SS}, and the temperature drops back to the case 1 value. The continuum front reaches the end of the condenser at 58 sec. Phase 2 lasts less than half as long as it does with the case 1 heating profile. In phase 3 the temperature rises in a manner similar to that of case 1, but the startup process is completed in only 172 sec.

Case 2 shows that the startup process can be shortened significantly without any temperature overshoot, if a heat pulse of magnitude greater than Q_{SS} is restricted to phase 1 and/or phase 2.

With case 3, a heat pulse identical to that of case 2 is applied, but after 160 sec when phase 3 is in effect. The curve of continuum front position in phase 2 is identical to that for case 1. However, a large temperature spike now develops, reaching 1755°F (957°C) when the heat pulse ends at 200 sec and the heat input rate drops back to Q_{SS}. The temperature then falls off exponentially, but does not approach the steady-state value until more than 300 sec have elapsed.

Thus, application of a heat pulse in phase 3 can result in a severe temperature overshoot and a significant increase in startup time.

In case 4, a ramp-type pulse is present at the start of the transient. The heat input rate rises linearly to $2Q_{SS}$ after 40 sec, then decreases linearly to Q_{SS} after 120 sec. For this case, the heating rate is elevated above Q_{SS} over portions of both phases 2 and 3.

Initial heating is at a considerably lower rate than for the other cases, with phase 1 continuing for 14–16 sec. A temperature peak of 1231°F (666°C) occurs at 40 sec, coinciding with the peak heating rate. The temperature then drops off slightly to 1203°F (651°C) at the end of phase 2, after 70 sec. A rapid rise to a second peak of 1517°F (825°C) occurs at 100 sec, followed by a gradual decrease in temperature to within 1° of the steady-state value at 196 sec. As occurred with case 3, a significant temperature overshoot develops in phase 3.

8.2.4. Transient Heat Transport Capability

Transient heat transport capability was established by comparing the actual heat transport rate with the various heat transport limits at selected times. Additionally, the vapor pressure and temperature drops, as well as the minimum liquid pressure, were examined.

The transient heat transport capability was evaluated for case 1 only, since this is the heating rate boundary condition that is applicable to example 1. Only phases 2 and 3 are of interest, as the heat pipe is nonfunctional in phase 1.

Calculational Approach. In evaluating the capillary pumping limit, the diameter of curvature corresponding to the actual heat transport rate was calculated instead of the limiting heat transport rate for capillary pumping. Associated vapor and liquid pressures then correspond to the actual heat transport rate.

The heat transport limit calculations in phase 2 were performed at the heat pipe vapor temperature T_V. Performance of the capillary pumping and entrainment limit calculations at the temperature T_2 behind the continuum front would yield lower, and probably more accurate, results.

In phase 2, the actual heat transport rate is equal to the sonic limit. In phase 3, it was calculated from Eq. (7.51). The volumetric thermal capacity of the heat pipe evaporator needed for this calculation was obtained from Eq. (7.5), and evaluated at the mean wall/wick temperature. The needed temperature gradient $\Delta T/\Delta t$ at a given time, where ΔT is the change in the mean evaporator wall/wick temperature, was obtained from the heat pipe temperature history for case 1.

Also, in performing the capillary pumping limit calculation in phase 2, the condenser length behind the continuum front was assumed to be adiabatic. The accuracy of this assumption was checked by comparing the heat removed by condensation behind the front (i.e., in segment 2) with the heat transport rate into the condenser. This comparison is briefly reviewed here.

In phase 2, the heat transport rate into segment 2 is equal to the sonic limit of 1.831 Btu/sec (1932 W). After 130 sec, the segment 2 length of 31.92 in. (0.811 m) is almost at its maximum value, the temperature is 1023°F (551°C), and the radiative heat rejection rate is 0.739 Btu/sec (789 W). Since the segment 2 temperature remains constant in segment 2 (see Figure 8.13), this is also the heat removal rate from the vapor stream.

The heat transport rate at the continuum front is then 1.831 – 0.739 = 1.092 Btu/sec (1152 W). The average heat transport rate in segment 2 is (1.831 + 1.092)/2 = 1.462 Btu/sec (1542 W), or 79.8% of the inlet heat transport rate. This percentage is higher during the earlier stages of phase 2.

For fully adiabatic flow, the average and inlet heat transport rates are equal, since there is then no heat transfer out of segment 2. Thus, the assumption of adiabatic heat transport is reasonable. It is also on the conservative side, since the higher average heat transport rate with fully adiabatic flow results in a lower capillary pumping limit.

Results. Results of the heat transport limit calculations are given in Figure 8.16 and Table 8.7.

The sonic, entrainment, and boiling limits, along with the actual heat transport rate, were calculated at selected times and plotted against time in Figure 8.16. The actual heat transport rate is equal to or below these heat transport limits throughout the transient. (In phase 2, the actual heat transport rate is equal to the sonic limit.) Thus, heat pipe operation during startup is not disrupted by entrainment, boiling, or a heat transport rate in excess of the sonic limit.

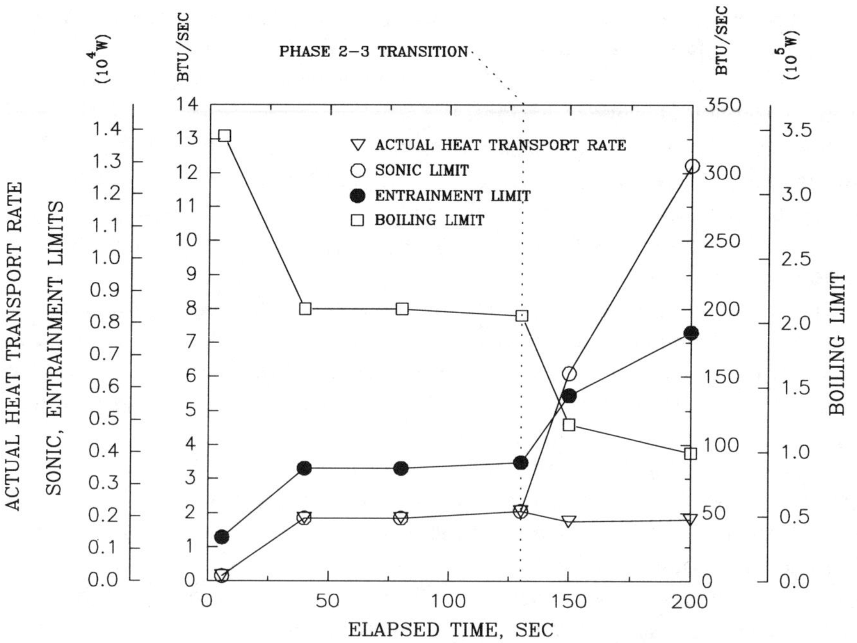

Figure 8.16. Sonic, entrainment, and boiling limits and actual heat transport rate for selected times during case 1 transient.

Capillary pumping characteristics of the heat pipe at selected times during the startup transient are presented in Table 8.7.

In phase 3, the diameter of curvature D_c exceeds the effective pore diameter D_p of the wick capillary pumping layer (60 microns), the vapor pressure and temperature drops ΔP_v and ΔT_v are small, and the minimum liquid pressure $P_{l\min}$ is positive (the liquid is under compression).

Table 8.7. Capillary pumping characteristics during case 1 transient (for design of Table 8.5)

Elapsed time, sec	D_c, microns	ΔP_v, %	ΔT_v, °F (°C)	$P_{l\min}$, psia (N/m^2)
		Phase 2		
6	2961	32.9	31.9 (17.7)	0.0114 (78.6)
40	88.0	37.0	51.3 (28.5)	−0.413 (−2848)
80	58.3	47.5	70.5 (39.2)	−0.892 (−6150)
130	48.0	54.1	84.1 (46.7)	−1.200 (−8274)
		Phase 3		
150	97.7	3.37	4.77 (2.65)	1.135 (7826)
200	95.2	0.95	1.51 (0.84)	3.292 (22,700)
After flow channel dimensions were doubled to 0.020 in. (0.000508 m)				
130	170.7	57.9	93.2 (51.8)	0.0431 (297)

The situation is considerably different in phase 2. As the elapsed time increases and the continuum front approaches the end of the condenser, D_c drops below D_p, vapor pressure and temperature drops are substantial, and the minimum liquid pressure becomes negative (the liquid is under tension).

The large but tolerable drops in vapor pressure and temperature in phase 2 reflect the fact that the heat transport rate is at the sonic limit over the entire length of condenser behind the continuum front. However, capillary pumping will be disrupted if $D_c < D_p$. Although a negative liquid pressure is probably tolerable, some operational uncertainty is associated with this condition.

Consequently, it was necessary to modify the heat pipe design once more, in order to increase both the diameter of curvature and the minimum liquid pressure in phase 2. The groove depth, groove width, and land width of the liquid flow channel were therefore doubled from 0.010 to 0.020 in. (0.000254 to 0.000508 m), reducing the liquid pressure drop.

The capillary pumping limit calculation was then repeated at 130 sec, the most critical transient time. Results are given in the last row of Table 8.7. The high vapor pressure and temperature drops remain, but the diameter of curvature has increased to 171 microns, and the liquid is no longer under tension.

The boiling limit is lower with the thicker liquid flow channel, but it still remains far above the actual heat transport rate. The overall heat pipe diameter also increases to 0.298 in. (0.00757 m), with a subsequent slight increase in the steady-state heat input and heat rejection rates. The thermal capacity of the heat pipe with the thicker wick increases by roughly 20%, slowing the heatup rate and extending the startup time.

These changes are relatively modest, and are not reevaluated here. However, for an actual design situation, recalculation of all transient heat pipe characteristics following the change in liquid channel dimensions would be warranted.

The foregoing results show that a heat pipe design based on steady-state conditions may require modification to insure adequate heat transport capability during startup. In the case under consideration, liquid flow channel dimensions were increased in order to reduce the liquid pressure drop, thereby maintaining adequate capillary pumping capability and minimum liquid pressure during the transient.

8.3. HEAT PIPE RECUPERATOR

This example considers the design of a heat pipe recuperator for recovery of heat in the exhaust gas stream of a gas turbine. The recovered heat is transferred to the air stream leaving the gas turbine compressor, raising its temperature prior to entering the combustor.

As with any heat pipe heat exchanger, the recuperator design requires that varying and sometimes conflicting requirements for heat exchange, pressure drop, structural adequacy, and heat pipe heat transport be satisfactorily resolved.

In this example, a nominal design is identified, and then examined for conformance with the design criteria. As will be seen, the design is found to be inadequate in several respects. The nature of the modifications that would be needed to correct these deficiencies is then discussed, but the numerous iterations required to arrive at a satisfactory design are not carried out. This approach provides useful insights into the capabilities and limitations of heat pipes that act as heat exchange elements as well as heat transport devices.

The recuperator design procedure is described first, followed by an account of the results obtained from a first iteration of the design process.

8.3.1. Design Procedure

In this section the recuperator configuration is presented, design criteria are given, the recuperator geometry is defined, and calculational methods for heat transfer coefficients and friction factors are described.

Recuperator Configuration. The recuperator consists of a radially oriented array of heat pipes spanning annular passages through which the exhaust gas and air streams flow in counterflow, as shown in Figure 8.17.

The heat pipes are arranged as a staggered array of circular tubes, as shown in the rectangular array of Figure 8.18. The heat pipe diameter is D_{hp}, the transverse spacing between tube centers is $x_t D_{hp}$ (normal to the flow direction), and the longitudinal spacing between centers is $x_l D_{hp}$ (along the flow direction). The transverse spacing is constant in the rectangular array, and is a linear function of the radius in the radially oriented array of the recuperator.

Design Criteria. The recuperator design was based on the criteria listed in Table 8.8.

Temperature Distribution. The temperature distribution along the recuperator length

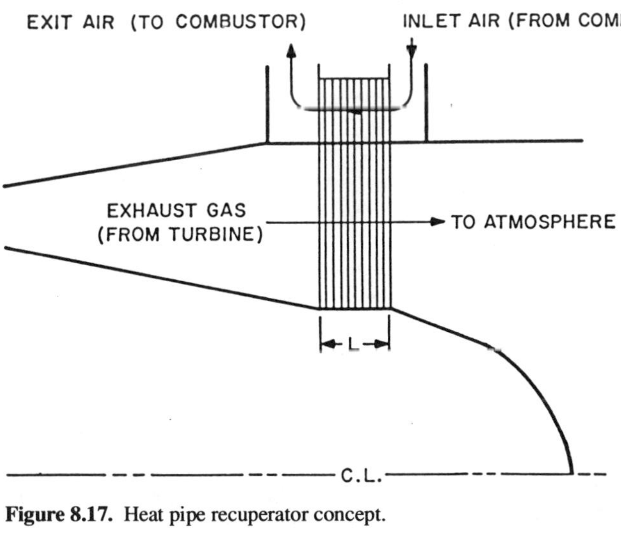

Figure 8.17. Heat pipe recuperator concept.

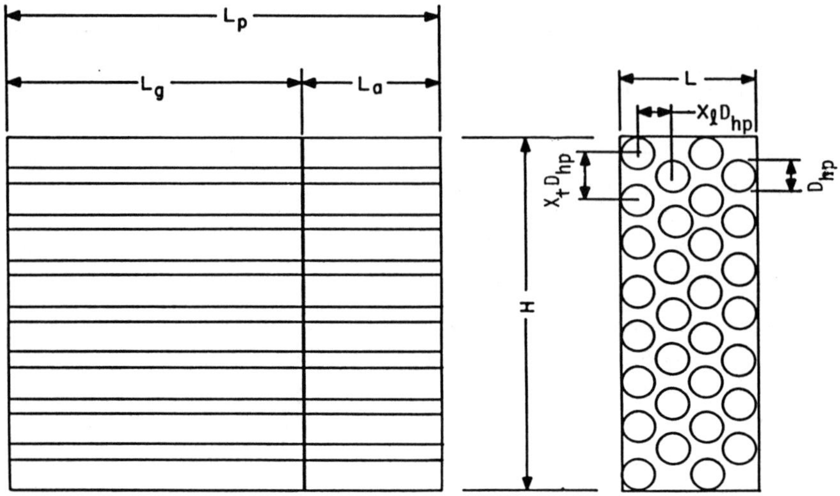

Figure 8.18. Arrangement of the heat pipe array.

Table 8.8. Design criteria for heat pipe recuperator

Heat exchanger effectiveness	50%
Air plus gas side pressure drops	6%
Air and gas flow rates	5 lb/sec (2.27 kg/sec)
Gas stream inlet temperature	1350°F (732°C)
Gas stream inlet pressure	15.5 psia (1.07×10^5 N/m^2)
Air stream inlet temperature	650°F (343°C)
Air stream inlet pressure	147 psia (1.01×10^6 N/m^2)

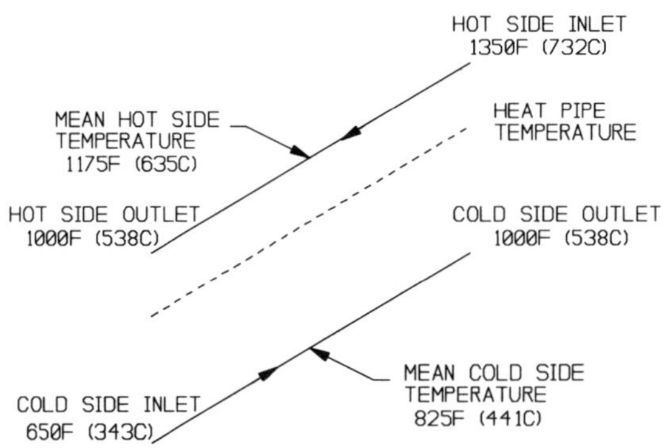

Figure 8.19. Temperature distribution along recuperator length.

for a heat exchanger effectiveness of 50% is as indicated in Figure 8.19. This smoothly varying temperature distribution is characteristic of an infinite number of heat pipe stages.

With a finite number of stages, the heat pipe temperature is nominally constant in each stage (row), but varies from stage to stage. The heat pipe temperature curve shown in Figure 8.19 is an approximation to the actual stepped temperature distribution. Actual values of the heat pipe temperature are determined as part of the recuperator design process.

At a given temperature, temperature-dependent properties of the combustion gas were assumed to be the same as those of air. In this example, gas and air properties are evaluated at the mean gas and air stream temperatures of 1175 and 825°F (635 and 441°C), and then assumed constant over the entire recuperator length. In a more precise analysis, the variation of fluid properties with temperature along the recuperator length would be taken into account.

Geometrical Relationships. The geometry of a radial heat pipe array is depicted in Figure 8.20. Geometrical relationships needed to calculate heat transfer coefficients and friction factors were derived from Figure 8.20, as well as from basic definitions. These relationships are now presented.

The mean transverse spacing between heat pipes, and the gas and air side frontal areas, are given by

$$x_{tm} = \frac{x_{ti}}{2}\left(2 + \frac{L_p}{R_i}\right)$$

$$A_{fg} - 2\pi\, R_i\, L_g \left(1 + \frac{L_g}{2\, R_i}\right)$$

$$A_{fa} = 2\pi\, (R_i + L_g)\, L_a \left[1 + \frac{L_a}{2\,(R_i + L_g)}\right]$$

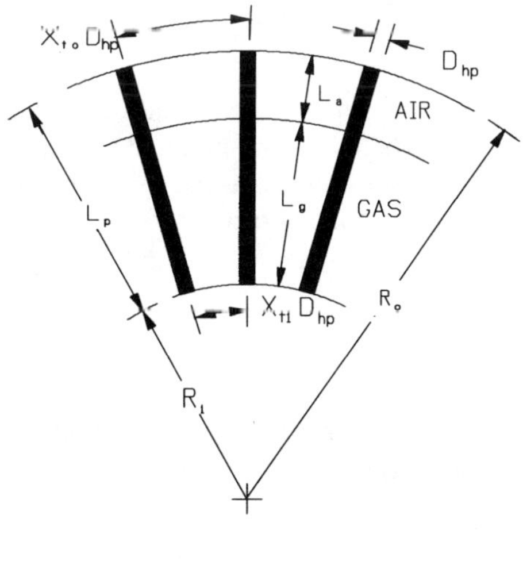

Figure 8.20. Geometry of a radial heat pipe array.

where R_i is the inner recuperator radius, x_{ti} is the transverse spacing in diameters at R_i, x_{tm} is the mean transverse spacing between heat pipes in diameters, L_g is the gas side flow passage width (also the heat pipe evaporator length), L_a is the air side flow passage width (also the heat pipe condenser length), A_{fg} is the gas side frontal area, and A_{fa} is the air side frontal area.

The equation for x_{tm} can also be used to obtain the mean transverse spacing for the heat pipe segments in the gas or air flow passages. R_i would then represent the inner radius of the flow passage in question, x_{ti} would be the transverse spacing at that radius, and L_p would be the flow passage width (L_g for the gas side and L_a for the air side).

The number of heat pipes per stage n_s is

$$n_s = \frac{2\,\pi\,R_i}{x_{ti}\,D_{hp}}$$

The recuperator core length L is given by

$$L = [1 + (s - 1)x_l]D_{hp}$$

where s is the number of stages (or rows).

If the number of stages s is large and the longitudinal spacing x_l is close to 1, the expression for the recuperator core length can be conservatively approximated by

$$L = x_l D_{hp} s$$

If s is at least 4 and x_l is not larger than 1.2, the simplified equation for core length yields results less than 5% greater than those obtained with the exact expression. Since these conditions are very likely to be satisfied in a heat pipe recuperator, the simplified equation for L is used here.

The heat transfer area A on either the gas or the air side is given by

$$A = \pi D_{hp} l n_s s$$

where l is either L_g or L_a.

On substituting $D_{hp}s$ from the previous equation into the equation for A,

$$A = \pi l n_s L/x_l$$

The hydraulic diameter for the gas or air side is defined as (Ref. 8.1)

$$D_h = 4A_{minf}L/A = 4\sigma_a A_f L/A$$

where A_{minf} is the minimum free flow area in a stage on the gas or air side, A_f is the frontal area on the gas or air side, and $\sigma_a = A_{minf}/A_f$. When the minimum free flow area is a function of the heat exchanger radius, it should be evaluated at the mean radius, using the mean transverse spacing x_{tm} for the flow passage in question.

On substituting A from the previous equation into the equation for D_h,

$$D_h = \frac{4\,\sigma_a\,A_f\,x_l}{\pi\,l\,n_s}$$

Heat Transfer Coefficients and Friction Factors. Data on heat transfer coefficients and friction factors for flow normal to an infinitely long rectangular array of staggered circular tubes are given in Ref. 8.1. The data are correlated by the equations

Table 8.9. Coefficients for Stanton number and friction factor

x_l	x_t 1.2	1.5	1.8	2.1	2.5
			C_h		
0.75	0.21	0.30	0.34	0.37	0.37
1.00	0.21	0.31	0.37	0.42	—
1.25	0.21	0.32	0.38	—	—
1.50	0.21	0.33	0.40	—	—
			C_f		
0.75	0.14	0.25	0.34	0.39	0.45
1.00	0.14	0.25	0.38	0.50	—
1.25	0.14	0.27	0.41	—	—
1.50	0.15	0.29	—	—	—

$$\text{St} = C_h \text{Re}^{-0.4} \text{Pr}^{-2/3}$$

$$f = C_f \text{Re}^{-0.18}$$

where Re is the Reynolds number, Pr is the Prandtl number, St is the Stanton number, f is the friction factor, and C_h and C_f are coefficients whose values are functions of x_t and x_l. The correlations are valid over a Reynolds number range of 300 to 15,000. If there are at least 15 tube rows, the Stanton number and heat transfer coefficient are at least 95% of the values for an infinite number of rows.

Values of C_h and C_f, as estimated from the data of Ref. 8.1, are given in Table 8.9. The dimensionless variables Re, Pr, and St are defined as follows:

$$\text{Re} = \frac{D_h G}{\mu}$$

$$\text{Pr} = \frac{c_p \mu}{k}$$

$$\text{St} = \frac{h}{G c_p}$$

where c_p is the specific heat at constant pressure, μ is the viscosity, k is the thermal conductivity, D_h is the hydraulic diameter, G is the mass velocity, and h is the heat transfer coefficient. Once St has been determined, h is readily found from the last equation.

The mass velocity G is defined as

$$G = \frac{W}{A_{\min f}} = \frac{W}{\sigma_a A_f}$$

where W is the mass flow rate, $A_{\min f}$ is the minimum free flow area in a row of the tube bank, A_f is the total frontal area, and $\sigma_a = A_{\min f}/A_f$.

From Figure 8.18,

$$\sigma_a = \frac{x_{tm} D_{hp} - D_{hp}}{x_{tm} D_{hp}} = \frac{x_{tm} - 1}{x_{tm}}$$

In this example, the mean transient spacing x_{tm} for the entire annular width of the recuperator (i.e., the heat pipe length L_p) was used in calculating σ_a, and in obtaining C_h and C_f from Table 8.9. In a more precise calculation, the mean transverse spacings for the gas and air flow passages would be used to establish C_h and C_f separately for the gas and air sides.

8.3.2. Design Calculations

The following major steps are involved in designing the heat pipe recuperator:

1. Select recuperator stage dimensions.
2. Calculate heat exchange and pressure drop characteristics.
3. Calculate heat pipe temperatures and heat transport capability.
4. Calculate stresses in heat pipe wall.
5. Repeat steps 1 through 4 as necessary to arrive at a viable design.

Because of the large number of interdependent variables, many iterations may be needed before all of the design criteria can be satisfied. In this example, steps 1 through 4 are carried out once, and the modifications that would be needed to obtain a viable design in subsequent calculations are discussed.

Stage Dimensions. The following stage dimensions were specified for initial evaluation:

Inner radius of recuperator R_i	10.0 in. (0.254 m)
Heat pipe length L_p	5.0 in. (0.127 m)
Evaporator length (gas side width L_g)	4.5 in. (0.114 m)
Condenser length (air side width L_a)	0.5 in. (0.0127 m)
Heat pipe diameter D_{hp}	0.12 in. (0.00305 m)
Longitudinal spacing $x_l D_{hp}$	$1.0D_{hp}$
Transverse spacing at inner radius $x_{ti} D_{hp}$	$1.2D_{hp}$

The mean transverse spacing $x_{tm} D_{hp}$ is then $1.5D_{hp}$, and $\sigma_a = 0.333$.

A large evaporator/condenser length ratio was selected to reduce the thermal resistance between the gas and the heat pipes relative to that between the heat pipes and the air. In this way, the minimum heat pipe temperature (at the air inlet end of the recuperator) can be made larger, with increased heat transport capability.

From Table 8.9, the values of the coefficients C_h and C_f are

$$C_h = 0.31 \qquad C_f = 0.25$$

Heat Exchange and Pressure Drop. Using the design criteria of Table 8.8, the equations in Section 8.3.1, equations for the stage and overall heat exchanger effectiveness from Sections 6.6.4 and 6.6.5, pressure drop equations in Section 6.6.3, and air tables

Table 8.10. Heat exchange and pressure drop characteristics of the recuperator

Parameter	Gas side	Air side
Frontal area, ft^2 (m^2)	2.41 (0.224)	0.322 (0.0299)
Mass velocity, lb/ft^2-sec (kg/m^2-sec)	6.22 (30.4)	46.58 (227)
Number of heat pipes per stage	436	436
Hydraulic diameter, in. (m)	0.0749 (0.001902)	0.0901 (0.00224)
Mean temperature, °F (°C)	1175 (635)	825 (441)
Viscosity, lb/sec-ft (N-sec/m^2)	2.63×10^{-5} (3.91×10^{-5})	2.27×10^{-5} (3.38×10^{-5})
Specific heat, Btu/lb-°F (J/kg-°K)	0.268 (1122)	0.257 (1076)
Density, lb/ft^3 (kg/m^3)	0.0256 (0.410)	0.309 (4.95)
Prandtl number	0.697	0.684
Reynolds number	1476	15,410
Stanton number	0.0213	0.00844
Heat transfer coefficient, Btu/hr-ft^2°F (W/m^2-°K)	127.8 (726)	363.7 (2065)
Friction factor	0.0672	0.0441
Heat transfer area per stage, ft^2 (m^2)	5.14 (0.478)	0.704 (0.0654)
Heat exchanger effectiveness per stage	0.1296	0.0523
Number of heat transfer units per stage*	0.1388	0.0541
Number of stages**	26	
Recuperator length, in. (m)	3.25 (0.0826)	
Pressure drop, %	11.99	3.43
Gas + air side pressure drop, %	15.42	

Note: *, Based on mean specific heat of 0.263 Btu/lb-°F (1101 J/kg-°K) for entire recuperator.
**, corresponds to overall heat exchanger effectiveness of 0.502.

(Ref. 8.2) to evaluate gas and air side properties at their respective mean temperatures, the heat exchange and pressure drop characteristics of the recuperator were obtained. This information, along with relevant supporting data, is given in Table 8.10.

The heat transfer requirement (an overall heat exchanger effectiveness of 50%) can be achieved with the initial design by using 26 heat pipe stages, with an overall length of 3.25 in. (0.0826 m). However, this design results in a gas plus air side pressure drop well in excess of the allowable 6%.

The most effective way to reduce the pressure drop is to reduce the mass velocity G by increasing the gas and/or air side flow area. This can be accomplished by using a larger inner radius R_i and/or increasing the heat pipe length L_p. These steps will also have an influence on the heat transfer calculations, possibly affecting the required number of heat pipe stages.

Heat Pipe Temperatures and Heat Transport. Results of heat pipe temperature and heat transport calculations are presented in this section.

A heat pipe wick of the two-layer, screened-grooved type was selected. Axial grooves extend along the heat pipe length, and are covered by a layer of 400 mesh screen. Potassium and cesium are candidate heat pipe fluids. The wall and wick are constructed from 304 stainless steel.

Table 8.11. Heat pipe geometry

Outer radius, in. (m)	0.060 (0.001524)
Vapor space radius, in. (m)	0.050 (0.001270)
Wall thickness, in. (m)	0.004 (0.000102)
Wick thickness, in. (m)	0.006 (0.000152)
Wick type	Screen-covered axial grooves
Liquid flow channel	Axial grooves
Groove depth, in. (m)	0.004 (0.000102)
Groove width, in. (m)	0.006 (0.000152)
Land width, in. (m)	0.004 (0.000102)
Capillary pumping layer	400 mesh screen
Thickness, in. (m)	0.002 (0.000051)
Porosity	0.6
Effective wick pore diameter, microns	60

The heat pipe geometry is given in Table 8.11. Table 8.12 provides information on heat pipe temperatures and heat transfer rates. In Table 8.13, actual and calculated heat transport limits are given, along with important design information on heat pipe pressures and temperature drops.

Heat pipe temperatures were calculated using the equations in Section 6.6.5. Heat transport limits were obtained using the methods described in Chapter 5.

It should first be noted that the boiling limit is far greater than the evaporator surface heat flux for all vapor temperatures of interest, indicating that boiling is not a problem in the recuperator heat pipes.

The following discussion pertains to the air inlet side of the recuperator, since this is where heat pipe design conditions are most critical.

The entrainment limit is well above the axial heat flux for potassium, but below the axial heat flux for cesium. The sonic limit exceeds the axial heat flux for both fluids, but to a greater extent with cesium. The proximity to the sonic limit results in an unacceptably high vapor temperature drop with potassium, and a marginally acceptable vapor temperature drop with cesium.

Capillary pumping capability, as measured by the diameter of curvature required for capillary pumping, is quite adequate for potassium heat pipes, barely adequate for cesium heat pipes with a horizontal orientation, but quite inadequate for those cesium heat pipes that are vertically oriented.

The potassium liquid of the vertically oriented heat pipes is slightly under tension

Table 8.12. Heat pipe temperatures and heat transport rates

Item	At air inlet	At gas inlet
Heat pipe temperature, °F (°C)	909 (487)	1245 (674)
Recuperator heat transfer rate, Btu/sec (W)	460 (485,300)	
Heat transport rate per heat pipe, Btu/sec (W)	0.0406 (42.8)	
Gas side heat flux, Btu/ft^2-sec (W/m^2)	3.44 (39,080)	
Air side heat flux, Btu/ft^2-sec (W/m^2)	31.0 (352,200)	

Table 8.13. Heat pipe heat transport characteristics

Characteristic				
Heat load, Btu/sec (W)			0.0406 (42.8)	
Axial heat flux, Btu/in^2-sec (W/m^2)			5.17 (8.46 × 10^6)	
Evaporator heat flux at outer surface, Btu/ft^2-sec (W/m^2)			3.44 (3.91 × 10^4)	
Wick pore diameter for capillary pumping, microns			60	
	At air inlet		At gas inlet	
	Potassium	Cesium	Potassium	Cesium
Vapor temperature, °F (°C)	909 (487)	909 (487)	1245 (674)	1245 (674)
Sonic limit, Btu/in^2-sec (10^6 W/m^2)	5.94 (9.72)	7.84 (12.83)	64.2 (105)	61.7 (101)
Entrainment limit, Btu/in^2-sec (10^6 W/m^2)	9.30 (15.2)	4.01 (6.56)	27.4 (44.8)	9.36 (15.3)
Boiling limit, Btu/ft^2-sec * (10^6 W/m^2)	3263 (37.1)	926 (10.5)	1209 (13.7)	211 (2.40)
Vapor pressure, psia (N/m^2)	0.49 (3380)	1.64 (11,310)	6.24 (43,030)	15.13 (104,300)
Vapor temperature drop, °F (°C)	60.9 (33.8)	7.51 (4.17)	0.88 (0.49)	0.33 (0.18)
Horizontal orientation				
Minimum liquid pressure, psia (N/m^2)	0.10 (690)	1.19 (8205)	6.05 (41,720)	14.65 (101,010)
Diameter of curvature, microns	126.3	66.6	211.3	64.1
Vertical orientation				
Minimum liquid pressure, psia (N/m^2)	–0.03 (–207)	0.90 (6206)	5.92 (40,820)	14.39 (99,220)
Diameter of curvature, microns	94.0	40.7	129.7	37.5

Note: *In calculating the boiling limit, the thermal conductivity of Haynes 25 was used; it is about 10% lower than that of stainless steel.

(under negative pressure). Although this condition may be acceptable, it should be eliminated if possible to avoid uncertainty about heat pipe performance during recuperator operation.

For both heat pipe fluids, the axial heat flux in the heat pipes will have to be reduced at the air inlet side. This is required for cesium in order to drop the axial heat flux below the entrainment limit and raise the diameter of curvature for capillary pumping above the wick pore diameter. A reduction in the axial heat flux well below the sonic limit is needed for the potassium heat pipes in order to reduce the high level of nonisothermality in the heat pipe vapor. Also, the resultant lower pressure drop in the potassium liquid and vapor should eliminate tension in the liquid of the vertically oriented heat pipes.

The axial heat flux can be reduced by increasing the heat pipe diameter, thereby increasing the vapor space area. Recuperator heat transfer and pressure drop will also be affected by the change in heat pipe diameter.

It is not necessary to increase the diameter of all the potassium heat pipes in the recuperator, since at some distance from the air inlet the vapor temperature will have increased sufficiently to raise the sonic limit well above the original axial heat flux. An excessive vapor temperature drop will then not be encountered in the heat pipes beyond that point. The rising vapor pressure associated with higher vapor temperatures will also eliminate the problem of tension in the potassium liquid in the later stages. Larger-diameter heat pipes would be used in perhaps the first eight or nine stages from the air inlet end.

For the cesium heat pipes, the diameter of curvature does not increase with temperature. Therefore, larger-diameter cesium heat pipes will be required over the entire recuperator length.

Heat Pipe Temperature Drop. Methods for calculating the effective thermal conductivity of the wick, and the temperature drop through the wick, are given in Section 6.5.1. Potassium and cesium thermal conductivities were obtained from Appendix C. The following values were used for the thermal conductivity of 304 stainless steel (Ref. 8.1):

Vapor Temperature, °F (°C)	Thermal Conductivity, Btu-hr-ft-°F (W/m-°K)
909 (487)	12 (20.8)
1245 (674)	14 (24.2)

Calculated temperature drops through the heat pipe wall and wick are given in Table 8.14. The total temperature drop is less than 1°F (0.6°C) in the evaporator (gas side) section, and between 7 and 9°F (4 and 5°C) in the condenser (air side) section.

The significant temperature drop through the condenser wall and wick, combined with the temperature drop that occurs in the heat pipe vapor, means that there is a smaller ΔT available for heat transfer from the heat pipes to the compressed air. As a result, the recuperator heat transfer rate will be less than the design requirement. An additional one or two stages might then be necessary to restore the design heat transfer rate.

Stress in Heat Pipe Wall. Thermal, hoop, and net stresses in the wall of the heat pipe evaporator and condenser sections were calculated at the air and gas inlets, using equations in Sections 6.4.1 and 6.4.2.

Table 8.14. Heat pipe wall and wick temperature drops at air inlet (vapor temperature = 909°F [487°C])

Temperature drop, °F (°C)	In evaporator		In condenser	
	Potassium	Cesium	Potassium	Cesium
Through wick	0.38 (0.21)	0.61 (0.34)	3.45 (1.92)	5.49 (3.10)
Through wall	0.36 (0.20)	0.36 (0.20)	3.24 (1.80)	3.24 (1.80)
Through wall + wick*	0.74 (0.41)	0.97 (0.54)	6.69 (3.72)	8.73 (4.85)

Note: *At the gas inlet end (vapor temperature = 1245°F [674°C]), the wall + wick temperature drop is slightly lower in the evaporator, about 0.5°F (0.3°C) lower in the condenser.

Table 8.15. Stresses in the heat pipe wall

Stress	At air inlet (heat pipe temperature = 909°F [487°C])		At gas inlet (heat pipe temperature = 1245°F [674°C])	
	Potassium	Cesium	Potassium	Cesium
	Evaporator (gas side)			
Thermal stress, psi (10^5 N/m^2)	71 (4.90)	71 (4.90)	61 (4.21)	61 (4.21)
Hoop stress, psi (10^5 N/m^2)	218 (15.0)	202 (13.9)	135 (9.31)	7 (0.483)
Maximum stress, psi (10^5 N/m^2)	289 (19.9)	273 (18.8)	196 (13.5)	68 (4.69)
	Condenser (air side)			
Thermal stress, psi (10^5 N/m^2)	641 (44.2)	641 (44.2)	550 (37.9)	550 (37.9)
Hoop stress, psi (10^5 N/m^2)	2124 (146)	2108 (145)	2042 (141)	1914 (132)
Maximum stress, psi (10^5 N/m^2)	2765 (191)	2749 (190)	2592 (179)	2464 (170)

Thermal stresses were calculated using the following nominal (room temperature) properties of stainless steel (Ref. 8.3):

Modulus of elasticity	28×10^6 psi (1.93×10^{11} N/m^2)
Coefficient of thermal expansion	9.9×10^{-6} in./in. per °F (1.78×10^{-5} m/m per °C)

In the evaporator, the thermal stress is compressive at the outer wall surface and tensile at the inner wall surface. The situation is reversed in the condenser. The hoop stress is always compressive because the external gas or air pressure is always greater than the vapor pressure of the heat pipe fluid. The maximum stress is compressive, and equal to the sum of the thermal and hoop stresses.

Results of the stress calculations are summarized in Table 8.15.

Stress is a maximum in the condenser wall, because of the higher air pressure and the greater wall temperature difference on the air side of the recuperator. Peak stresses occur at the air inlet end. Stresses are not significantly different for potassium and cesium heat pipes.

For stainless steel, a stress level of 4000 to 10,000 psi (2.76×10^7 to 6.90×10^7 N/m^2) at 1250°F (677°C) is required to produce a creep rate of 0.1%/1000 hr (Ref. 8.3). A higher stress is required to produce this creep rate at 900°F (482°C).

Since the maximum stress in the stainless steel heat pipe walls is less than 3000 psi (2.07×10^7 N/m^2), the initial recuperator design should be capable of operating for at least several thousand hours without excessive creep.

8.3.3. Comments on the Heat Pipe Recuperator Design

This example demonstrates that the design of a heat pipe recuperator is influenced by a large number of variables, most of which are strongly interrelated. The most critical aspect of heat pipe design generally occurs at the air inlet, where the vapor temperature and pressure of the heat pipes are at a minimum. Achievement of an optimal design may require numerous design iterations.

8.4. REFERENCES

8.1. Kays, W. M., and London, A. L., *Compact Heat Exchangers,* McGraw-Hill, New York, 1964, pp. 7, 127, 128, 232.
8.2. Eckert, E. R. G., and Drake, R. M., Jr., *Heat and Mass Transfer,* 2nd Ed., McGraw-Hill, New York, 1959, p. 504.
8.3. Baumeister, T. (Ed.), *Standard Handbook for Mechanical Engineers,* 7th Ed., McGraw-Hill, New York, 1967, pp. 5-14, 6-10.

APPENDIX

A

NOMENCLATURE

The various symbols assigned to variables and parameters have been defined as they are introduced in the body of the book. They are summarized here for convenience of reference.

Because of the large number of variables and parameters, it has been necessary on occasion to represent more than one with the same symbol. For example, A denotes both the heat transfer area on the gas or air side of a recuperator and a constant in the equation relating pressure and temperature of a saturated vapor. In such cases, symbol meanings will be clear from the context in which they are employed.

Symbols for temperature generally denote the absolute temperature (°R or °K). However, in situations in which temperature differences are involved or the temperature is used to evaluate a temperature-dependent property, temperature symbols may also denote relative temperature (°F or °C). For example, if the heat pipe vapor temperature T_v is being used to denote the temperature at which heat radiates away from the heat pipe condenser, it would be expressed in absolute units. If T_v were being used in an expression for the wick temperature drop, or if the surface tension at T_v were required, relative as well as absolute units could be employed.

A.1. GREEK LETTER SYMBOLS

α	ratio of vapor and liquid flow passage areas, A_v/A_l
α_b	constant in Eq. (5.139) for boiling limit parameter, angle between acceleration of gravity and L_b

α_{dc}	distributive wick area ratio in condenser, A_v/A_{ldc}
α_{de}	distributive wick area ratio in evaporator, A_v/A_{lde}
α_e	constant in Eq. (5.130) for entrainment limit parameter
α_g	angle between gravity vector and L_b
α_s	constant in Eq. (5.116) for sonic limit
α_{te}	coefficient of thermal expansion
α_{wall}	fraction of A_{ww} occupied by wall
$\alpha_{\text{wall}m}$	fraction of A_{wwm} occupied by wall
β	boiling limit parameter, defined by Eq. (5.137)
β_b	angle between external acceleration vector and L_b
β_p	k_{ep}/k_l
β_s	k_{es}/k_l
γ	critical pressure ratio defined by Eq. (7.38)
γ_1	t_{wick1}/r_v
γ_2	t_{wick2}/r_v
γ_l	Q_{ll}/Q
γ_v	Q_{vl}/Q
δ	angle defined by Eq. (5.48)
δ_1	arbitrarily small quantity
δ_g	geometrical parameter in Eq. (6.2)
ΔP	pressure drop or pressure difference
ΔP_b	radial pressure difference across cylinder wall for buckling
ΔP_c	capillary pressure in tube or wick pores
ΔP_f	frictional pressure drop (pressure drop due to frictional shear force)
ΔP_{ft}	frictional pressure drop for turbulent flow
ΔP_{fl}	frictional pressure drop for laminar flow
ΔP_l	liquid pressure drop
ΔP_{la}	liquid pressure drop due to gravity/external acceleration
ΔP_{lf}	frictional pressure drop in liquid
ΔP_{lfa}	total liquid frictional pressure drop with artery wick
ΔP_{lfd}	liquid frictional pressure drop in distribution wick
ΔP_v	vapor pressure drop
ΔP_{vf}	frictional pressure drop in vapor
ΔP_{vl}	maximum pressure difference between heat pipe vapor and its liquid
ΔP_{vm}	momentum pressure drop in vapor
Δt	time increment
ΔT	change in mean temperature of evaporator (segment 1), fluid temperature change, temperature increment

ΔT_{hp}	maximum possible temperature drop per stage on hot side of heat pipe heat exchanger
ΔT_i	temperature drop across liquid-vapor interface
ΔT_l	cumulative temperature drop through liquid layers of wick with series heat flow
ΔT_m	mean temperature difference between hot side and cold side fluids of heat exchanger
$\Delta T_{\max}$	maximum possible temperature difference in heat exchanger = $T_{hi} - T_{ci}$, maximum steady-state temperature drop across heat pipe wall and wick
ΔT_{pc}	maximum possible temperature rise per stage on cold side of heat pipe heat exchanger
ΔT_s	cumulative temperature drop through solid layers of wick with series heat flow
ΔT_v	vapor temperature drop
ΔT_{vc}	vapor temperature drop at sonic limit
ΔT_{vs}	drop in saturation vapor temperature along evaporator
ΔT_w	wick temperature drop
ΔT_{wall}	temperature drop through heat pipe wall
ΔT_{wb}	wick temperature drop at inception of boiling
Δx	increment in position of continuum front
ε	heat exchanger effectiveness, overall heat exchanger effectiveness
ε_c	heat exchanger effectiveness for heat transfer between heat pipe and cold side fluid
ε_{cf}	heat exchanger effectiveness for counterflow heat exchanger
ε_h	heat exchanger effectiveness for heat transfer between hot side fluid and heat pipe
$\varepsilon_{\max}$	largest possible value of overall heat exchanger effectiveness
ε_{pa}	ratio of peak to average heat flux on outer surface of segment 1
ε_{pf}	heat exchanger effectiveness for parallel flow heat exchanger
ε_{po}	ratio of pressure drop in porous and open liquid flow channels
ε_s	surface emissivity
ε_{st}	heat exchanger effectiveness in stages of heat pipe heat exchanger when hot and cold side stage effectivenesses are equal
ζ	$\log_{10} P_v$
η	dimensionless heat transfer area
θ	angle between heat pipe or flow passage axis and gravity vector
θ_c	angle between axis of spherical shell and clamp point
θ_i	ratio of actual interface heat transfer coefficient to $h_{i\max}$
θ_r	angle between centrally applied load and support ribs

κ	heat of vaporization
κ_{21}	ratio of thermal conductivities in layers of two-layer wick, k_2/k_1
κ_m	mean heat of vaporization
κ_{sl}	ratio of solid and liquid thermal conductivities
λ	mean free path
λ_w	characteristic wick dimension for calculation of entrainment limit
μ	fluid viscosity
μ_l	viscosity of heat pipe liquid
μ_v	vapor viscosity
μ_{vm}	mean vapor viscosity averaged over $L_e + L_{aa}$
ξ	axial heat flux, Q/A_v
ξ_a	axial heat flux at evaporator exit
ξ_c	axial heat flux at capillary pumping limit
ξ_d	heat flow rate per parallel flow path through distribution wick per unit vapor space flow area
ξ_e	axial heat flux at entrainment limit
$\xi_e{}^*$	axial heat flux at entrainment limit based on vapor shear stress criterion
ξ_s	axial heat flux at sonic limit
ρ	density
ρ_l	liquid density
ρ_{la}	liquid density at initial nonoperating (ambient) temperature
ρ_{lo}	liquid density at heat pipe operating temperature
ρ_m	mean density of wall/wick in evaporator
ρ_s	density of solid wick material
ρ_v	vapor density
ρ_{ve}	vapor density at exit of evaporator or adiabatic section
ρ_{vi}	vapor density at beginning of evaporator
ρ_{vs}	vapor density at sonic limit
ρ_{wall}	density of wall
σ	surface tension
σ_a	ratio of minimum free flow area to frontal area in heat exchanger core
σ_r	Stefan-Boltzmann radiation constant
τ_c	cold side dimensionless temperature
τ_h	hot side dimensionless temperature
τ_s	shear stress at periphery of fluid flow passage
υ	Poisson's ratio

ϕ	liquid volume fraction in wick, angle between heat pipe axis and external acceleration vector
ϕ_1	porosity or volume fraction of layer 1 (liquid flow channel) in two-layer wick
ϕ_{1v}	fraction of total volume occupied by section 1 (Figure 6.9)
ϕ_2	porosity or volume fraction of layer 2 (capillary pumping layer) in two-layer wick
ϕ_e	entrainment limit parameter, defined by Eq. (5.129)
ϕ_l	porosity of liquid flow channel
ϕ_{l1}	liquid volume fraction in section 1 (Figure 6.9)
ϕ_{l2}	liquid volume fraction in section 2 (Figure 6.9)
ϕ_{l2w}	liquid volume fraction of wick in section 2 (Figure 6.9)
ϕ_{ls}	contact angle between liquid and solid surfaces
ϕ_m	ratio of average to peak heat flux on outer surface of evaporator
Ψ	unit thermal capacity ratio = ratio of smaller to larger unit thermal capacity rates
ω	dT_2/dT_v
Ω	momentum factor
Ω_c	momentum factor in condenser
Ω_e	momentum factor in evaporator
Ω_{po}	momentum factor for Poiseuille flow
Ω_{tc}	ratio of cold and hot side unit thermal capacity rates = $W_c c_c/W_h c_h$

A.2. ENGLISH LETTER SYMBOLS

a	cumulative heat transfer area at a point along heat exchanger length
a_c	constant in Eqs. (7.48), (7.48c), and (7.46c–7.47c)
$a_f - f_f, i_f, j_f$	fluid coefficients defined by Eqs. (5.49)–(5.56)
A	constant in Eq. (5.100) for saturated vapor, heat transfer area on gas or air side
A_b	constant in Eq. (5.139) for boiling limit parameter
A_c	cold side heat transfer area, condenser surface area
A_{minf}	minimum free flow area in row of tube bank or stage of heat pipe heat exchanger
A_c'	total cold side heat transfer area for conventional heat exchanger
A_{cs}	cross-sectional area of flow passage
A_e	evaporator surface area, constant in Eq. (5.130) for entrainment limit parameter
A_f	frontal area of heat exchanger core

A_{fa}	air side frontal area
A_{fg}	gas side frontal area
A_h	hot side heat transfer area, total hot side heat transfer area in heat pipe heat exchanger
A_h'	total hot side heat transfer area for conventional heat exchanger
A_l	cross-sectional area of liquid flow passage
A_{ldc}	area for flow of liquid through distribution wick in condenser
A_{lde}	area for flow of liquid through distribution wick in evaporator
A_{lv}	heat transfer area at liquid-vapor interface in evaporator or condenser
A_o	outer surface area of heat pipe segment
A_{ref}	reference heat transfer area
A_s	constant in Eq. (5.116) for sonic limit
A_t	total heat transfer area (hot plus cold side) in heat pipe heat exchanger
A_t'	total heat transfer area on hot or cold side for conventional heat exchanger when $A_h' = A_c'$
A_{tc}	total cold side heat transfer area in heat pipe heat exchanger
A_{ts}	total heat transfer area for equal hot and cold side heat exchanger effectivenesses
A_v	cross-sectional area of vapor flow passage
A_{ve}	vapor space flow area at exit of evaporator or adiabatic section
$A_{\text{wick}e}$	outer wick surface area in evaporator
A_{ww}	cross-sectional area of wall and wick at x
A_{wwm}	mean cross-sectional area of wall and wick
b	distance between center of hemispherical shell and support ribs
b_c	constant in Eqs. (7.49) and (7.49c)
B	constant in Eq. (5.100) for saturated vapor
B_c	inner periphery of cooling jacket
B_{cs}	boundary of cross-section (wetted periphery)
B_l	periphery of liquid flow channel
B_{mc}	mean periphery in conductive layer
B_p	periphery of heat pipe
B_v	periphery of vapor flow passage
c	specific heat
c_{1m}	mean specific heat of wall/wick in evaporator
c_c	specific heat of cold side fluid
c_h	specific heat of hot side fluid
c_l	specific heat of liquid in wick
c_p	specific heat at constant pressure
c_s	specific heat of solid material in wick
c_{wall}	specific heat of wall

C	thermal conductance
C_f	coefficient used to calculate friction factor for flow normal to tubes
C_{g1}	geometrical correction factor for coordinate system
C_{g2}	geometrical correction factor for radial variation of surface heat flux
C_h	coefficient used to calculate Stanton number for flow normal to tubes
C_{po}	friction coefficient for Poiseuille flow
C_t	thermometric thermal capacity
$C_t(T_2)$	thermometric thermal capacity at temperature T_2
$C_t(T_3)$	thermometric thermal capacity at temperature T_3
C_{tl}	thermometric thermal capacity of liquid material in wick
C_{ts}	thermometric thermal capacity of solid material in wick
$C_{t\mathrm{wall}}$	thermometric thermal capacity of wall material
C_v	volumetric thermal capacity
C_{v1}	volumetric thermal capacity of segment 1
C_{v2}	volumetric thermal capacity of segment 2
C_{v3}	volumetric thermal capacity of segment 3
CF	pressure drop correction factor for convective cooling of heat pipe
d_1	groove depth
d_i	inner diameter of artery
d_o	outer diameter of artery
d_p	thickness of "liquid land"
dP_{fp}	differential frictional pressure drop along length dx of porous flow channel
dP_{fl}	differential laminar frictional pressure drop along length dx of open flow channel
D	diameter of flow passage, inner diameter
D_c	diameter of curvature of liquid-vapor interface in wick pores
D_{cb}	minimum diameter of curvature for no boiling
D_h	hydraulic diameter
D_{hl}	hydraulic diameter of liquid flow channel
D_{hld}	hydraulic diameter of distribution wick liquid flow passage
D_{hp}	heat pipe diameter
D_{hv}	hydraulic diameter of vapor flow passage
D_p	effective wick pore diameter
D_{pe}	maximum pore diameter for no entrainment
D_s	spacing between wires
D_v	diameter of vapor flow passage
D_w	wire diameter
E	modulus of elasticity
E_h	ratio of cold and hot side heat exchanger effectivenesses $= \varepsilon_c/\varepsilon_h$

f	friction factor
f_c	condensation coefficient
f_l	friction factor for laminar flow
f_t	friction factor for turbulent flow
F_a	axial force
F_g	geometrical factor relating surface and axial heat fluxes [see Eq. (5.150)]
F_s	axial stagnation force on hemispherical shell
F_{su}	unsupported load on hemispherical shell
FM	ratio of frictional and momentum pressure drops in vapor
FM_l	value of FM for laminar flow
FM_t	value of FM for turbulent flow
g	acceleration of gravity
$g(\gamma)$	function of γ defined by Eq. (5.13)
G	mass velocity = flow rate per unit of minimum free flow area
G_p	geometry-dependent parameter [see Eq. (5.64)]
G_v	axial mass flow rate per unit cross section of vapor flow passage
G_{vs}	mass flow rate through vapor passage periphery per unit surface area
G_{vsm}	mean mass flow rate through vapor passage periphery per unit surface area
h	height, heat transfer coefficient
h_c	heat transfer coefficient of cold side fluid
h_c'	cold side heat transfer coefficient for conventional heat exchanger
h_{cc}	convective heat transfer coefficient
h_h	heat transfer coefficient of hot side fluid
h_h'	hot side heat transfer coefficient for conventional heat exchanger
h_i	interface heat transfer coefficient
$h_{i\max}$	upper limit on interface heat transfer coefficient
H_v	minimum vapor space dimension
i	superscript denoting conditions at end of *i*th time increment, subscript denoting *i*th conductive layer for heat transfer
I	function defined by Eq. (5.158a)
k	thermal conductivity
k_{1m}	mean wall/wick thermal conductivity of segment 1 (evaporator)
k_e	effective thermal conductivity of wick
k_{e1}	effective thermal conductivity of section 1 (Figure 6.9)

k_{e1w}	effective thermal conductivity of wick in section 1 (Figure 6.9)
k_{e2}	effective thermal conductivity of section 2 (Figure 6.9)
k_{e2l}	effective thermal conductivity of two-layer wick
k_{e2w}	effective thermal conductivity of wick in section 2 (Figure 6.9)
k_{eo}	overall effective thermal conductivity of sections 1 and 2 (Figure 6.9)
k_{ep}	effective thermal conductivity of wick with parallel heat flow
k_{es}	effective thermal conductivity of wick with series heat flow
k_{ew}	effective thermal conductivity of open grooved wick
$k_{e\text{wall}}$	effective thermal conductivity of wall section of open grooved wick
k_l	thermal conductivity of liquid component of wick
k_{l1}	effective thermal conductivity of layer 1 (grooved part) of covered grooved wick
k_{l2}	effective thermal conductivity of layer 2 (screened or sintered fiber layer) of covered grooved wick
k_s	ratio of specific heats at constant pressure and constant volume, thermal conductivity of solid component of wick
k_w	mean thermal conductivity of wick
K	friction factor for porous medium ($1/K$ = permeability)
K_c	entrance loss coefficient
K_e	exit loss coefficient
l	thickness of combined wall and wick, L_g or L_a
L	total length of flow passage, length of heat exchanger or recuperator core
L_a	length of heat pipe adiabatic section, air side flow passage width
L_b	straight-line distance between lowest and highest points of heat pipe
L_c	length of heat pipe condenser, width of cold side duct in heat pipe heat exchanger
$L_c \quad x$	length of segment 3
L_{cl}	distance from condenser end of heat pipe over which flow is laminar
L_{cyl}	length of cylinder subjected to axial load
L_d	length of parallel flow path between arteries in distribution wick
L_e	length of heat pipe evaporator (segment 1)
L_{el}	distance from evaporator end of heat pipe over which flow is laminar
L_g	gas side flow passage width (heat pipe evaporator length)
L_h	width of hot side duct in heat pipe heat exchanger
L_p	heat pipe length
L_s	length of heat pipe segment
M	molecular weight
Ma	Mach number
Ma_e	Mach number at exit of evaporator or adiabatic section

Ma_m	mean Mach number along $L_e + L_a$
n	subscript denoting current value of variable, number of gs of external acceleration, number of stages in heat pipe heat exchanger
n_s	number of heat pipes per stage
N	number of arteries, number of equally spaced arteries
NTU	number of heat transfer units
NTU′	number of heat transfer units for conventional heat exchanger
NTU_c	number of heat transfer units for heat transfer between heat pipe and cold side fluid
NTU_h	number of heat transfer units for heat transfer between hot side fluid and heat pipe
NTU_s	number of heat transfer units in heat pipe heat exchanger when hot and cold side heat exchanger effectivenesses for each stage are equal
P	fluid pressure
P_1	pressure at inlet of heat exchanger core
P_2	pressure at exit of heat exchanger core, vapor pressure in segment 2
P_a	atmospheric pressure
P_{cf}	vapor pressure corresponding to continuum flow temperature T_{cf}
P_f	fluid factor defined by Eq. (5.154)
P_g	geometrical factor defined by Eq. (5.155)
P_{ht}	liquid heat transport parameter
P_l	liquid pressure
$P_{l\min}$	minimum liquid pressure
P_m	mean fluid pressure in heat exchanger core
P_s	stagnation pressure
P_{sa}	actual stagnation pressure
P_v	vapor pressure, vapor pressure in segment 1 (evaporator)
P_{vc}	exit vapor pressure at sonic limit
P_{ve}	exit vapor pressure
P_{vi}	vapor pressure of liquid at liquid-vapor interface, initial (maximum) vapor pressure
$P_{v\min}$	minimum vapor pressure
P_{vs}	pressure of saturated vapor
P_{vsi}	pressure of saturated vapor at beginning of evaporator
P_{vw}	vapor pressure of liquid at wall-wick interface
Pr	Prandtl number
q	surface heat flux
q_1	constant heat flux on evaporator outer surface
q_{1av}	average heat flux in wall/wick interior
q_{1m}	average surface heat flux entering evaporator

q_{1max}	maximum surface heat flux entering evaporator
q_a	surface heat flux at inner wick surface of evaporator
q_{ah}	heat flux on lateral wing surface due to aerodynamic heating
q_{am}	average surface heat flux over inner wick surface of evaporator
q_{amax}	maximum surface heat flux leaving evaporator
q_b	boiling limit = surface heat flux for incipient boiling at outer wick surface
q_{bi}	surface heat flux for incipient boiling at liquid-vapor interface
q_{bmax}	maximum surface heat flux at incipient boiling
q_{bo}	surface heat flux for incipient boiling with negligible interface thermal resistance
q_c	average surface heat flux over L_c
q_{cl}	average surface heat flux over L_{cl}
q_e	heat flux on surface of heat pipe evaporator
q_{elm}	average surface heat flux over L_{el}
q_{em}	average surface heat flux over L_e
q_i	surface heat flux at liquid-vapor interface
q_l	heat flux through liquid layers of wick with parallel heat flow
q_m	average surface heat flux
q_{max}	maximum surface heat flux at outer wick diameter
q_r	average radial heat flux
q_s	heat flux through solid layers of wick with parallel heat flow
q_{sb}	maximum possible surface heat flux
q_{sl}	surface heat flux corresponding to axial heat flux at sonic limit ξ_s
q_v	heat flux based on inner surface area of wick
Q	heat transport or heat transfer rate, heat transfer rate through outer sur face of heat pipe segment
Q_1	heat transfer rate to segment 1
Q_2	heat transfer rate to segment 2
Q_3	heat transfer rate to segment 3
Q_a	axial heat transport rate at evaporator exit
Q_b	boiling limit = heat transport rate at incipient boiling
Q_c	capillary pumping limit
Q_d	heat transport rate in parallel flow path between arteries in distribution wick
Q_e	entrainment limit
Q_{ht}	actual heat transport rate in phase 3
Q_l	maximum heat transport rate at which fluid flow remains laminar
Q_{ll}	maximum heat transport rate at which liquid flow remains laminar
Q_{max}	maximum possible heat transport or heat transfer rate
Q_r	rate at which heat is radiated from condenser
Q_s	sonic limit
Q_{ss}	steady-state heating rate

Q_{vl}	maximum heat transport rate at which vapor flow remains laminar
r	radius
r_c	minimum radius of curvature of liquid-vapor interface in wick pores
r_{cb}	radius of curvature at incipient boiling
r_i	inner radius of wall or wick
r_m	mean wall radius
r_n	nucleation site radius, smallest equivalent nucleation site radius
r_{na}	actual nucleation site radius
r_{ne}	effective nucleation site radius, defined by Eq. (5.138)
r_o	outer radius
r_p	wick pore radius
r_v	vapor space radius = radius of wick inner surface
r_w	outer radius of heat pipe wall or wick
R	resistance to liquid flow through wick
R_1	resistance to flow through liquid flow channel
R_2	resistance to liquid flow through capillary pumping layer
R_g	gas constant of heat pipe vapor
R_i	inner recuperator radius
R_t	total thermal resistance
R_u	universal gas constant
Re	Reynolds number
Re_a	axial Reynolds number
Re_{as}	axial Reynolds number at sonic limit
Re_r	radial Reynolds number
Re_{re}	radial Reynolds number in evaporator
s_a	axial stress
s_{ab}	axial stress for buckling of cylinder
s_b	bending stress
s_h	hoop stress
s_r	tensile/compressive stress in ribs
s_s	shear stress
s_{th}	thermal stress
S_c	stress for 0.5% or 1% creep in 1000 hr
S_r	stress for rupture in 1000 hr
S_u	ultimate tensile strength
S_y	yield strength at 0.2% offset (strain)
St	Stanton number
t	elapsed time, thickness of conductive layer
t_1	thickness of liquid flow channel, annular thickness
t_2	thickness of capillary pumping layer

t_d	thickness of distribution wick
t_e	end cap thickness
t_h	height
t_l	cumulative thickness of liquid layers of wick
t_{max}	elapsed time of heat pipe transient
t_r	rib thickness
t_s	cumulative thickness of solid layers of wick
t_w	width
t_{wall}	wall thickness
t_{wick}	wick thickness
t_{wick1}	thickness of layer 1 of two-layer wick
t_{wick2}	thickness of layer 2 of two-layer wick
t_{wicko}	optimum wick thickness for maximum surface heat flux at incipient boiling
T	temperature
$T(Q_{max})$	temperature at Q_{max}
T_1	mean temperature of segment 1 (evaporator), fluid temperature at inlet of heat exchanger core
T_2	mean temperature of segment 2, fluid temperature at exit of heat ex changer core
T_3	mean temperature of segment 3
T_a	initial or ambient temperature
T_c	temperature of cold side fluid, cooler gas, heat pipe condenser
T_{cf}	continuum flow temperature = lowest temperature for continuum flow in heat pipe vapor
T_{ci}	inlet temperature of cold side fluid or coolant
T_{co}	outlet temperature of cold side fluid or coolant
T_e	radiation equilibrium temperature
T_h	temperature of hotter gas, temperature of hot side fluid
T_{hi}	inlet temperature of hot side fluid
T_{ho}	outlet temperature of hot side fluid
T_i	liquid temperature at liquid-vapor interface
T_{im}	mean of absolute temperature at liquid-vapor interface and of saturated vapor
T_{ls}	saturation temperature corresponding to liquid pressure
T_m	mean fluid temperature in heat exchanger core, mean wick temperature
T_{om}	mean temperature at outer surface of heat pipe evaporator
T_p	heat pipe operating temperature
T_{pl}	lower limit on heat pipe temperature in heat exchanger
T_{pu}	upper limit on heat pipe temperature in heat exchanger
T_r	recovery temperature of high-velocity air stream
T_{ref}	reference temperature
T_s	temperature of saturated vapor

T_s'	saturation temperature corresponding to liquid pressure in heat pipe wick
T_{sb}	vapor temperature corresponding to q_{sb}, minimum vapor temperature at which boiling is possible
T_v	heat pipe vapor temperature, temperature of inner wick surface
T_{ve}	vapor temperature at evaporator exit
T_{vi}	vapor temperature at evaporator end of heat pipe
T_{vm}	mean absolute vapor temperature
T_{vs}	temperature of saturated vapor
T_{vsi}	temperature of saturated vapor at beginning of evaporator
T_{vsm}	mean temperature of saturated vapor entering flow passage over evaporator length
T_w	wick temperature at wick-wall interface
$T_{wo\max}$	maximum temperature of outer wall surface in evaporator
U	unit thermal conductance
U'	unit thermal conductance for conventional heat exchanger
U_c	unit thermal conductance based on cold side area
U_h	unit thermal conductance based on hot side area
UA	overall thermal conductance = $U_h A_h = U_c A_c$
v	volume
v_1	specific volume at inlet of heat exchanger core
v_2	specific volume at exit of heat exchanger core
v_f	fractional change in liquid volume when temperature drops from T_p to T_a
v_m	mean specific volume in heat exchanger core
v_{wall}	volume of wall
v_{wick}	volume of wick
V	average axial velocity of fluid over flow passage cross section
V_a	axial fluid velocity at radial distance y from flow passage wall
V_{aa}	average axial fluid velocity over flow length
V_i	mean axial velocity at flow passage inlet ($x = 0$)
V_{ll}	maximum axial velocity at which liquid flow is laminar
$V_{l\max}$	maximum mean axial velocity of liquid
V_m	average axial velocity of fluid over flow length L
$V_{\max}$	maximum axial flow velocity
V_o	mean axial velocity at flow passage exit ($x = L$)
V_r	radial velocity of fluid
V_{rm}	average radial velocity at vapor passage periphery
V_s	sonic velocity
V_{se}	sonic velocity at exit of evaporator or adiabatic section
V_{si}	sonic velocity at beginning of evaporator

V_v	vapor velocity
V_{ve}	vapor velocity at exit of evaporator or adiabatic section
V_{vl}	maximum mean axial velocity at which vapor flow is laminar
V_{vm}	mean vapor velocity averaged over $L_e + L_a$
V_{vr}	radial velocity of vapor
$V_{v\max}$	maximum mean axial velocity of vapor
w	surface mass flux (radial mass flow rate per unit wall area)
w_g	width of liquid-filled grooves
w_l	width of land between grooves
$w_{\max}$	maximum surface mass flux
w_o	axial displacement
W	centerline distance between ribs of flat plate heat pipe structure, maximum mass flow rate in heat pipe, mass flow rate
W_1	mass flow rate through liquid flow channel
W_2	liquid mass flow rate through capillary pumping layer
W_c	mass flow rate of cold side fluid
W_h	mass flow rate of hot side fluid
W_v	mass flow rate of vapor
Wc	unit thermal capacity rate
$(Wc)_{\text{high}}$	larger unit thermal capacity rate
$(Wc)_{\text{low}}$	smaller unit thermal capacity rate
x	continuum front position, segment 2 length, distance along heat pipe or flow passage
x_1	temperature coefficient in Eq. (7.15)
x_2	temperature coefficient in Eq. (7.16)
x_l	longitudinal spacing between heat pipe stages in diameters
x_t	transverse spacing between heat pipes in diameters
x_{ti}	transverse spacing between heat pipes in diameters at R_i
x_{tm}	mean transverse spacing between heat pipes in diameters
y	radial distance
y_1	variable in Eq. (5.106), defined by Eq. (5.107)
y_w	maximum wall deflection
z	l/r_v
z_e	correction factor in Eq. (5.123) for entrainment limit, defined by Eq. (5.124)
Z	calculational parameter in Eqs. (5.58a) and (5.58b)

APPENDIX

B

CONVERSION FACTORS (English units to SI units)

Length

1 in. = 0.0254 m
1 ft = 0.3048 m

Volume

1 in.^{3} = 1.6387×10^{-5} m^3
1 ft^3 = 0.02832 m^3

Mass

1 lb = 0.4536 kg

Density

1 lb/ft^3 = 16.0185 kg/m^3
1 lb/in.^{3} = 2.768×10^4 kg/m^3

Force

1 lb = 4.4482 N

Power

1 Btu/sec = 1055 W
1 Btu/hr = 0.2931 W

Heat of vaporization

1 Btu/lb = 2326 J/kg

Specific heat

1 Btu/lb-°R = 4187 J/kg-°K

Absolute viscosity

1 lb/sec-ft = 1.4882 N-sec/m^2

Thermal conductivity

1 Btu/hr-ft-°R = 1.7307 W/m-°K

Pressure

1 lb/in.^{2} (psi) = 6895 N/m^2

Energy

1 Btu = 1055 J

Temperature

°F = 1.8°C + 32
°R = °F + 459.67
°K = °C + 273.15
°R = 1.8°K
Δ°F = Δ°R
Δ°C = Δ °K

Surface tension

1 lb/ft = 14.593 N/m

Heat transfer coefficient

1 Btu/hr-ft^2-°R = 5.678 W/m^2-°K

Heat flux

1 Btu/ft^2-sec = 1.136×10^4 W/m^2
1 Btu/in^2-sec = 1.636×10^6 W/m^2

Physical constants

Acceleration of gravity

32.174 ft/sec^2 = 9.8067 m/sec^2

Molar gas constant

1.9857 Btu/°R-lb mole = 8.3144 J/°K-g mole

Stefan-Boltzmann constant

0.476×10^{-12} Btu/ft^2-sec-°R^4 =
5.67×10^{-8} W/m^2-°K^4

B.1. REFERENCES

Keenan, J. H., Chao, J., and Kaye, J., *Gas Tables,* 2nd Ed. (English Units), Wiley, New York, 1980, pp. 182–185.

Collier, J. G., *Convective Boiling and Condensation,* 2nd Ed., McGraw-Hill, New York, 1981, pp. 427–429.

APPENDIX

C

MATERIAL PROPERTIES

In this appendix, properties used in the design and analysis of heat pipes are presented for various materials. Data are given for the heat pipe fluids cesium, potassium, sodium, lithium, and water, and for the wall/wick materials Haynes 188 and TZM molybdenum.

Data for the heat pipe fluids as a function of vapor temperature are contained in Tables C.1–C.4. Relevant property data are given in Table C.1. Values of terms in equations for fluid coefficients are given in Table C.2. (Fluid coefficients are combinations of fluid properties used in the capillary pumping limit equation.) The fluid coefficient equations for which these terms apply are presented following Table C.2. Equations for calculating additional properties of the heat pipe fluids are given in Tables C.3 and C.4. The data in Tables C.2–C.4 are derived from those of Table C.1.

Temperature-dependent property data for TZM molybdenum and Haynes 188 may be found in Tables C.5–C.7. Table C.5 contains equations for properties needed to calculate heat pipe thermal characteristics. Strength and strength-related properties are given in Tables C.6 and C.7.

Properties of sintered fibers and screens relevant to the design of heat pipe wicks are given in Table C.8.

The equations for material properties as a function of temperature are estimated to correlate with tabular data within 10%. Data on material strength and wick properties are subject to considerable variability, depending on such factors as heat treatment, cold working, and fabrication techniques. The data are considered suitable for preliminary design and analysis purposes.

C.1. REFERENCES

C.1. Silverstein, C. C., "A Feasibility Study of Heat-Pipe-Cooled Leading Edges for Hypersonic Cruise Aircraft," CR-1857, NASA, Washington, D.C., November 1971, pp. 70, 134–135.

C.2. Collier, J. G., *Convective Boiling and Condensation,* 2nd Ed., McGraw-Hill, New York, 1981, p. 430.

C.3. "Haynes Alloy No. 188," Cabot Wrought Products Division, Cabot Corporation, Kokomo, Indiana, 1983.

C.4. Briggs, J. Z., and Barr, R. Q., "Arc-Cast Molybdenum TZM Alloy: Properties and Applications," *High Temperatures–High Pressures,* Vol. 3, 1971, pp. 363–409.

Table C.1. Properties of heat pipe fluids (Refs. C.1, C.2)

Temperature, °F (°C)	P_v, psia (10^4 N/m^2)	$\rho_v \times 10^3$, lb/ft^3 (kg/m^3)	ρ_l, lb/ft^3 (kg/m^3)	μ_v, lb/hr-ft (10^{-5} N-sec/m^2)	μ_l, lb/hr-ft (10^{-4} N-sec/m^2)	k_l, Btu/hr-ft-°F (W/m-°K)	c_l, Btu/lb-°F (J/kg-°K)	κ, Btu/lb (10^5 J/kg)	$\sigma \times 10^3$, lb/ft (N/m)
Cesium (Melting point, 83°F [28°C]; molecular weight, 132.9)									
900 (482)	1.6 (1.10)	28 (448)	99.5 (1594)	0.0576 (2.38)	0.470 (1.943)	10.8 (18.69)	0.057[a] (239)	228 (5.30)	3.56 (52.0)
1100 (593)	6.6 (4.55)	65 (1041)	95.2 (1525)	0.0617 (2.55)	0.420 (1.736)	9.97 (17.26)		218 (5.07)	3.15 (46.0)
1300 (704)	20.0 (13.8)	152 (2435)	91.0 (1458)	0.0657 (2.72)	0.377 (1.558)	9.18 (15.89)		209 (4.86)	2.74 (40.0)
1500 (816)	48.0 (33.1)	360 (5767)	86.5 (1386)	0.0692 (2.86)	0.345 (1.426)	8.38 (14.51)		199 (4.63)	2.32 (33.9)
1700 (927)	97. (66.9)	680 (10,890)	82.0 (1314)	0.0726 (3.00)	0.320 (1.323)	7.60 (13.15)		189 (4.40)	1.92 (28.0)
Potassium (Melting point, 146°F [63°C]; molecular weight, 39.1)									
1100 (593)	2.4 (1.66)	5.6 (89.7)	44.0 (705)	0.0622 (2.57)	0.375 (1.550)	20.3 (35.1)	0.185 (775)	881 (20.5)	5.27 (76.9)
1300 (704)	8.8 (6.07)	18.0 (288)	42.3 (678)	0.0660 (2.73)	0.332 (1.372)	18.6 (32.2)	0.189 (791)	846 (19.68)	4.77 (69.6)
1500 (816)	24.5 (16.89)	50.0 (801)	40.5 (649)	0.0697 (2.88)	0.302 (1.248)	16.9 (29.2)	0.195 (816)	813 (18.91)	4.28 (62.5)
1700 (927)	56. (38.6)	107. (1714)	38.7 (620)	0.0730 (3.02)	0.280 (1.157)	15.3 (26.5)	0.205 (858)	782 (18.19)	3.80 (55.5)
1900 (1038)	110. (75.9)	205. (3284)	37.0 (593)	0.0763 (3.15)	0.260 (1.075)	13.8 (23.9)	0.215 (900)	752 (17.49)	3.28 (47.9)
Sodium (Melting point, 208°F [98°C]; molecular weight, 23.0)									
1300 (704)	2.1 (1.45)	2.7 (43.3)	48.6 (779)	0.0500 (2.07)	0.444 (1.835)	34.6 (59.9)	0.301 (1260)	1742 (40.5)	9.05 (132)
1500 (816)	7.2 (4.96)	8.8 (141)	47.2 (756)	0.0532 (2.20)	0.392 (1.620)	33.0 (57.1)	0.304 (1273)	1695 (39.4)	8.30 (121)
1700 (927)	20.8 (14.34)	22.0 (352)	45.8 (734)	0.0562 (2.32)	0.356 (1.472)	31.6 (54.7)	0.310 (1298)	1648 (38.3)	7.50 (109)
1900 (1038)	48. (33.1)	50.0 (801)	44.5 (713)	0.0591 (2.44)	0.326 (1.348)	30.5 (52.8)	0.318 (1331)	1606 (37.4)	6.75 (98.5)
2100 (1149)	100. (69.0)	96.0 (1538)	43.3 (694)	0.0620 (2.56)	0.300 (1.240)	29.5 (51.1)	0.330 (1382)	1564 (36.4)	6.00 (87.6)
Lithium (Melting point, 367°F [186°C]; molecular weight, 6.94)									
2000 (1093)	2.1 (1.45)	0.6 (9.6)	27.2 (436)	0.0424 (1.753)	0.500 (2.07)	29.1 (50.4)	0.986 (4128)	8700 (202)	19.0 (277)
2200 (1204)	5.5 (3.79)	1.47 (23.6)	26.7 (428)	0.0445 (1.840)	0.460 (1.902)	29.4 (50.9)	0.983 (4116)	8530 (198)	17.8 (260)
2400 (1316)	13.0 (8.96)	3.35 (53.7)	26.3 (421)	0.0470 (1.943)	0.432 (1.786)	29.7 (51.4)	0.980 (4103)	8370 (195)	16.6 (242)
2600 (1427)	27.2 (18.75)	6.7 (107)	25.8 (413)	0.0491 (2.03)	0.412 (1.703)	30.0 (51.9)	0.976 (4087)	8210 (191)	15.4[b] (225)[b]
2800 (1538)	54. (37.2)	12.2 (195)	25.4 (407)	0.0510 (2.11)	0.398 (1.645)	30.3 (52.4)	0.973 (4074)	8070 (188)	14.2[b] (207)[b]
3000 (1649)	97. (66.9)	21.0 (336)	25.0 (401)	0.0530 (2.19)	0.388 (1.604)	30.6 (53.0)	0.970 (4061)	7930 (183)	13.0[b] (190)[b]

Table C.1. (continued)

Temperature, °F (°C)	P_v, psia (10^4 N/m^2)	$\rho_v \times 10^3$, lb/ft^3 (kg/m^3)	ρ_l, lb/ft3 (kg/m^3)	μ_v, lb/hr-ft (10^{-5} N-sec/m^2)	μ_l, lb/hr-ft (10^{-4} N-sec/m^2)	k_l, Btu/hr-ft-°F (W/m-°K)	c_l, Btu/lb-°F (J/kg-°K)	κ, Btu/lb (10^5 J/kg)	$\sigma \times 10^3$, lb/ft (N/m)
Water (Melting point, 32°F [0°C]; molecular weight, 18.0)									
104 (40)	1.07 (0.74)	3.19 (51.1)	61.88 (991)	0.0235 (0.97)	1.582 (6.54)	0.365 (0.632)	0.998 (4179)	1037 (24.1)	4.77 (69.6)
140 (60)	2.89 (1.99)	8.12 (130)	61.35 (983)	0.0254 (1.05)	1.130 (4.67)	0.377 (0.652)	0.999 (4183)	1016 (23.6)	4.53 (66.1)
176 (80)	6.87 (4.74)	18.30 (293)	60.64 (971)	0.0273 (1.13)	0.860 (3.56)	0.387 (0.670)	1.002 (4195)	995 (23.1)	4.29 (62.6)
212 (100)	14.69 (10.13)	37.29 (597)	59.77 (957)	0.0292 (1.21)	0.685 (2.83)	0.394 (0.682)	1.007 (4216)	973 (22.6)	4.03 (58.8)
248 (120)	28.79 (19.85)	69.98 (1121)	58.82 (942)	0.0310 (1.28)	0.559 (2.31)	0.397 (0.687)	1.013 (4241)	949 (22.1)	3.76 (54.9)
284 (140)	52.40 (36.1)	122.7 (1965)	57.77 (925)	0.0328 (1.36)	0.469 (1.94)	0.398 (0.689)	1.023 (4283)	925 (21.5)	3.48 (50.8)
320 (160)	89.62 (61.8)	203.3 (3257)	56.59 (907)	0.0346 (1.43)	0.406 (1.68)	0.395 (0.684)	1.035 (4334)	898 (20.9)	3.19 (46.6)
356 (180)	145.4 (100.3)	321.5 (5150)	55.34 (887)	0.0363 (1.50)	0.359 (1.48)	0.391 (0.677)	1.051 (4401)	868 (20.2)	2.89 (42.2)

Notes: [a]For temperatures greater than 570°F (299°C).

[b]Estimated values.

Examples when power of 10 appears in variable or unit (cesium at 900°F):

$\rho_v = 28 \times 10^{-3}$ lb/ft^3

$\mu_v = 2.38 \times 10^{-5}$ N-sec/m^2

Variables in Table C.1

P_v	vapor pressure	k_l	liquid thermal conductivity
ρ_v	vapor density	c_l	liquid specific heat
ρ_l	liquid density	κ	heat of vaporization
μ_v	vapor viscosity	σ	surface tension
μ_l	liquid viscosity		

Table C.2. Constants in fluid coefficient equations

Constant	Cesium	Potassium	Sodium	Lithium	Water
a_1	45.80	67.67	118.9	262.9	47.48
a_2	−0.0174	−0.210	−0.0324	−0.0509	−0.0643
b_1	3.448×10^{-4}	2.597×10^{13}	9.22×10^{16}	3.469×10^{21}	53.63
b_2	−0.00352	5.957	−6.975	−8.008	−3.28
c_1	4.896×10^{10}	1.718×10^{15}	9.162×10^{18}	2.2×10^{23}	12,030
c_2	−4.504	−6.044	−7.183	−8.333	−3.498
d_1	3.393	1.632×10^{-8}	1.175×10^{-7}	3.434×10^{-8}	4.47×10^{-6}
d_2	-2.50×10^{-5}	-5.60×10^{-13}	−0.363	−0.299	−1.013
e_1	4.512×10^{8}	1.538×10^{13}	2.12×10^{16}	1.935×10^{21}	109.6
e_2	−4.506	−6.042	−6.999	−8.329	−3.498
f_1	3.025×10^{-9}	4.443×10^{-10}	9.92×10^{-14}	4.494×10^{-15}	3.545
f_2	7.08×10^{-4}	0.000613	7.186×10^{-11}	5.448×10^{-12}	0.00195
i_1	0.466	1.908	0.654	1.151	0.339
i_2	2.80×10^{-5}	0.000120	0.223	0.326	0.2108
j_1	5.814	9654	4.764×10^{4}	1.024×10^{5}	3.478×10^{4}
j_2	-2.16×10^{-3}	−0.9527	−1.034	−0.857	−1.359

Temperature-Dependent Equations for Fluid Coefficients (English Units; T_v = Vapor Temperature, °F)

Lithium, sodium	Potassium	Cesium	Water
$a_f = a_1 = a_2 T_v$	$a_f = a_1 = a_2 T_v$	$a_f = a_1 = a_2 T_v$	$a_f = a_1 = a_2 T_v$
$b_f = b_1 T_v^{b_2}$	$b_f = b_1 T_v^{b_2}$	$b_f = b_1 e^{b_2 T_v}$	$b_f = b_1 T_v^{b_2}$
$c_f = c_1 T_v^{c_2}$	$c_f = c_1 T_v^{c_2}$	$c_f = c_1 T_v^{c_2}$	$c_f = c_1 T_v^{c_2}$
$d_f = d_1 T_v^{d_2}$	$d_f = d_1 + d_2 T_v$	$d_f = d_1 + d_2 T_v$	$d_f = d_1 T_v^{d_2}$
$e_f = e_1 T_v^{e_2}$	$e_f = e_1 T_v^{e_2}$	$e_f = e_1 T_v^{e_2}$	$e_f = e_1 T_v^{e_2}$
$f_f = f_1 T_v + f_2$	$f_f = f_1 e^{f_2 T_v}$	$f_f = f_1 e^{f_2 T_v}$	$f_f = f_1 e^{f_2 T_v}$
$i_f = i_1 T_v^{i_2}$	$i_f = i_1 + i_2 T_v$	$i_f = i_1 + i_2 T_v$	$i_f = i_1 T_v^{i_2}$
$j_f = j_1 T_v^{j_2}$	$j_f = j_1 T_v^{j_2}$	$j_f = j_1 + j_2 T_v$	$j_f = j_1 T_v^{j_2}$

	English units	SI units
a_f	psi-micron	a_f'(N-micron/m^2) = $6895 a_f$
b_f	lb-sec-in./Btu	b_f'(N-m/W) = $b_f/9330$
c_f	lb-in^2-sec^2/Btu^2	c_f'(N-m^2/W^2) = $2.578 \times 10^{-9} c_f$
d_f	lb-in.-sec-/Btu	d_f'(N-m/W) = $d_f/9330$
e_f	lb-in^2-sec^2/Btu^2	e_f'(N-m^2/W^2) = $2.578 \times 10^{-9} e_f$
f_f	lb-in^2-sec^2/Btu^2	f_f'(N-m^2/W^2) = $2.578 \times 10^{-9} f_f$
i_f	Btu/in.-sec	i_f'(W/m) = $4.154 \times 10^4 i_f$
j_f	Btu/in.-sec	j_f'(W/m) = $4.154 \times 10^4 j_f$
	Heat pipe dimensions, in.	Heat pipe dimensions, m
	Axial heat flux, Btu/in^2-sec	Axial heat flux, W/m^2

Note: To calculate fluid coefficients in SI units (primed values), first calculate values in English units, and then convert to SI units:

Example: Find d_f' in SI units for sodium at 1000°C.

$T_v = 1.8(1000) + 32 = 1832$°F

From Table C.2, $d_1 = 1.175 \times 10^{-7}$, $d_2 = -0.363$

$d_f = 1.175 \times 10^{-7} (1832)^{-0.363} = 7.684 \times 10^{-9}$ lb-in.-sec/Btu

$d_f' = 7.684 \times 10^{-9}/9330 = 8.236 \times 10^{-13}$ N-m/W

Table C.3. Properties of liquid lithium and sodium

$y = x_0 + x_1T + x_2T^2 + x_3T^3 + x_4T^4$ (T = temperature, °F)

y	x_0	x_1	x_2	x_3	x_4	Temperature range, °F
			Lithium			
Thermal conductivity, Btu/hr-ft-°F	20.26	0.0149	3.268×10^{-6}	0	0	
Heat of vaporization, Btu/lb	10,227	−0.770	0	0	0	
Density, lb/in.^{3}	0.019	0	0	0	0	<367
	0.01939	-2.008×10^{-6}	0	0	0	≥367
Specific heat, Btu/lb-°F	0.79	0	0	0	0	<367
	186	0	0	0	0	$367 \le T \le 368$
	1	0	0	0	0	368
Thermometric thermal capacity, Btu/in.^{3}*	−1.351	0.01501	0	0	0	<367
	−1294	3.61	−0.000187	0	0	$367 \le T \le 368$
	0.628	0.01939	-1.004×10^{-6}	0	0	>368
			Sodium			
Thermal conductivity, Btu/hr-ft-°F	20.26	0.0149	-3.268×10^{-6}	0	0	
Heat of vaporization, Btu/lb	2029	−0.2225	0	0	0	
Density, lb/in.^{3}	0.0346	0	0	0	0	<209
	0.0345	-4.798×10^{-6}	0	0	0	≥209
Specific heat, Btu/lb-°F	0.3089	0	0	0	0	<208
	48.68	0	0	0	0	$208 \le T \le 209$
	0.3469	-8.125×10^{-5}	3.521×10^{-8}	0	0	>209
Thermometric thermal capacity, Btu/in.^{3}*	−0.962	0.01069	0	0	0	<208
	−349	1.684	0	0	0	$208 \le T \le 209$
	0.536	0.01197	-2.234×10^{-6}	5.35×10^{-10}	-4.223×10^{-14}	>209

Notes: *Reference temperature = 90°F (32°C).

To calculate a property in SI units, first calculate it in English units, and then convert to SI units:

Example: Find the density of liquid sodium at 1000°C.

$T = 1.8(1000) + 32 = 1832$°F

Density $= 0.0345 - 4.798 \times 10^{-6}(1832) = 0.0257$ lb/in.$^3 \times 2.768 \times 10^4 = 711.4$ kg/m^3

Table C.4. Properties of liquid potassium, cesium, and water

$y = x_0 + x_1 T$ (T = temperature, °F)

y	x_0	x_1	Temperature range, °F
	Potassium		
Thermal conductivity, Btu/hr-ft-°F	29.21	−0.00815	1100–1900
Heat of vaporization, Btu/lb	1056	0.161	1100–1900
Density, lb/in.^{3}	0.0311	-5.09×10^{-6}	1100–1900
Specific heat, Btu/lb-°F	0.1408	3.80×10^{-5}	1100–1900
Thermometric thermal capacity, Btu/in.^{3}*	−0.2971	0.00462	1100–1900
	Cesium		
Thermal conductivity, Btu/hr-ft-°F	14.38	−0.00400	900–1700
Heat of vaporization, Btu/lb	272	−0.0485	900–1700
Density, lb/in.^{3}	0.0690	-1.267×10^{-5}	900–1700
Specific heat, Btu/lb-°F	0.057	—	900–1700
Thermometric thermal capacity, Btu/in.^{3}*	0.2258	0.00300	900–1700
	Water		
Thermal conductivity, Btu/hr-ft-°F	0.372	6.39×10^{-5}	104–392
Heat of vaporization, Btu/lb	1115	−0.690	104–392
Density, lb/in.^{3}	0.0378	-1.615×10^{-5}	104–392
Specific heat, Btu/lb-°F	0.9688	0.000205	104–356
Thermometric thermal capacity, Btu/in.^{3}*	−3.034	0.0347	104–356

Notes: *Reference temperature = 90°F (32°C).

To calculate a property in SI units, first calculate it in English units, and then convert to SI units:

Example: Find the heat of vaporization of water at 120°C.

$T = 1.8(120) + 32 = 248$°F

Heat of vaporization = 1115 − 0.690(248) = 944 Btu/lb = 944(2326) = 2.195×10^6 J/kg

Table C.5. Properties of TZM molybdenum and Haynes 188 (Refs. C.3, C.4)

$y = x_0 + x_1 T + x_2 T^2 + x_3 T^3$ (T = temperature, °F)

y	x_0	x_1	x_2	x_3
	TZM Molybdenum			
Density, lb/in.^{3}	0.3618	-5.017×10^{-6}	0	0
Specific heat, Btu/lb-°F	0.0593	7.33×10^{-6}	0	0
Thermal conductivity, Btu/hr-ft-°F	74.17	−0.00833	0	0
Thermometric thermal capacity, Btu/in.^{3}*	−1.945	0.0215	1.718×10^{-6}	-1.226×10^{-11}
	Haynes 188			
Density, lb/in.^{3}	0.324	0	0	0
Specific heat, Btu/lb-°F	0.0955	2.53×10^{-5}	0	0
Thermal conductivity, Btu/hr-ft-°F	6.06	0.00542	0	0
Thermometric thermal capacity, Btu/in.^{3}*	−2.814	0.0309	4.099×10^{-6}	0

Notes: *Reference temperature = 90°F (32°C).

To calculate a property in SI units, first calculate it in English units, and then convert to SI units:

Example: Find the thermal conductivity of Haynes 188 at 800 °C.

$T = 1.8(800) + 32 = 1472$ °F

Thermal conductivity = 6.06 + 0.00542(1472) = 14.04 Btu/hr-ft-°F = 14.04(1.731) = 24.30 W/m-°K

Table C.6. Strength and strength-related properties of Haynes 188 (Ref. C.3)

Temperature, °F (°C)	E, 10^6psi (10^9 N/m^2)	$10^6\alpha_{te}$, in./in. per °F (m/m per °K)	S_u, 1000 psi (10^6 N/m^2)	S_y, 1000 psi (10^6 N/m^2)	S_c, 1000 psi (10^6 N/m^2)	S_r, 1000 psi (10^6 N/m^2)
70 (21)*	33.6 (232)	6.4 (11.5)	137.2 (946)	63.7 (464)	—	—
1000 (538)	27.6 (190)	8.2 (14.8)	108.5 (748)	42.0 (290)	—	—
1200 (649)	26.2 (180)	8.6 (15.5)	103.3 (712)	39.7 (274)	—	—
1400 (760)	24.9 (172)	9.0 (16.2)	89.9 (620)	38.9 (268)	11.7 (81)	23.0 (159)
1500 (816)	—	—	—	—	7.8 (54)	15.0 (103)
1600 (871)	23.6 (163)	9.4 (16.9)	60.0 (414)	35.9 (248)	5.0 (34)	9.4 (65)
1700 (927)	—	—	—	—	3.1 (21)	5.5 (38)
1800 (982)	22.3 (154)	9.9 (17.8)	35.2 (243)	19.0 (131)	1.8 (12)	2.4 (17)
2000 (1093)	21.0 (145)	10.3 (18.5)	18.7 (129)	9.3 (64)	—	—

Notes: *For E, temperature is 86°F (30°C).

E = modulus of elasticity, α_{te} = mean coefficient of thermal expansion from 70°F (21°C) to indicated temperature, S_u = ultimate tensile strength, S_y = yield strength at 0.2% offset, S_c = stress for 0.5% creep/1000 hr, S_r = stress for rupture in 1000 hr.

Table C.7. Strength and strength-related properties of TZM molybdenum (Ref. C.4)

Temperature, °F (°C)	E, 10^6psi (10^9 N/m^2)	$10^6\alpha_{te}$, in./in. per °F (m/m per °K)	S_u, 1000 psi (10^6 N/m^2)	S_y, 1000 psi (10^6 N/m^2)	S_c, 1000 psi (10^6 N/m^2)	S_r, 1000 psi (10^6 N/m^2)
70 (21)	41.2 (284)	2.80 (5.04)	115 (793)	103 (710)	—	—
1472 (800)	32.6 (225)	3.20 (5.76)	97 (668)	—	—	—
1652 (900)	—	—	—	—	55.3 (381)	62.4 (430),
1832 (1000)	31.2 (215)	3.50 (6.30)	83 (572)	—	45.4 (313)	52.5 (362)
2192 (1200)	28.4 (196)	3.80 (6.84)	63 (434)	—	—	—
2552 (1400)	24.1 (166)	4.25 (7.65)	36 (248)	—	—	—
2912 (1600)	17.0 (117)	4.80 (8.64)	17 (117)	—	—	—
3272 (1800)	9.9 (68.3)	5.40 (9.72)	12 (82.7)	—	—	—

Notes: E = modulus of elasticity, α_{te} = mean coefficient of thermal expansion from 70°F (21°C) to indicated temperature, S_u = ultimate tensile strength, S_y = yield strength at 0.2% offset, S_c = stress for 1% creep/1000 hr, S_r = stress for rupture in 1000 hr.

E, S_u, and S_y are mean values. Actual values could vary from the mean by ±12% or more, depending on size, shape, cold work, and heat treatment of body subjected to stress.

Table C.8. Properties of porous wick materials (Ref. C.1)

Type	Mean fiber or wick diameter, in. (10^{-5} m)	Porosity	Friction factor, 10^6 in^{-2} (10^9 m^{-2})	Mean pore diameter, microns[a]
Sintered fiber	0.0013 (3.30)	0.887	1.18 (1.83)	190
Sintered fiber	0.0013 (3.30)	0.807	3.30 (5.12)	129
50 Mesh sintered screen	0.009 (22.9)	0.625	0.974 (1.510)	609
100 Mesh sintered screen	0.0045 (11.43)	0.585	4.21 (6.53)	263
200 Mesh sintered screen	0.0022 (5.59)	0.651	8.35 (12.94)	126
400 Mesh sintered screen[b]	0.0011 (2.79)	0.600	16.7 (25.9)	60

Notes: [a]Based on capillary rise experiments with water, using oxidized samples. Perfect wetting assumed.
[b]Data for 400 mesh screen are estimated, based on extrapolation of other data in this table.

APPENDIX

D

HEAT EXCHANGER RELATIONSHIPS

Equations for the heat exchanger effectiveness and temperature distribution in counterflow and parallel flow heat exchangers are derived in this appendix. The derivations are based on methods developed in Ref. D.1. Results are presented in dimensionless form.

D.1. DIMENSIONLESS VARIABLES

The following dimensionless variables are used in the derivations:

Cold side temperature:

$$\tau_c = \frac{T_c - T_{ci}}{\Delta T_{\max}}$$

Hot side temperature:

$$\tau_h = \frac{T_h - T_{ci}}{\Delta T_{\max}}$$

Temperature difference between hot and cold sides:

$$\tau_h - \tau_c = \frac{T_h - T_c}{\Delta T_{\max}}$$

Hot side inlet temperature:

$$\tau_{hi} = \frac{T_{hi} - T_{ci}}{\Delta T_{\max}} = 1$$

Hot side outlet temperature:

$$\tau_{ho} = \frac{T_{ho} - T_{ci}}{\Delta T_{\max}}$$

Cold side inlet temperature:

$$\tau_{ci} = \frac{T_{ci} - T_{ci}}{\Delta T_{\max}} = 0$$

Cold side outlet temperature:

$$\tau_{co} = \frac{T_{co} - T_{ci}}{\Delta T_{\max}}$$

Heat exchanger effectiveness:

$$\varepsilon = \frac{T_{co} - T_{ci}}{\Delta T_{\max}} \qquad W_c c_c = (Wc)_{\text{low}}$$

$$\varepsilon = \frac{T_{hi} - T_{ho}}{\Delta T_{\max}} \qquad W_h c_h = (Wc)_{\text{low}}$$

Heat transfer area:

$$\eta = \frac{U\,a}{(Wc)_{\text{low}}}$$

Number of heat transfer units:

$$\text{NTU} = \frac{U\,A}{(Wc)_{\text{low}}}$$

Thermal capacity ratio:

$$\psi = \frac{(Wc)_{\text{low}}}{(Wc)_{\text{high}}} \leq 1$$

where T_h is the hot side temperature at a given location in the heat exchanger, T_{hi} is the temperature at the hot side inlet, T_{ho} is the temperature at the hot side outlet, T_c is the cold side temperature at a given location in the heat exchanger, T_{ci} is the temperature at the cold side inlet, T_{co} is the temperature at the cold side outlet, $\Delta T_{max} = T_{hi} - T_{ci}$, a is the hot or cold side heat transfer area between the hot side inlet and a given location in the heat exchanger, A is the total hot or cold side heat transfer area, U is the overall unit thermal conductance corresponding to the hot or cold side heat transfer area, W_h is the hot side mass flow rate, W_c is the cold side mass flow rate, c_h is the hot side specific heat at constant pressure, c_c is the cold side specific heat at constant pressure, $(Wc)_{low}$ is the smaller of $W_h c_h$ and $W_c c_c$, and $(Wc)_{high}$ is the larger of $W_h c_h$ and $W_c c_c$.

When a is equal to the total hot or cold side heat transfer area A, η is equal to NTU.

Relationships will now be derived for counterflow and parallel heat exchange under the following conditions:

- Thermal capacity ratio $\Psi < 1$, $W_c c_c < W_h c_h$
- Thermal capacity ratio $\Psi < 1$, $W_h c_h < W_c c_c$

- Thermal capacity ratio $\Psi = 0$, $W_c c_c < W_h c_h = \infty$
- Thermal capacity ratio $\Psi = 0$, $W_h c_h < W_c c_c = \infty$
- Thermal capacity ratio $\Psi = 1$, $W_c c_c = W_h c_h$

D.2. COUNTERFLOW HEAT EXCHANGE

D.2.1. $\Psi < 1$, $W_c c_c < W_h c_h$

A representative heat exchanger temperature distribution is shown qualitatively in Figure D.1. Both dimensional and dimensionless variables are indicated.

Heat Exchanger Effectiveness. The heat exchanger effectiveness is found by considering the heat transfer rate and the temperature differential between the hot side and cold side fluids.

The differential heat transfer rate between the hot and cold side fluids at a given location a can be written in the following forms:

$$dQ = -W_c c_c dT_c = -W_h c_h dT_h \tag{D.1}$$

$$dQ = Uda(T_h - T_c) \tag{D.2}$$

The minus signs in Eq. (D.1) appear because both the hot and cold side fluid temperature gradients along the area axis are negative. Equation (D.1) can be recast in the form

$$d\,(T_h - T_c) = \frac{d\,Q}{W_c c_c}\left(1 - \frac{W_c c_c}{W_h c_h}\right) \tag{D.3}$$

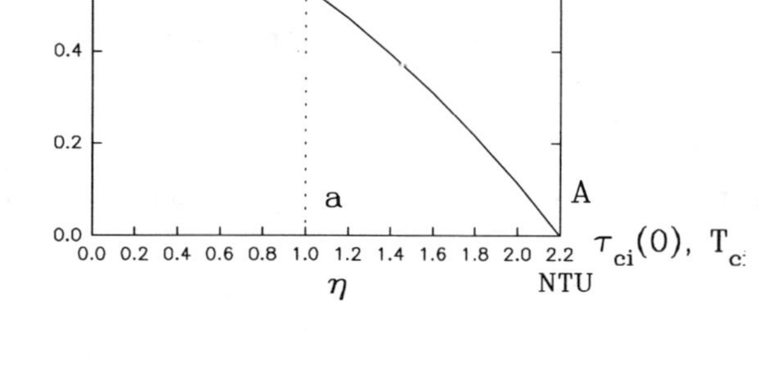

Figure D.1. Temperature distribution in counterflow heat exchanger ($W_c c_c = 0.5 W_h c_h$, $\varepsilon = 0.8$).

Upon substituting Eq. (D.2) for dQ, Eq. (D.3) can be written

$$\frac{d\,(T_h - T_c)}{T_h - T_c} = \frac{U\,da}{W_c c_c}\left(1 - \frac{W_c c_c}{W_h c_h}\right) \tag{D.4}$$

Equation (D.4) can be expressed in dimensionless form as

$$\frac{d\,(\tau_h - \tau_c)}{\tau_h - \tau_c} = (1 - \psi)\,d\,\eta \tag{D.5}$$

Upon integrating Eq. (D.5) from the hot side inlet to the cold side inlet, and noting that η = NTU when $a = A$,

$$\frac{\tau_{ho}}{1 - \tau_{co}} = e^{(1-\psi)\,\mathrm{NTU}} \tag{D.6}$$

Now, equating the rate of heat transfer from the hot side fluid to that of the cold side fluid,

$$W_h c_h (T_{hi} - T_{ho}) = W_c c_c (T_{co} - T_{ci}) \tag{D.7}$$

Equation (D.7) can be written in dimensionless form as

$$\tau_{ho} = 1 - \Psi \tau_{co} \tag{D.8}$$

Substituting this value of τ_{ho} into Eq. (D.6), and noting that

$$\varepsilon = \frac{T_{co} - T_{ci}}{\Delta T_{\max}} = \tau_{co}$$

we obtain

$$\frac{1 - \psi\,\varepsilon}{1 - \varepsilon} = e^{(1-\psi)\,\mathrm{NTU}} \tag{D.9}$$

Solving this equation for ε,

$$\varepsilon = \frac{1 - e^{(1-\psi)\,\mathrm{NTU}}}{\psi - e^{(1-\psi)\,\mathrm{NTU}}} \tag{D.10}$$

Multiplying the numerator and denominator of Eq. (D.10) by $-e^{-(1-\psi)\,\mathrm{NTU}}$, Eq. (D.10) becomes

$$\varepsilon = \frac{1 - e^{-(1-\psi)\,\mathrm{NTU}}}{1 - \psi\, e^{-(1-\psi)\,\mathrm{NTU}}} \tag{D.11}$$

Equation (D.11) relates the heat transfer effectiveness ε to the number of heat transfer units NTU and the unit thermal capacity ratio Ψ.

Temperature Distribution. If Eq. (D.5) is integrated from $\eta = 0$ to η, we obtain

$$\frac{\tau_h - \tau_c}{1 - \tau_{co}} = e^{(1-\psi)\,\eta} \tag{D.12}$$

Upon equating the heat transfer rate from the hot side fluid to the cold side fluid between $\eta = 0$ and η,

$$W_h c_h (T_{hi} - T_h) = W_c c_c (T_{co} - T_o) \tag{D.13}$$

In dimensionless form, Eq.(D.13) becomes

$$1 - \tau_h = \Psi(\tau_{co} - \tau_c) \tag{D.14}$$

Solving for τ_c from Eq. (D.14),

$$\tau_c = \tau_{co} + \frac{\tau_h - 1}{\psi} \tag{D.15}$$

It follows from Eq. (D.15) that

$$\tau_h - \tau_c = \tau_h \left(\frac{\psi - 1}{\psi} \right) + \frac{1 - \psi\,\tau_{co}}{\psi} \tag{D.16}$$

Substituting $\tau_h - \tau_c$ from Eq. (D.16) into Eq. (D.12), letting $\tau_{co} = \varepsilon$ (see Section D.1), and solving for τ_h,

$$\tau_h = \frac{1}{1 - \psi} \left[1 - \psi\,\varepsilon - \psi\,(1 - \varepsilon)\,e^{(1-\psi)\,\eta} \right] \tag{D.17}$$

Equation (D.17) represents the hot side dimensionless temperature distribution in the heat exchanger.

Substituting Eq. (D.17) for τ_h into Eq. (D.15), letting $\tau_{co} = \varepsilon$, and simplifying, we obtain for the cold side dimensionless temperature distribution

$$\tau_c = \varepsilon + \left(\frac{1 - \varepsilon}{1 - \psi} \right) [1 - e^{(1-\psi)\,\eta}] \tag{D.18}$$

It may be shown that, for $\eta = 0$, $\tau_h = 1$ and $\tau_c = \varepsilon$. Also, for $\eta = \text{NTU}$, $\tau_h = 1 - \psi\varepsilon$ and $\tau_c = 0$.

D.2.2. $\Psi < 1$, $W_h c_h < W_c c_c$

Heat Exchanger Effectiveness. The heat exchanger effectiveness for this case is exactly the same as that given by Eq. (D.11).

Temperature Distribution. The hot and cold side temperature distributions are similar in shape to those for which $W_c c_c < W_h c_h$, but are inverted and displaced according to the relations

$$\tau_h(\eta) = 1 - \tau_c(\text{NTU} - \eta)_{\text{Eq. (D.18)}} \tag{D.19}$$

$$\tau_c(\eta) = 1 - \tau_h(\text{NTU} - \eta)_{\text{Eq. (D.17)}} \tag{D.20}$$

That is, τ_h at location η is equal to $1 - \tau_c$, where τ_c is evaluated using Eq. (D.18) at $\eta = \text{NTU} - \eta$. Similarly, τ_c at location η is equal to $1 - \tau_h$, where τ_h is evaluated using Eq. (D.17) at $\eta = \text{NTU} - \eta$.

D.2.3. $\Psi = 0$, $W_c c_c < W_h c_h = \infty$

This case corresponds to a constant-temperature, condensing vapor on the hot side of the heat exchanger, with an effective c_h of ∞.

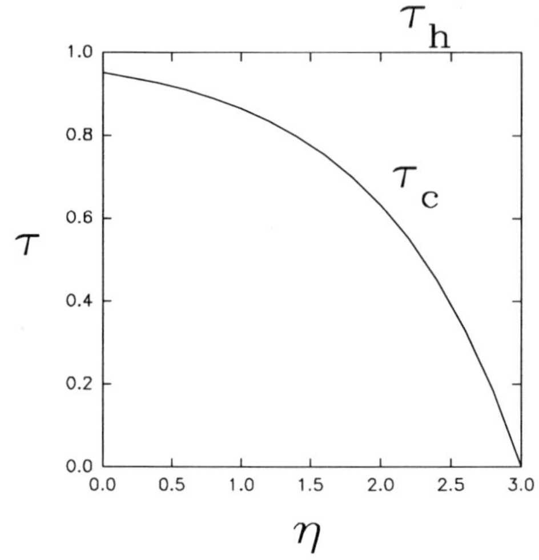

Figure D.2. Temperature distribution in counterflow heat exchanger ($W_c c_c < W_h c_h = \infty$, NTU = 3).

Heat Exchanger Effectiveness. Letting $\Psi = 0$ in Eq. (D.11),

$$\varepsilon = 1 - e^{-\mathrm{NTU}} \tag{D.21}$$

Temperature Distribution. Letting $\Psi = 0$ in Eqs. (D.17) and (D.18), and substituting Eq. (D.21) for ε, the temperature distributions take the form

$$\tau_h = 1 \tag{D.22}$$

$$\tau_c = 1 - e^{-(\mathrm{NTU}-\eta)} \tag{D.23}$$

Equations (D.22) and (D.23) are plotted in Figure D.2 for NTU = 3.

D.2.4. $\Psi = 0$, $W_h c_h < W_c c_c = \infty$

This case corresponds to a constant-temperature, vaporizing liquid on the cold side of the heat exchanger, with an effective c_c of ∞.

Heat Exchanger Effectiveness. The heat exchanger effectiveness is the same as for the case of Section D.2.3, and is given by Eq. (D.21).

Temperature Distribution. The temperature distributions are evaluated from Eqs. (D.19) and (D.20), letting $\Psi = 0$ and using Eq. (D.21) for ε. Then,

$$\tau_h = e^{-\eta} \tag{D.24}$$

$$\tau_c = 0 \tag{D.25}$$

Equations (D.24) and (D.25) are plotted in Figure D.3 for NTU = 3.

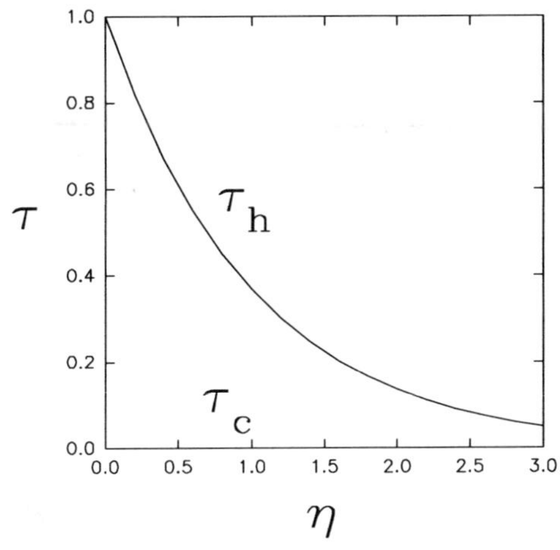

Figure D.3. Temperature distribution in counter-flow heat exchanger ($W_h c_h < W_c c_c = \infty$, NTU = 3).

D.2.5. $\Psi = 1, W_c c_c = W_h c_h$

This case corresponds to heat exchange between hot and cold side fluids of equal thermal capacity, a situation that is approximated in a gas turbine recuperator.

If Ψ is set equal to 1 in the previously derived equations for ε, τ_h, and τ_c in Section D.2.1, these equations become indeterminate. Although the indeterminacies can be evaluated mathematically, the equations will be rederived here from physical considerations. Reference should be made to the qualitative temperature distribution of Figure D.4.

Heat Exchanger Effectiveness. Equation (D.3) now takes the form

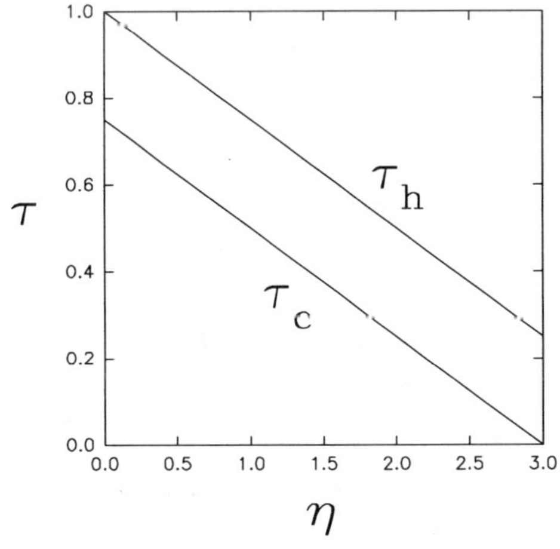

Figure D.4. Temperature distribution in counter-flow heat exchanger ($W_c c_c = W_h c_h$, NTU = 3).

$$d(\tau_h - \tau_c) = 0 \tag{D.26}$$

Upon integrating Eq. (D.26) along the entire heat exchanger,

$$\tau_h - \tau_c = 1 - \tau_{co} = \tau_{ho} = 1 - \varepsilon \tag{D.27}$$

In Eq. (D.27), the integration constant is equal to both $1 - \tau_{co}$ and τ_{ho}. From Section D.1, $\tau_{co} = \varepsilon$.

Combining Eqs. (D.1) and (D.2),

$$dQ = Uda(T_h - T_c) = -Wc\,dT_c = -Wc\,dT_h$$

This equation can be written in dimensionless form as

$$(\tau_h - \tau_c)\,d\eta = -d\tau_c = -d\tau_h \tag{D.28}$$

Substituting Eq. (D.27) for $\tau_h - \tau_c$,

$$(1 - \varepsilon)\,d\eta = -d\tau_c = -d\tau_h \tag{D.29}$$

Integrating this equation along the entire heat exchanger,

$$(1 - \varepsilon)\mathrm{NTU} = \tau_{co} = \varepsilon$$

Upon solving for ε,

$$\varepsilon = \frac{\mathrm{NTU}}{\mathrm{NTU} + 1} \tag{D.30}$$

This is the desired relation between ε and NTU for counterflow heat exchange, when $\Psi = 1$.

Temperature Distribution. The temperature distributions are found by integrating Eq. (D.29) from $\eta = 0$ to η. Then,

$$(1 - \varepsilon)\eta = \tau_{co} - \tau_c = \varepsilon - \tau_c = 1 - \tau_h \tag{D.31}$$

Solving this equation for τ_c,

$$\tau_c = \varepsilon - (1 - \varepsilon)\eta \tag{D.32}$$

Substituting Eq. (D.30) for ε,

$$\tau_c = \frac{\mathrm{NTU} - \eta}{\mathrm{NTU} + 1} \tag{D.33}$$

Equation (D.33) represents the cold side temperature distribution for counterflow heat exchange, when $\Psi = 1$.

From Eq. (D.31),

$$\tau_h = 1 - \varepsilon + \tau_c \tag{D.34}$$

Substituting ε from Eq. (D.30) and τ_c from Eq. (D.33) in Eq. (D.34),

$$\tau_h = \frac{\mathrm{NTU} + 1 - \eta}{\mathrm{NTU} + 1} \tag{D.35}$$

Equation (D.35) represents the hot side temperature distribution for counterflow heat exchange, when $\Psi = 1$.

D.3. PARALLEL FLOW HEAT EXCHANGE

Heat exchanger relations for parallel flow will now be considered. As was done for counterflow heat exchange, the following cases are evaluated:

- $\Psi < 1$, $W_c c_c < W_h c_h$
- $\Psi < 1$, $W_h c_h < W_c c_c$
- $\Psi = 0$, $W_c c_c < W_h c_h = \infty$
- $\Psi = 0$, $W_h c_h < W_c c_c = \infty$
- $\Psi = 1$, $W_c c_c = W_h c_h$

D.3.1. $\Psi < 1, W_c c_c < W_h c_h$

A representative temperature distribution for this case is shown qualitatively in Figure D.5.

Heat Exchanger Effectiveness. As with counterflow, the differential heat transfer rate between the hot and cold side fluids can be written

$$dQ = W_c c_c dT_c = -W_h c_h dT_h \tag{D.36}$$

$$dQ = U da(T_h - T_c) \tag{D.2}$$

Since the cold side temperature gradient along the heat exchanger is now positive, there is no minus sign before the cold side differential heat transfer rate in Eq. (D.36). Equation (D.2) is applicable to parallel flow as well as counterflow.

Performing the same manipulations on Eqs. (D.36) and (D.2) as were carried out on Eqs. (D.1) and (D.2) for counterflow, and expressing the results in dimensionless form,

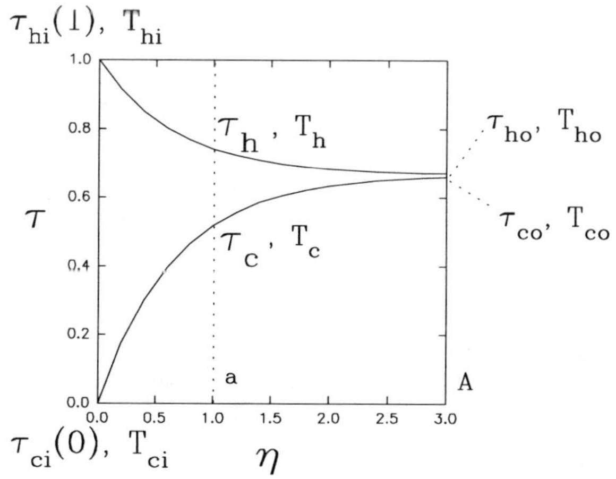

Figure D.5. Temperature distribution in parallel flow heat exchanger ($W_c c_c = 0.5 W_h c_h$, NTU = 3).

$$\frac{d\,(\tau_h - \tau_c)}{\tau_h - \tau_c} = -\,(1 + \psi)\,d\eta \tag{D.37}$$

Upon integrating this equation over the entire heat exchanger,

$$\tau_{ho} - \tau_{co} = e^{-(1 + \Psi)\mathrm{NTU}} \tag{D.38}$$

If the total hot and cold side heat transfer rates are equated and converted to dimensionless form, the same equation is obtained for τ_{ho} as was the case for counterflow. Therefore,

$$\tau_{ho} = 1 - \Psi\tau_{co} \tag{D.8}$$

Using this equation, we may write

$$\tau_{ho} - \tau_{co} = 1 - (1 + \Psi)\tau_{co} = 1 - (1 + \Psi)\varepsilon \tag{D.39}$$

Substituting Eq. (D.39) into Eq. (D.38) and solving for ε,

$$\varepsilon = \frac{1 - e^{-(1+\psi)\mathrm{NTU}}}{1 + \psi} \tag{D.40}$$

Equation (D.40) is the desired expression for ε as a function of NTU and Ψ.

Temperature Distribution. The temperature distributions are found by carrying out the integration of Eq. (D.37), and performing a heat balance between $\eta = 0$ and η instead of over the entire heat exchanger. It then follows that

$$\tau_h - \tau_c = e^{-(1 + \Psi)\eta} \tag{D.41}$$

$$\tau_h = 1 - \Psi\tau_c \tag{D.42}$$

Substituting Eq. (D.42) into Eq. (D.41) and solving for τ_c,

$$\tau_c = \frac{1 - e^{-(1+\psi)\eta}}{1 + \psi} \tag{D.43}$$

Then, substituting τ_c from Eq. (D.43) into Eq. (D.42) and solving for τ_h,

$$\tau_h = \frac{1 + \psi\, e^{-(1+\psi)\eta}}{1 + \psi} \tag{D.44}$$

Equations (D.43) and (D.44) represent the desired form for the cold and hot side temperature distributions in parallel flow when $W_c c_c < W_h c_h$.

D.3.2. $\Psi < 1$, $W_h c_h < W_c c_c$

Heat Exchanger Effectiveness. It may be shown that Eq. (D.40) for ε applies for either $W_c c_c < W_h c_h$ or $W_h c_h < W_c c_c$.

Temperature Distribution. The dimensionless temperature distributions take a somewhat different form for this case, as is now shown.

On equating the hot and cold side heat transfer rates between $\eta = 0$ and η,

$$W_h c_h (T_{hi} - T_h) = W_c c_c (T_c - T_{ci})$$

The dimensionless form of this equation is

$$\Psi(1 - \tau_h) = \tau_c \tag{D.45}$$

Solving for τ_h in Eq. (D.45),

$$\tau_h = 1 - \tau_c/\Psi \tag{D.46}$$

Equation (D.41) for the dimensionless temperature difference between the hot and cold side fluids is valid for either $W_c c_c < W_h c_h$ or $W_h c_h < W_c c_c$.

Upon substituting τ_h from Eq. (D.46) into Eq. (D.41) and solving for τ_c,

$$\tau_c = \left(\frac{\psi}{1+\psi}\right)\left[1 - e^{-(1+\psi)\eta}\right] \tag{D.47}$$

Now if τ_c from Eq. (D.47) is substituted into Eq. (D.46),

$$\tau_h = \frac{\psi + e^{-(1+\psi)\eta}}{1+\psi} \tag{D.48}$$

Equations (D.47) and (D.48) represent the desired cold and hot side temperature distributions for parallel flow when $W_h c_h < W_c c_c$.

D.3.3. $\Psi = 0$, $W_c c_c < W_h c_h = \infty$

This case corresponds to parallel flow heat exchange in which the hot side fluid is a constant-temperature, condensing vapor with an effective c_h of ∞.

Heat Exchanger Effectiveness. By letting $\Psi = 0$ in Eq. (D.40),

$$\varepsilon = 1 - e^{-\text{NTU}} \tag{D.49}$$

Temperature Distribution. By letting $\Psi = 0$ in Eqs. (D.44) and (D.43),

$$\tau_h = 1 \tag{D.50}$$

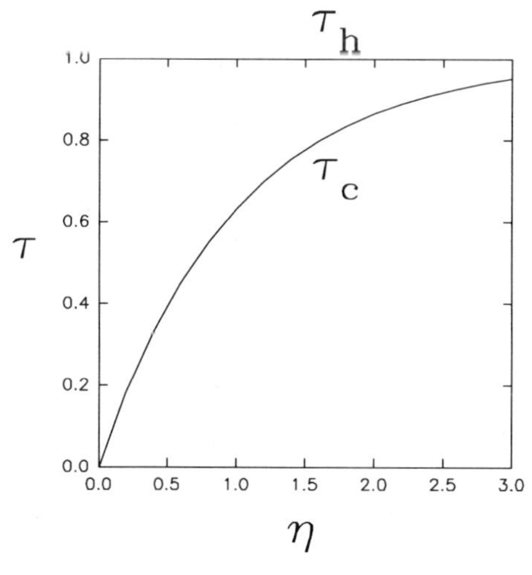

Figure D.6. Temperature distribution in parallel flow heat exchanger ($W_c c_c < W_h c_h = \infty$, NTU = 3).

$$\tau_c = 1 - e^{-\eta} \tag{D.51}$$

Equations (D.50) and (D.51) are plotted in Figure D.6, for NTU = 3.

D.3.4. $\Psi = 0, W_h c_h < W_c c_c = \infty$

This case corresponds to parallel flow heat exchange in which the cold side fluid is a constant-temperature, vaporizing liquid with an effective c_c of ∞.

Heat Exchanger Effectiveness. The heat exchanger effectiveness for this case is also given by Eq. (D.49).

Temperature Distribution. By letting $\Psi = 0$ in Eqs. (D.48) and (D.47),

$$\tau_h = e^{-\eta} \tag{D.52}$$

$$\tau_c = 0 \tag{D.53}$$

Equations (D.52) and (D.53) are plotted in Figure D.7, for NTU = 3.

D.3.5. $\Psi = 1, W_c c_c = W_h c_h$

This case represents heat exchange in parallel flow in which the thermal capacities of the hot and cold side fluids are equal. This situation is approximated in a gas turbine recuperator when the gas and air streams both flow in the same direction.

Unlike the case of heat exchange with counterflow, letting $\Psi = 1$ with parallel flow does not lead to indeterminacies in the ε, τ_h, and τ_c equations. The following results can then be determined directly from Eqs. (D.40), (D.44), and (D.43).

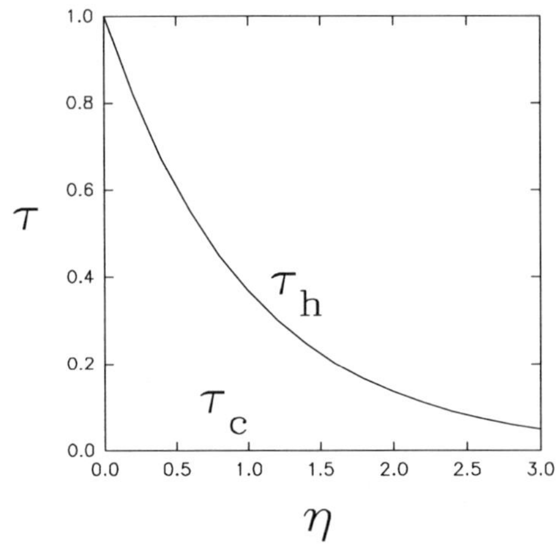

Figure D.7. Temperature distribution in parallel flow heat exchanger ($W_h c_h < W_c c_c = \infty$, NTU = 3).

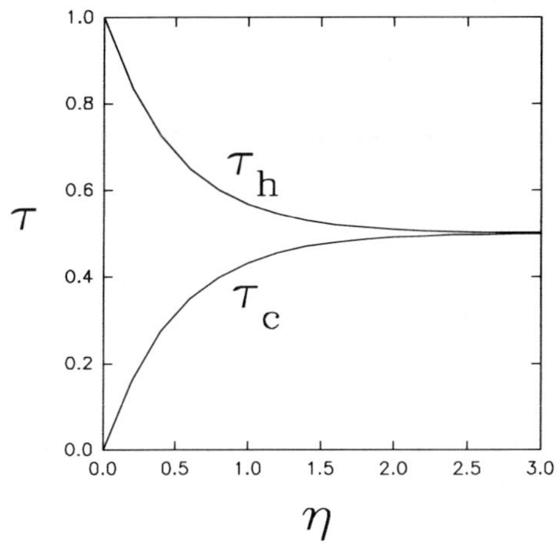

Figure D.8. Temperature distribution in parallel flow heat exchanger ($W_c c_c = W_h c_h$, NTU = 3).

Heat Exchanger Effectiveness.

$$\varepsilon = \frac{1 - e^{-2\text{NTU}}}{2} \tag{D.54}$$

Note that the maximum heat exchanger effectiveness is limited to 0.5.

Temperature Distribution.

$$\tau_h = \frac{1 + e^{-2\eta}}{2} \tag{D.55}$$

$$\tau_c = \frac{1 - e^{-2\eta}}{2} \tag{D.56}$$

Equations (D.55) and (D.56) are plotted in Figure D.8 for a parallel flow heat exchanger with NTU = 3.

D.4. REFERENCE

D.1. Kays, W. M., and London, A. L., *Compact Heat Exchangers,* McGraw-Hill, New York, 1964, pp. 264–266.

APPENDIX

E

LIQUID FLOW THROUGH CAPILLARY PUMPING LAYER

In a two-layer wick, liquid condensate returning to the evaporator has a choice of two parallel flow paths. The first is the liquid flow channel, which is specifically designated for the return liquid flow. The second is the capillary pumping layer.

In most circumstances the resistance to liquid flow is much greater in the capillary pumping layer, and most of the liquid flow passes through the designated liquid flow channel. Then, for practical purposes the capillary pumping layer may be ignored when calculating the axial resistance to liquid flow through the wick and the axial liquid frictional pressure drop.

However, this is not always the case. For example, in a covered grooved wick with a relatively thick capillary pumping layer, a significant fraction of the liquid flow may move through the capillary pumping layer. Overall resistance to liquid flow is then reduced, resulting in a higher recirculation rate of the heat pipe fluid and a higher capillary pumping limit.

In this appendix, expressions are derived for the liquid frictional pressure drop when the resistance through the capillary pumping layer is not much greater than that through the liquid flow channel.

E.1. LIQUID PRESSURE DROP IN TWO-LAYER WICK

Reference is made to the wick cross section of Figure E.1. The main liquid flow channel is designated by subscript 1, and the capillary pumping layer by subscript 2. The

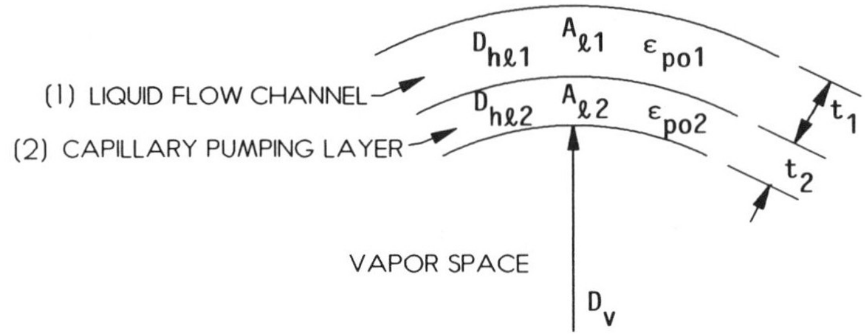

Figure E.1. Geometry of a two-layer wick.

hydraulic diameter is D_{hl}, the flow cross-sectional area is A_l, the flow channel thickness is t, and the ratio of frictional pressure drops in porous and open flow channels is ε_{po}.

Liquid flow is assumed to be laminar in both the main liquid flow channel and the capillary pumping layer. Since flow channels 1 and 2 are in parallel, they undergo the same pressure drop, and the total mass flow rate is the sum of that through each channel. We may then write for the length dx at any point x along the liquid flow path through the wick,

$$\Delta P_l = R_1 W_1 = R_2 W_2 = \text{RW} \tag{E.1}$$

$$W = W_1 + W_2 \tag{E.2}$$

where ΔP_l is the pressure drop, R_1 is the flow resistance of channel 1, R_2 is the flow resistance of channel 2, R is the total flow resistance, W_1 is the mass flow rate through channel 1, W_2 is the mass flow rate through channel 2, and W is the total mass flow rate.

Solving for the flow rates in Eq. (E.1), substituting the results in Eq. (E.2), and solving for R_1/R,

$$R_1/R = 1 + R_1/R_2 \tag{E.3}$$

If the flow resistance R_2 is infinite, $W_2 = 0$, $W = W_1$, and $R = R_1$. We may then write

$$\Delta P_{lo} = R_1 W \tag{E.4}$$

where ΔP_{lo} is the liquid pressure drop through the wick length dx when $R_2 = \infty$.

Dividing Eq. (E.1) by Eq. (E.4), with the mass flow rate W being the same for both situations, we obtain

$$\Delta P_l = \Delta P_{lo}/(1 + R_1/R_2) \tag{E.5}$$

Although Eq. (E.5) was derived for a differential flow length dx, it applies as well for the entire heat pipe length if R_1/R_2 is constant and independent of x. This is the case for laminar flow, as is shown subsequently.

ΔP_{lo} then represents the liquid frictional pressure drop term in the capillary pumping limit equation when $R_2 = \infty$ (the basis on which the capillary pumping limit equation was derived). When R_2 is finite and not much greater than R_1, ΔP_l from Eq. (E.5) should be used instead of ΔP_{lo} in the capillary pumping equation. That is, the liquid frictional term in the pressure drop equation should be divided by $1 + R_1/R_2$.

Introducing Eq. (4.34) for the differential laminar liquid frictional pressure drop at

x, and Eq. (4.35) for the ratio of laminar differential pressure drops in porous and open flow channels with the same cross section and flow rate, and using Eq. (E.1), R_1/R_2 may be written as

$$\frac{R_1}{R_2} = \left(\frac{\varepsilon_{po1}}{\varepsilon_{po2}}\right)\left(\frac{D_{hl2}}{D_{hl1}}\right)^2\left(\frac{A_{l2}}{A_{l1}}\right) \tag{E.6}$$

Since Eq. (E.6) is independent of length, it applies for the ratio of flow resistances over the entire heat pipe length, as well as for any differential length at x. Thus, as stated earlier, Eq. (E.5) is also applicable to the liquid pressure drop along the entire wick length.

Equation (E.6) will now be used to obtain explicit expressions for R_1/R_2 for two-layer wicks with a porous capillary pumping layer and various liquid flow channel configurations. In the following equations, K_1 and K_2 are the friction factors when channels 1 and 2 are porous.

E.2. FLOW RESISTANCE RATIO IN TWO-LAYER WICKS

E.2.1. Open Annular Liquid Flow Channel

For an open flow channel, $\varepsilon_{po} = 1$. Therefore, using Eq. (4.35),

$$\frac{\varepsilon_{po1}}{\varepsilon_{po2}} = \frac{32}{K_2 D_{hl2}^2} \tag{E.7}$$

Assuming that the wick thickness is small compared to the vapor space dimensions, D_{hl1} and A_{l2}/A_{l1} are well represented by the equations

$$D_{hl1} = 2t_1 \tag{E.8}$$

$$\frac{A_{l2}}{A_{l1}} = \frac{t_2/t_1}{1 + (2\,t_2/D_v)} \tag{E.9}$$

where D_v is the mean vapor space dimension. For a circular heat pipe, D_v is the vapor space diameter. For a rectangular heat pipe with vapor space width W_v and height H_v, $D_v = (W_v + H_v)/2$.

Upon substitution of Eqs. (E.7), (E.8), and (E.9) in Eq. (E.6),

$$\frac{R_1}{R_2} = \left(\frac{8}{K_2\,t_1^{\,2}}\right)\left[\frac{t_2/t_1}{1 + (2\,t_2/D_v)}\right] \tag{E.10}$$

E.2.2. Porous Annular Liquid Flow Channel

Using Eq. (4.35),

$$\frac{\varepsilon_{po1}}{\varepsilon_{po2}} = \left(\frac{K_1}{K_2}\right)\left(\frac{D_{hl1}}{D_{hl2}}\right)^2 \tag{E.11}$$

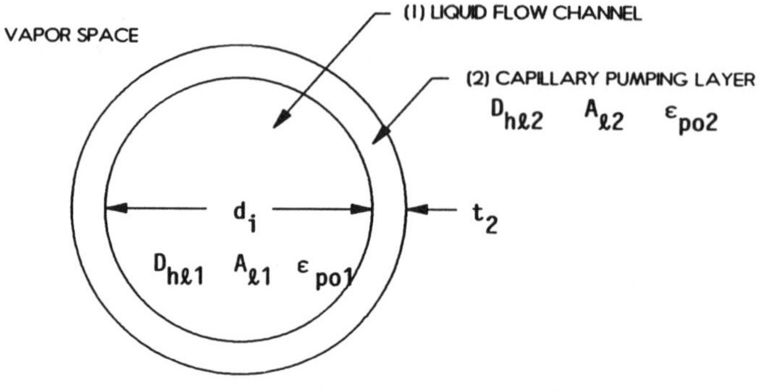

Figure E.2. Geometry of an artery.

Substituting Eqs. (E.11), (E.8), and (E.9) into Eq. (E.6),

$$\frac{R_1}{R_2} = \left(\frac{K_1}{K_2}\right)\left[\frac{t_2/t_1}{1 + (2t_2/D_v)}\right] \tag{E.12}$$

E.2.3. Grooved Annular Liquid Flow Channel

If the groove width is w_g and the width of land between grooves is w_l, then it may be shown that

$$D_{hl1} \approx \frac{2\,t_1}{1 + (t_1/w_g)} \tag{E.13}$$

$$\frac{A_{l2}}{A_{l1}} \approx \left[\frac{t_2/t_1}{1 + (2t_2/D_v)}\right]\left(1 + \frac{w_l}{w_g}\right) \tag{E.14}$$

Equation (E.7) for an open liquid flow channel also applies for a grooved flow channel. Therefore, upon substituting Eqs. (E.7), (E.13), and (E.14) in Eq. (E.6),

$$\frac{R_1}{R_2} = \left(\frac{8}{K_2\,t_1^2}\right)\left[\frac{t_2/t_1}{1 + (2t_2/D_v)}\right]\left(1 + \frac{t_1}{w_g}\right)^2\left(1 + \frac{w_l}{w_g}\right) \tag{E.15}$$

As w_g becomes very large and $w_l \to 0$, Eq. (E.15) becomes identical to Eq. (E.10) for an open annular liquid flow channel.

E.2.4. Artery

In an artery the liquid flow channel is an open cylinder with diameter d_i, surrounded by a porous capillary pumping layer of thickness t_2, as shown in Figure E.2.

Then

$$\mathrm{D}_{hl1} = d_i \tag{E.16}$$

$$A_{l2}/A_{l1} = 4[1 + (t_2/d_i)](t_2/d_i) \tag{E.17}$$

Upon substituting Eqs. (E.16), (E.17), and (E.7) into Eq.(E.6),

$$\frac{R_1}{R_2} = \left(\frac{128}{K_2\, d_i^2}\right)\left(1 + \frac{t_2}{d_i}\right)\left(\frac{t_2}{d_i}\right) \tag{E.18}$$

Usually, R_1/R_2 for an open artery will be negligibly small unless $t_2/d_i > 0.2$.

The various equations for R_1/R_2 were derived on the assumption of laminar flow in an open liquid flow channel. In the absence of other information, they may be considered as first approximations to R_1/R_2 when flow in an open liquid flow channel is turbulent.

INDEX